WELDING

Principles and Applications

EIGHTH EDITION

Study Guide/Lab Manual

WELDING
Principles and Applications
EIGHTH EDITION
Study Guide/Lab Manual
Larry Jeffus

CENGAGE
Learning·

Australia • Brazil • Mexico • Singapore • United Kingdom • United States

Welding: Principles and Applications
Eighth Edition

Larry Jeffus

SVP, GM Skills & Global Product Management: **Dawn Gerrain**

Product Director: **Matt Seeley**

Product Team Manager: **Erin Brennan**

Associate Product Manager: **Nicole Sgueglia**

Senior Director, Development: **Marah Bellegarde**

Senior Product Development Manager: **Larry Main**

Senior Content Developer: **Sharon Chambliss**

Product Assistant: **Maria Garguilo**

Vice President, Marketing Services: **Jennifer Ann Baker**

Marketing Director: **Michele McTighe**

Marketing Manager: **Jonathan Sheehan**

Marketing Coordinator: **Andrew Ouimet**

Senior Production Director: **Wendy Troeger**

Production Director: **Andrew Crouth**

Senior Content Project Manager: **Betsy Hough**

Senior Art Director: **Benjamin Gleeksman**

Software Development Manager: **Pavan K. Ethakota**

Cover Image(s): Larry Jeffus

For product information and technology assistance, contact us at
Cengage Learning Customer & Sales Support, 1-800-354-9706

For permission to use material from this text or product,
submit all requests online at **www.cengage.com/permissions.**
Further permissions questions can be e-mailed to
permissionrequest@cengage.com

Library of Congress Control Number: 201594388

Book Only ISBN: 978-1-3054-9470-1

Cengage Learning
20 Channel Center Street
Boston, MA 02210
USA

Cengage Learning is a leading provider of customized learning solutions with employees residing in nearly 40 different countries and sales in more than 125 countries around the world. Find your local representative at **www.cengage.com.**

Cengage Learning products are represented in Canada by Nelson Education, Ltd.

To learn more about Cengage Learning, visit **www.cengage.com**
Purchase any of our products at your local college store or at our preferred online store **www.cengagebrain.com**

Notice to the Reader

Printed in the United States of America
Print Number: 01   Print Year: 2016

CONTENTS

INTRODUCTION

Many job opportunities exist for welders who have demonstrated their ability to make acceptable welds in various positions using the major welding processes. In addition to welding skills, the welder must have a basic understanding of the science of welding, including knowing how various types of welds are made, how varying the different aspects of the welding setup affects the welds, and how adjusting welding factors can be used to obtain a better weld or overcome specific conditions.

To gain welding skills, you will perform the practices outlined in *Welding: Principles and Applications*, Eighth Edition. You will be performing welds on different types and thicknesses of metal in different positions using each of the major welding processes. In addition, you will be making a variety of cuts using several different cutting processes. With practice, you will gain enough skill and confidence in your abilities to complete certification with a variety of welding processes for the various welding positions.

Another important part of becoming a good welder is to gain a broad understanding of welding. Why do metals react as they do to heat? Why is one welding process preferred over another for a particular job? Why does the act of modifying one or more of the welding factors result in a better or different weld? The purpose of this guide is to help you answer these questions by developing a firm understanding of the theory of welding through study and practice.

This *Study Guide/Lab Manual* is designed to focus your studies and assist in organizing your learning of welding principles and applications. Once the reading assignments in the text are completed, the instructor will ask you to complete the appropriate section in this guide. The sections consist of a combination of multiple-choice questions, lists, matching items, identification exercises, and completion questions. (Not all types of questions appear in each chapter of the guide.) Do not guess the answers to the questions. If you do not know the answer, leave the space blank. You can then identify the topics that you need to review again until you understand the information. Ask your instructor for help if you continue to have difficulty with any of the welding principles. You will be a better welder if you combine welding skills with a firm understanding of the how and why of welding.

WELDING

Principles and Applications

EIGHTH EDITION

Study Guide/Lab Manual

Chapter 1

Introduction to Welding

■ PRACTICE 1-1

Name _____ Date _____

Class _____ Instructor _____ Grade _____

OBJECTIVE: After completing this practice, you should be able to identify items that are manufactured using various welding processes.

EQUIPMENT AND MATERIALS NEEDED FOR THIS PRACTICE

Paper and pencil.

INSTRUCTIONS

Answer the following questions by looking around your school, home, and community to identify items manufactured with the following welding process.

1. List four items that are manufactured using the oxyfuel welding, brazing, or cutting process.

    Process   Item

    a. _____ , _____

    b. _____ , _____

    c. _____ , _____

    d. _____ , _____

2. List four items that are manufactured using the shielded metal arc welding process.

    a. _____

    b. _____

    c. _____

    d. _____

3. List four items that are manufactured using the gas metal arc welding process.

   a. _____

   b. _____

   c. _____

   d. _____

4. List four items that are manufactured using the flux cored arc welding process.

   a. _____

   b. _____

   c. _____

   d. _____

5. List four items that are manufactured using the gas tungsten arc welding process.

   a. _____

   b. _____

   c. _____

   d. _____

INSTRUCTOR'S COMMENTS _____

_____

_____

■ PRACTICE 1-2

Name _____ Date _____

Class _____ Instructor _____ Grade _____

OBJECTIVE: After completing this practice, you should be able to identify those jobs available in your local newspaper's employment section that require welding skills.

EQUIPMENT AND MATERIALS

Paper and pencil.

INSTRUCTIONS

Using your local newspaper, make a list of the jobs listed in the help wanted section that might require welding skills.

a. _____

b. _____

c. _____

d. _____

e. _____

f. _____

g. _____

h. _____

i. _____

j. _____

INSTRUCTOR'S COMMENTS _____

_____

_____

CHAPTER 1: INTRODUCTION TO WELDING QUIZ

Name _____   Date _____

Class _____   Instructor _____   Grade _____

INSTRUCTIONS: Carefully read Chapter 1 in the text and answer the following questions.

A. IDENTIFICATION

In the space provided, identify the items shown on the illustration.

   1. Identify the welding process shown.

_____

   2. Identify the welding process shown.

_____

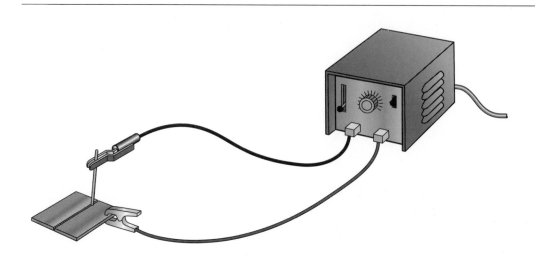

3. Identify the welding process shown.

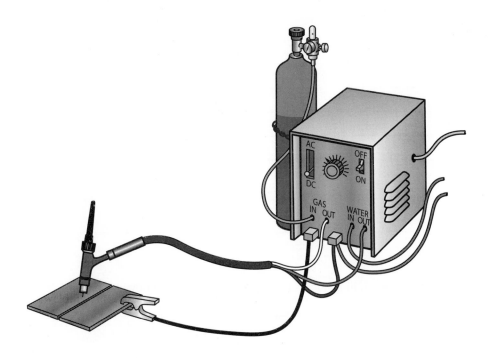

4. Identify the welding process shown.

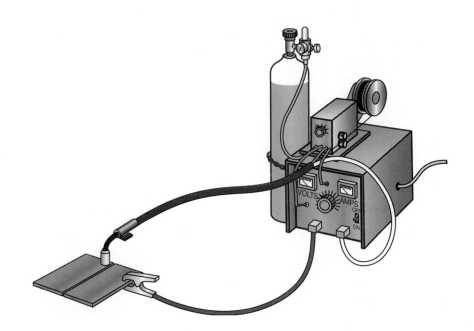

B. SHORT ANSWER

5. Using Table 1-1 on page 9 in the text give the full name for the following welding or allied process abbreviations.

   a. AB _____

   b. DS _____

   c. PEW _____

   d. MAC _____

   e. PAW _____

   f. FOW _____

   g. OFC-A _____

   h. SAW _____

   i. IW _____

   j. TW _____

6. What were the early work implements that were fastened together by individuals?

   _____

7. When did forge welding become popular? _____

8. What is the correct term for stick welding? _____

9. Convert 12 in. to mm _____

10. Convert 10°F to °Celsius _____

11. Convert 0°Kelvin to °C _____

12. Convert 50 sq ft to sq yd _____

13. Convert 8 yards to ft _____

14. Convert 40,000 meters to kilometers _____

15. Convert 460°Rankine to °F _____

C. ESSAY

Provide complete answers for all of the following questions.

16. Describe the type of jobs the following workers may do.

    a. welders _____

    b. tack welders _____

    c. welding operator _____

    d. welder's helper _____

    e. welder assemblers or welding fitters _____

    f. welding inspectors _____

    g. welding shop supervisor _____

    h. welding salespersons _____

    i. welding shop owner _____

    j. welding engineer_____

17. List and explain the various considerations that must be used when selecting the best joining process to be used for a particular welding job.

_____

_____

_____

18. How does the American Welding Society define a weld?

_____

_____

19. What is welding?

_____

_____

D. DEFINITIONS

20. Define the following terms:

    qualification _____

    certification _____

    coalescence_____

    OFC _____

    OFW_____

    TB _____

    SMAW_____

    GTAW_____

    GMAW _____

    FCAW_____

    manual _____

    semiautomatic _____

    machine _____

    automated _____

INSTRUCTOR'S COMMENTS _____

_____

_____

Chapter 2

Safety in Welding

1. A welder will only weld with proper eye protection: One of the primary concerns in welding is the protection of the eyes.

2. Welders will always wear protective clothing: Protective clothing for your personal protection will be required for all welding.

3. Welders may only weld with proper ventilation: Welding fumes can be hazardous to the welding operator. Fumes will vary depending on the type of welding or cutting operation along with the type of filler metal, fluxes, coating, and the base metal being welded on.

4. Welders must never weld on any type of vessel or tank that has contained any type of flammable material: Welding or cutting should not be performed on drums, barrels, tanks, or other containers until they have been cleaned thoroughly, eliminating all flammable materials and all substances (such as detergents, solvents, greases, tars, or acids), which might produce flammable, toxic, or explosive vapors when heated.

5. Welders are not to lubricate pressure regulators: Do not use oil, grease, or any type of hydrocarbon product on any torch, regulator, or cylinder fitting. Oil and grease in the presence of oxygen may burn with explosive force.

6. Welders must always use acetylene cylinders in the vertical (upright) position: Acetylene cylinders that have been lying on their side must stand upright for four hours or more before they can be used safely.

7. Welders must always use less than 15 psig acetylene working pressure: Acetylene becomes unstable at pressures higher than 30 psig and could explode. For safety, the red line on the acetylene working pressure gauge is at 15 psig, which is halfway to the point of explosion.

8. Welders will always secure cylinders: Cylinders must be secured with a chain or other device so that they cannot be knocked over accidentally.

9. Welders are to grind only iron, steel, or stainless steel on a grinding stone designed for ferrous material. The stone will become glazed (the surface clogs with metal) and may explode due to frictional heat buildup if nonferrous metals are ground.

10. Welders are always to keep a clean workstation: Always keep a safe clean work area.

11. Welders must always protect others from the welding arc's light: Welding curtains must always be used to protect other workers in an area that might be exposed to the welding light.

12. Welders must protect their hearing: Damage to your hearing caused by high sound levels may not be detected until later in life, and the resulting loss in hearing is nonrecoverable. It will not get better with time. Each time you are exposed to high levels of sound your hearing will become worse.

13. Welders must always guard against problems caused by welding on used metal: Extreme caution must be taken to avoid the fumes produced when welding is done on dirty, painted, plated, or used metal. Any chemicals that might be on the metal will become mixed with the welding fumes, and this combination can be extremely hazardous. All metal must be cleaned and ground to bare metal before welding to avoid this potential problem.

14. Welders must avoid electrical shock: Welding cables must never be spliced within 10 ft (3 m) of the electrode holder. All extension cords and equipment power cords must be in good repair. Never use equipment not properly grounded.

15. Welders must not carry flammable items in their pockets: There is no safe place to carry butane lighters and matches while welding or cutting. They may catch fire or explode if they are subjected to welding heat or sparks. Butane lighters may explode with the force of 1/4 stick of dynamite. Matches can erupt into a ball of fire. Both butane lighters and matches must always be removed from the welder's pockets and placed a safe distance away before any work is started.

16. Welders must always be well informed about safe and proper equipment usage: Before operating any power equipment for the first time, you must read the manufacturer's safety and operating instructions and should be assisted by someone who has experience with the equipment. Be sure your hands are clear before the equipment is started. Always turn the power off and lock it off before working on any part of the equipment.

■ PRACTICE 2-1

Name _____ Date _____

Class _____ Instructor _____ Grade _____

OBJECTIVES: After completing this practice, you should be able to explain:

- Why welding curtains are used in a welding environment
- What can happen to your hearing, short-term and long-term, if proper hearing protection is not used
- Why caution must be taken to avoid the fumes produced when welding is done on dirty, painted, plated, or used metal
- Why professional welders NEVER carry butane or propane lighters or matches on their person when welding or cutting
- When and why it is important to read and understand the manufacturer's safety and operating instructions on any piece of equipment

EQUIPMENT AND MATERIALS NEEDED FOR THIS PRACTICE

Paper and pencil.

INSTRUCTIONS

Look carefully at Chapter 2, "Safety in Welding." Reread all the figure captions, shaded caution areas, and tables, then answer the following questions in short answer form.

1. Why are welding curtains used in a welding environment?

_____

_____

_____

2. What can happen if you do not use proper hearing protection?

_____

_____

_____

_____

3. What is the one way to avoid potentially hazardous fumes when welding on metal that has been painted or has any grease, oil, or chemicals on its surface?

_____

_____

_____

_____

4. Why do professional welders never carry butane or propane lighters or matches on their person?

_____

_____

_____

_____

_____

5. When is it important to read and understand the manufacturer's safety and operating instructions regarding a piece of equipment?

_____

_____

_____

_____

_____

INSTRUCTOR'S COMMENTS _____

_____

_____

CHAPTER 2: WELDING SAFETY QUIZ

Name _____ Date _____

Class _____ Instructor _____ Grade _____

Instructions: Before working with any type of welding or cutting operation, you must be knowledgeable about the following. Carefully read Chapter 2 in the text and answer each question.

A. SHORT ANSWER

Write a brief answer in the space provided that will answer the question or complete the statement.

1. Burns can be caused by _____ types of light rays.

2. Arc welding produces _____, _____, and _____ types of light.

3. _____ protection must be worn in the welding shop at all times.

4. When grinding, chipping, or when overhead work is being done, _____ and _____ should be worn.

5. The two types of ear protection used in the welding shop are _____ and _____.

6. Welders should recognize that fumes of any type should not be _____.

7. Forced ventilation is required when welding on metals that contain or are coated with

    _____, _____, _____, _____, _____, _____, _____, and _____.

8. Acetylene cylinders that have been lying on their side must stand upright for _____ or more before they are used.

9. When moving cylinders, the _____ should be replaced.

10. Highly combustible materials should be _____ or more away from welding.

11. List the guidelines you should consider when selecting the following clothes for welding.

a. Shirts

_____

b. Pants

_____

c. Boots

_____

d. Caps

_____

B. IDENTIFICATION

Identify the items shown in the illustration by placing the letter under the appropriate number.

12. Safety glasses with side shields

13. Full face shield

A

B

14. Welding helmet

C

D

15. Typical respirator

16. Power air-purifying respirator

E

C. ESSAY
Provide complete answers for all of the following questions.

17. Describe first-, second-, and third-degree burns and the first aid that would be administered for each type of burn.

_____

_____

18. Why should eye burns be checked by a professional?

_____

_____

19. List the areas mentioned in this chapter that should not be used for cylinder storage.

_____

_____

20. What should be done with a leaking cylinder if the leak cannot be stopped?

_____

_____

21. Why must an acetylene cylinder that was stored horizontally be set in a vertical position for four hours before it is used?

_____

_____

22. What is an MSDS?

_____

_____

23. Describe how a person should lift heavy objects.

_____

_____

24. What type of fire extinguishers are used in your shop? Where are they located?

_____

_____

25. Why is it so important to keep your work area clean?

_____

_____

26. How should a grinding stone be inspected for cracks?

_____

_____

27. What type of ventilation is used in your shop and is it adequate?

_____

_____

28. List two conditions that would require forced ventilation in a weld shop.

a. _____

b. _____

29. Describe how inert gases such as argon and carbon dioxide can be hazardous.

_____

_____

D. DEFINITIONS

30. Define the following terms:

infrared light _____

ultraviolet light _____

forced ventilation _____

natural ventilation _____

valve protection cap _____

acetone _____

INSTRUCTOR'S COMMENTS _____

_____

_____

Chapter 3

Shielded Metal Arc Equipment, Setup, and Operation

Name _____ Date _____

Class _____ Instructor _____ Grade _____

OBJECTIVE: After completing this practice, you will be able to accurately calculate the amperage when the knob is at various settings.

EQUIPMENT AND MATERIALS NEEDED FOR THIS PRACTICE

1. Paper and pencil.

2. Amperage ranges as given for this practice (or from a machine in your shop).

INSTRUCTIONS

Estimate the amperage when the knob is at the 1/4, 1/2, and 3/4 settings.

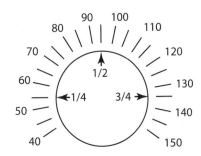

INSTRUCTOR'S COMMENTS _____

_____

_____

■ PRACTICE 3-2

Name _____   Date _____

Class _____   Instructor _____   Grade _____

OBJECTIVES: After completing this practice, you will be able to accurately calculate the amperages when the knob is set at 1, 4, 7, 9, 2.3, 5.7, and 8.5.

EQUIPMENT AND MATERIALS NEEDED FOR THIS PRACTICE

1. Paper and pencil or a calculator.

2. Amperage ranges as given for this practice (or from a machine in your shop).

INSTRUCTIONS

Calculate the amperages for each of the following knob settings: 1, 4, 7, 9, 2.3, 5.7, and 8.5.

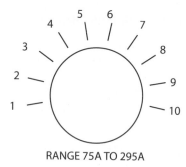

RANGE 75A TO 295A

INSTRUCTOR'S COMMENTS _____

_____

_____

■ PRACTICE 3-3

Name _____ Date _____

Class _____ Instructor _____ Grade _____

OBJECTIVE: After completing this practice you should be able to read a duty cycle chart.

EQUIPMENT AND MATERIALS NEEDED FOR THIS PRACTICE

1. Paper and pencil.

2. The duty cycle chart shown in the textbook as Figure 3-34 on page 66.

INSTRUCTIONS

1. Determine the maximum welding amperage for Welder 1.

2. Determine the percent duty cycle at the maximum amperage for Welder 1.

3. Determine the maximum welding amperage at 100% duty cycle for Welder 1.

4. Determine the maximum welding amperage for Welder 2.

5. Determine the percent duty cycle at the maximum amperage for Welder 2.

6. Determine the maximum welding amperage at 100% duty cycle for Welder 2.

7. Determine the maximum welding amperage for Welder 3.

8. Determine the percent duty cycle at the maximum amperage for Welder 3.

9. Determine the maximum welding amperage at 100% duty cycle for Welder 3.

INSTRUCTOR'S COMMENTS _____

_____

_____

■ PRACTICE 3-4

Name _____ Date _____

Class _____ Instructor _____ Grade _____

OBJECTIVE: After completing this practice you should be able to determine welding lead sizes.

EQUIPMENT AND MATERIALS NEEDED FOR THIS PRACTICE

1. Paper and pencil.

2. Table 3-6, shown on textbook page 67.

INSTRUCTIONS

1. Determine the minimum copper welding lead size for a 200 amp welder with 100 ft (30 m) leads.

2. Determine the minimum copper welding lead size for a 125 amp welder with 225 ft (69 m) leads.

3. Determine the maximum length aluminum welding lead that can carry 300 amps.

INSTRUCTOR'S COMMENTS _____

_____

_____

■ PRACTICE 3-5

Name _____ Date _____

Class _____ Instructor _____ Grade _____

OBJECTIVE: After completing this practice you should be able to properly repair electrode holders.

EQUIPMENT AND MATERIALS NEEDED FOR THIS PRACTICE

1. Manufacturer's instructions for your type of electrode holder.

2. Replacement parts for your type of electrode holder.

3. A set of hand tools.

INSTRUCTIONS

Before starting any work, make sure that the power to the welder is off and locked off or the welding lead has been removed from the machine.

1. Remove the electrode holder from the welding cable.

2. Remove the jaw insulating covers.

3. Replace the jaw insulating covers with new ones.

4. Reconnect the electrode holder to the welding cable.

5. Reconnect the welding cable to the welder (if it is not already connected) and turn on the welding power.

6. Make a weld to ensure that the repair was made correctly.

INSTRUCTOR'S COMMENTS _____

_____

_____

CHAPTER 3: SHIELDED METAL ARC WELDING EQUIPMENT, SETUP, AND OPERATION QUIZ

Name _____ Date _____

Class _____ Instructor _____ Grade _____

Instructions: Carefully read Chapter 3 in the text and complete the following questions.

A. SHORT ANSWER

Write a brief answer in the space provided that will answer the question or complete the statement.

1. From the illustrations below, which one of the two is classified as straight polarity (DCSP or DCEN)? Select "A" or "B."

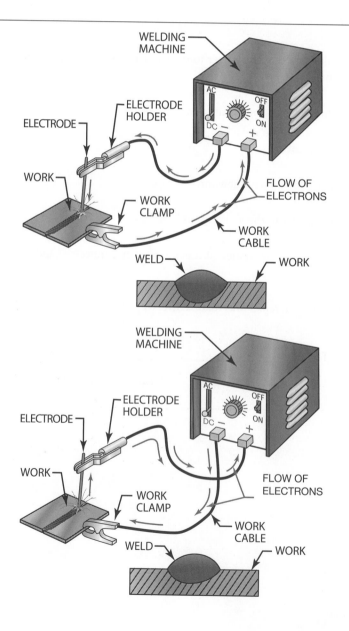

2. What is the minimum cable size that should be used if the length of the copper cable is 100 ft with a maximum of 200 amperes? Use the chart on the next page.

_____

_____

3. What is the minimum cable size that should be used if the length of the aluminum welding cable is 100 ft with a maximum of 200 amperes? Use the same chart as you used in question 2.

_____

_____

Amperes			Copper Welding Lead Sizes								
			100	150	200	250	300	350	400	450	500
ft	m											
50	15		2	2	2	2	1	1/0	1/0	2/0	2/0
75	23		2	2	1	1/0	2/0	2/0	3/0	3/0	4/0
100	30		2	1	1/0	2/0	3/0	4/0	4/0		
125	38		2	1/0	2/0	3/0	4/0				
150	46		1	2/0	3/0	4/0					
175	53		1/0	3/0	4/0						
200	61		1/0	3/0	4/0						
250	76		2/0	4/0							
300	91		3/0								
350	107		3/0								
400	122		4/0								

Amperes			Aluminum Welding Lead Sizes								
			100	150	200	250	300	350	400	450	500
ft	m											
50	15		2	2	1/0	2/0	2/0	3/0	4/0		
75	23		2	1/0	2/0	3/0	4/0				
100	30		1/0	2/0	4/0						
125	38		2/0	3/0							
150	46		2/0	3/0							
175	53		3/0								
200	61		4/0								
225	69		4/0								

Length of Cable

4. Why is a smaller cable sometimes spliced into the end of a large welding cable? What is the maximum length of the cable splice?

_____

_____

5. Name the item shown below.

_____

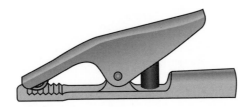

6. What can cause a properly sized electrode holder to overheat?

_____

7. Name the item shown below.

_____

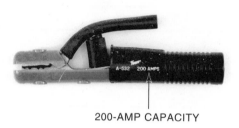

200-AMP CAPACITY

8. Why should you occasionally touch the work clamp?

_____

9. What is the temperature of the SMAW arc in degrees F and degrees C?

_____

10. List the parts of the SMAW process and electrode.

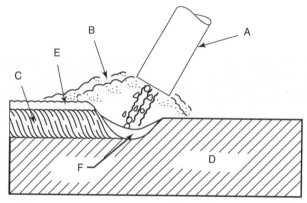

A. _____

B. _____

C. _____

D. _____

E. _____

F. _____

11. What is the minimum thickness of metal normally welded using the SMAW process?

   a. 8 gauge

   b. 10 gauge

   c. 12 gauge

   d. 16 gauge

12. What is the estimated amperage at which the following knob is set? (10 to 120 ampere range)

    a. 46 amperes

    b. 54 amperes

    c. 60 amperes

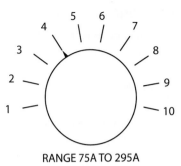

RANGE 75A TO 295A

13. List three items that should be routinely checked on portable welding equipment.

    a. _____

    b. _____

    c. _____

14. Why are transformer and transformer-rectifier welding machines classified as the industry standard?

    _____

    _____

    _____

    _____

15. Why must a welding machine periodically be blown out with compressed air?

    _____

    _____

16. What type of welding is done with a constant current welding machine?

    _____

    _____

17. What is the difference between a generator welder and an alternator welder?

    _____

    _____

B. MATCHING

In the space provided to the left, write the letter from Column B that best answers the question or completes the statement in Column A.

Column A   Column B

_____ 18. What is the measurement of electrical pressure?   a. duty cycle

_____ 19. What is the measurement of the amount of heat in the arc?   b. operating voltage

_____ 20. What is the measurement of the magnitude of electron flow?   c. voltage

_____ 21. What reduces the arc resistance, making the arc more stable? d. arc stabilizers

_____ 22. What is the voltage at the electrode before striking an arc? e. tap type

_____ 23. What is the voltage at the arc during welding?   f. step-down transformer

_____ 24. What takes high-voltage, low-amperage current and changes   g. amperage
            it into a low-voltage, high-amperage current?
                                                                         h. open circuit voltage
_____ 25. What is the percentage of time a welding machine can be
            used continuously?   i. wattage

_____ 26. What type of welding machine allows the selection of   j. rectifiers
            higher current settings by tapping into the secondary
            coil at higher turn values?

_____ 27. What is used to change AC to DC?

C. FILL IN THE BLANK

Fill in the blank with the correct word. Answers may be more than one word.

28. SMAW is a widely used welding process because of its _____,

    _____, and _____.

29. _____ to the flow of electrons (electricity) produces heat.

30. _____ controls the maximum gap the electrons can jump to form the arc.

31. _____ controls the size of the arc.

32. _____ is a measurement of the amount of electrical energy or power in the arc.

33. The quantity of heat in a material is a function of both its _____ and its

    _____.

34. In _____the total welding current (watts) remains the same.

35. A transformer with more turns of wire in the primary winding than in the secondary winding is known as a _____ transformer.

36. Alternating welding current can be converted to direct current by using a series of _____.

INSTRUCTOR'S COMMENTS _____

_____

_____

Chapter 4

Shielded Metal Arc Welding of Plate

Lens Filter Recommendation	
Shielded Metal Arc Welding Operation	Suggested Shade #
Shielded Metal Arc Welding, up to 5/32-in. electrodes	10
Shielded Metal Arc Welding, 3/16-in. to 1/4-in. electrodes	12

Welding Current Adjustment

The selection of the proper welding current depends on the electrode size, plate thickness, welding position, and the welder's skill. Electrodes of the same size can be used with a higher current in the flat position than they can in the vertical or overhead position. Since several factors affect the current requirements, data provided by welding equipment and electrode manufacturers should be used. For initial settings, use the table below.

Electrode Size	Classification					
	E6010	E6011	E6012	E6013	E7016	E7018
3/32 in.	40–80	50–70	40–90	40–85	75–105	70–110
1/8 in.	70–130	85–125	75–130	70–120	100–150	90–165
5/32 in.	100–165	130–160	120–200	130–160	140–190	125–220

After setting the current on the welding machine, you must master the skill of starting the arc.

Two methods are used for starting any arc, the striking or brushing method, Figure 4-1, and the tapping method, Figure 4-2.

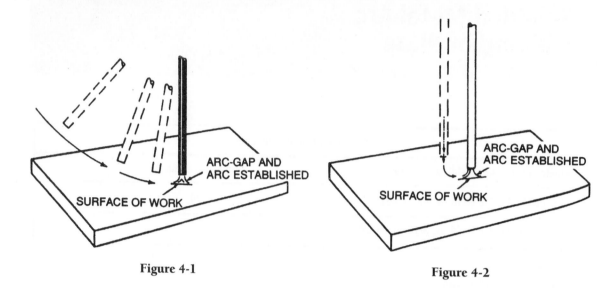

Figure 4-1   **Figure 4-2**

In the striking or brushing method, the electrode is brought to the surface of the work in a lateral motion similar to striking a match. As soon as the electrode touches the surface, the electrode is raised to establish the arc, Figure 4-1. The arc length or gap between the electrode and the work should be approximately equal to the diameter of the electrode. When the proper arc length is obtained, a sharp crackling sound is heard.

In the tapping method, the electrode is held in a vertical position to the work and tapped or bounced on the work surface, Figure 4-2. Upon contact, the electrode is raised approximately to the diameter of the electrode to establish the proper arc length.

If the electrode is raised too slowly with either of the above arc starting methods, the electrode will adhere and stick to the base metal. If this occurs, the electrode can usually be freed by a quick sideway twist to snap the end of the electrode from the base metal. However, if twisting does not dislodge the electrode, remove the holder from the electrode quickly. Allow the electrode to briefly cool, then free the electrode from the base metal by bending it back and forth. When using this method, use extreme caution because the electrode may still be hot.

WORKSTATION CHECKLIST

OBJECTIVE: You will demonstrate your knowledge of inspecting an SMA welding station.

In the space provided to the left, place an (S) for Satisfactory if the item is in good working condition or an (R) for Repair if the site needs repair.

_____ 1. Check to see that the power to the welding machine has been turned off (if not, turn it off before starting).

_____ 2. Inspect your welding helmet for holes, cracks, and proper headband adjustment.

_____ 3. Inspect your filter lens for cracks and for the correct shade, and make sure the lens cover is clean and that there are not any light leaks around the lens.

_____ 4. Make sure you are wearing proper clothing for welding: long-sleeved shirt, leather gloves, safety glasses, leather shoes or boots, and so forth. Refer to Chapter 2 in the text if you have any questions about welding apparel.

  5. Inspect the electrode lead and work lead for:

_____   a. proper connections between the electrode holder and cable

_____   b. proper connections between the work lead and the ground clamp

_____   c. broken insulation, exposed wire over the full length of the electrode and work lead

_____   d. tight connections at both the electrode and work lugs on the welding machine

_____ 6. Proper ventilation. Refer to Chapter 2 in the text if you have questions about proper ventilation.

_____ 7. Inspect the welding area for the safety of others who may be working around you. Check for holes in curtains, cracks in walls, or even doors that are to be closed while you are welding. Always be concerned with the safety of others.

_____ 8. Inspect for flammable materials in the work area such as paper, rags, wood, oil, and other flammable liquids.

_____ 9. Make sure you have pliers, a wire brush, and chipping hammer that are in good working condition.

OPERATING INSTRUCTIONS FOR SMAW EQUIPMENT

1. Check to see that your work area is safe for welding.

2. Select the proper electrode for the work to be done.

3. Set the welding machine on the correct current and polarity for the electrode you have chosen. Some welding machines will have a switch for the polarity while others only have welding cables that must sometimes be disconnected from the positive and negative poles of the machine and then reconnected again. Remember that on reverse polarity the electrode holder is connected to the positive lug of the welding machine and the work (ground) clamp is connected to the negative lug of the welding machine. When using straight polarity, the cables are the opposite, the electrode holder is connected to the negative lug of the welding machine and work (ground) clamp is connected to the positive lug of the welding machine. If you are using an AC welding machine, there is no straight or reverse polarity, therefore, the AC welding machine does not have a positive or negative lug. See Chapter 3 of the text for AC welding current information.

4. Connect the work clamp (ground) to the part that is to be welded. In some instances, you may wish to connect the clamp to a metal table on which the part to be welded can make direct contact.

5. Place the range selector switch in the desired position. Check the electrode amperage settings for the desired settings.

6. Place the amperage adjustment control to the desired setting.

7. Turn on the welding machine's power switch.

8. Place the electrode in the electrode holder.

9. Start welding.

10. Readjust the amperage adjustment control if necessary.

SMAW TROUBLESHOOTING	
Welding Problem	**How to Correct**
Arc Blow	1. adjust electrode angle 2. move ground clamp 3. use AC current instead of DC 4. inspect the part to see if it has become magnetized and if so, demagnetize (explain how to do this.)
Brittle Welds	1. use proper pre and post-heat on the metal to be welded 2. use low-hydrogen electrodes 3. make sure parts being welded are not cooled too quickly, as by dipping in water
Cracks in Weld	1. reduce welding speed, use an electrode that produces a more convex bead 2. use low-hydrogen electrodes 3. use pre and post-heat on the weld
Distortion	1. reduce current 2. use chill plates 3. increase welding speed 4. clamp or fix parts being welded 5. weld thick sections first
Incomplete Penetration	1. increase amperage 2. use a larger root opening 3. decrease the electrode diameter 4. reduce the welding speed
Inferior Appearance	1. use correct electrode 2. use electrodes that have not been wet 3. use the correct polarity 4. increase or decrease current 5. manipulate the electrode differently
Porosity	1. use dry electrodes 2. do not weld on wet metal 3. clean paint, grease, oil, etc., from the metal being welded 4. shorten arc length 5. use low-hydrogen electrodes
Slag Inclusions	1. increase current 2. decrease welding speed 3. do not allow the welding pool to get ahead of the arc 4. change polarities 5. if welding multipass welds, always chip and wire brush between passes
Spatter	1. decrease current 2. shorten arc length 3. weld on dry metal with dry electrodes
Undercutting	1. decrease current 2. reduce welding speed 3. shorten arc length 4. change electrode angle

■ PRACTICE 4-1

Name _____ Date _____ Electrode Used _____

Class _____ Instructor _____ Grade _____

OBJECTIVE: After completing this practice, you should be able to safely set up a shielded metal arc welding workstation.

EQUIPMENT AND MATERIALS NEEDED FOR THIS PRACTICE

1. Properly set-up and adjusted arc welding machine.

2. Proper safety protection (welding hood, safety glasses, wire brush, chipping hammer, leather gloves, pliers, long-sleeved shirt, long pants, and leather boots or shoes). Refer to Chapter 2 in the text for more specific safety information.

INSTRUCTIONS

1. Demonstrate the use of proper general work clothing, special protective clothing, eye and ear protection to prevent burns and other possible injuries.

2. Explain what material safety data sheets are and what types of information they contain.

3. Describe the ventilation provided in the area.

4. Point out electrical safety provisions.

5. List proper workstation cleanup procedures.

INSTRUCTOR'S COMMENTS _____

_____

_____

■ PRACTICE 4-2

Name _____   Date _____   Electrode Used _____

Class _____   Instructor _____   Grade _____

OBJECTIVE: After completing this practice, you should be able to make a straight stringer bead in the flat position using an E6010 or E6011 series electrode, an E6012 or E6013 series electrode, and an E7016 or E7018 series electrode.

EQUIPMENT AND MATERIALS NEEDED FOR THIS PRACTICE

1. Properly set-up and adjusted arc welding machine.

2. Proper safety protection (welding hood, safety glasses, wire brush, chipping hammer, leather gloves, pliers, long-sleeved shirt, long pants, and leather boots or shoes). Refer to Chapter 2 in the text for more specific safety information.

3. Arc welding electrodes with a 1/8-in. (3-mm) diameter.

4. One piece of mild steel plate, 6 in. (152 mm) long by 1/4 in. (6 mm) thick.

INSTRUCTIONS

1. Starting at one end of the plate, make a straight weld the full length of the plate. Watch the molten weld pool, not the end of the electrode. As you become more skillful, it is easier to watch the molten weld pool.

2. Repeat the beads with all three (F) groups of electrodes until you have consistently good quality beads that are the same height and width with good tie-in of the weld bead edges to the base metal.

3. Cool, chip, and inspect the bead for defects after completing it.

4. Turn off the welding machine and clean up your work area when you are finished welding.

INSTRUCTOR'S COMMENTS _____

_____

_____

■ PRACTICE 4-3

Name _____ Date _____ Electrode Used _____

Class _____ Instructor _____ Grade _____

OBJECTIVE: After completing this practice, you should be able to make stringer beads in the vertical up position using E6010 or E6011 electrodes, E6012 or E6013 electrodes, and E7016 or E7018 electrodes.

EQUIPMENT AND MATERIALS NEEDED FOR THIS PRACTICE

1. Properly set-up and adjusted arc welding machine.

2. Proper safety protection (welding hood, safety glasses, wire brush, chipping hammer, leather gloves, pliers, long-sleeved shirt, long pants, and leather boots or shoes). Refer to Chapter 2 in the text for more specific safety information.

3. Arc welding electrodes with a 1/8-in. (3-mm) diameter.

4. One piece of mild steel plate, 6 in. (152 mm) long by 1/4 in. (6 mm) thick.

INSTRUCTIONS

1. Start with the plate at a 45° angle. This technique is the same as that used to make a vertical weld. However, a lower level of skill is required at 45°, and it is easier to develop your skill. After the welder masters the 45° angle, the angle is increased by a few degrees successively until a vertical position is reached.

2. Before the molten metal drips down the bead, the back of the molten weld pool will start to bulge out and away from the base metal. When this happens, increase the speed of travel and the weave pattern.

3. Cool, chip, and inspect each completed weld for defects.

4. Repeat the beads as necessary with all three (F) groups of electrodes until consistently good quality beads are obtained in this position.

5. Turn off the welding machine and clean up your work area when you are finished welding.

INSTRUCTOR'S COMMENTS _____

_____

_____

■ PRACTICE 4-4

Name _____   Date _____   Electrode Used _____

Class _____   Instructor _____   Grade _____

OBJECTIVE: After completing this practice, you should be able to weld horizontal stringer beads using E6010 or E6011 electrodes, E6012 or E6013 electrodes, and E7016 or E7018 electrodes.

EQUIPMENT AND MATERIALS NEEDED FOR THIS PRACTICE

1. Properly set-up and adjusted arc welding machine.

2. Proper safety protection (welding hood, safety glasses, wire brush, chipping hammer, leather gloves, pliers, long-sleeved shirt, long pants, and leather boots or shoes). Refer to Chapter 2 in the text for more specific safety information.

3. Arc welding electrodes having a 1/8-in. (3-mm) diameter.

4. One piece of mild steel plate, 6 in. (152 mm) long by 1/4 in. (6 mm) thick.

INSTRUCTIONS

1. Start practicing these welds with the plate at a slight angle.

2. Strike the arc on the plate and build the molten weld pool.

3. The "J" weave pattern is recommended in order to deposit metal on the plate so that it can support the bead.

4. As you acquire more skill, gradually increase the plate angle until it is vertical and the weld is horizontal.

5. Cool, chip, and inspect the weld for uniformity and defects.

6. Repeat the welds with all three (F) groups of electrodes until you can consistently make welds free of defects.

7. Turn off the welding machine and clean up your work area when you are finished welding.

INSTRUCTOR'S COMMENTS _____

_____

_____

■ PRACTICE 4-5

Name _____   Date _____   Electrode Used _____

Class _____   Instructor _____   Grade _____

OBJECTIVE: After completing this practice, you should be able to weld a square butt joint in the flat position using E6010 or E6011 electrodes, E6012 or E6013 electrodes, and E7016 or E7018 electrodes.

EQUIPMENT AND MATERIALS NEEDED FOR THIS PRACTICE

1. Properly set-up and adjusted arc welding machine.

2. Proper safety protection (welding hood, safety glasses, wire brush, chipping hammer, leather gloves, pliers, long-sleeved shirt, long pants, and leather boots or shoes). Refer to Chapter 2 in the text for more specific safety information.

3. Arc welding electrodes with a 1/8-in. (3-mm) diameter.

4. Two or more pieces of mild steel plate, 6 in. (152 mm) long by 1/4 in. (6 mm) thick.

INSTRUCTIONS

1. Hold the plates together tightly.

2. Tack weld the plates together.

3. Chip the tacks before you start to weld.

4. The zigzag weave pattern works well on this joint.

5. Strike the arc and establish a molten pool directly in the joint.

6. Continue the weld along the joint.

7. Cool, chip, and inspect the weld for uniformity and defects.

8. Repeat the welds with all three (F) groups of electrodes until you can consistently make welds free of defects.

9. Turn off the welding machine and clean up your work area when you are finished welding.

INSTRUCTOR'S COMMENTS _____

_____

_____

■ PRACTICE 4-6

Name _____   Date _____   Electrode Used _____

Class _____   Instructor _____   Grade _____

OBJECTIVE: After completing this practice, you should be able to weld a vertical (3G) up-welded square butt joint using E6010 or E6011 electrodes, E6012 or E6013 electrodes, and E7016 or E7018 electrodes.

EQUIPMENT AND MATERIALS NEEDED FOR THIS PRACTICE

1. Properly set-up and adjusted arc welding machine.

2. Proper safety protection (welding hood, safety glasses, wire brush, chipping hammer, leather gloves, pliers, long-sleeved shirt, long pants, and leather boots or shoes). Refer to Chapter 2 in the text for more specific safety information.

3. Arc welding electrodes with a 1/8-in. (3-mm) diameter.

4. Two or more pieces of mild steel plate, 6 in. (152 mm) long by 1/4 in. (6 mm) thick.

INSTRUCTIONS

1. Hold the plates together tightly.

2. Tack weld the plates together.

3. Start with the plates at a 45° angle, then increase the angle as skill develops.

4. The "C," "J," or square weave pattern works well for this joint.

5. After completing the weld, cool, chip, and inspect the weld for uniformity and defects.

6. Repeat the welds using all three (F) groups of electrodes until you can consistently make welds free of defects.

7. Turn off the welding machine and clean up your work area when you are finished welding.

INSTRUCTOR'S COMMENTS _____

_____

_____

■ PRACTICE 4-7

Name _____ Date _____ Electrode Used _____

Class _____ Instructor _____ Grade _____

OBJECTIVE: After completing this practice, you will be able to weld a horizontal square butt joint in the 2G position using E6010 or E6011 electrodes, E6012 or E6013 electrodes, and E7016 or E7018 electrodes.

EQUIPMENT AND MATERIALS NEEDED FOR THIS PRACTICE

1. Properly set-up and adjusted arc welding machine.

2. Proper safety protection (welding hood, safety glasses, wire brush, chipping hammer, leather gloves, pliers, long-sleeved shirt, long pants, and leather boots or shoes). Refer to Chapter 2 in the text for more specific safety information.

3. Arc welding electrodes with a 1/8-in. (3-mm) diameter.

4. Two or more pieces of mild steel plate, 6 in. (152 mm) long by 1/4 in. (6 mm) thick.

INSTRUCTIONS

1. After the plates are tack welded together, place them on the welding table so that the weld bead will be in the horizontal position.

2. Start practicing these welds at a slight angle.

3. Strike the arc on the bottom plate and build the molten weld pool until it bridges the gap.

4. The "J" weave pattern works well on this joint.

5. When the 6-in. (152-mm) long weld is completed, cool, chip, and inspect it for uniformity and soundness.

6. Repeat the welds as needed for all these groups of electrodes until you can consistently make welds free of defects.

7. Turn off the welding machine and clean up your work area when you are finished welding.

INSTRUCTOR'S COMMENTS _____

_____

_____

■ PRACTICE 4-8

Name _____ Date _____ Electrode Used _____

Class _____ Instructor _____ Grade _____

OBJECTIVE: After completing this practice, you should be able to weld an edge joint in the flat position using E6010 or E6011 electrodes, E6012 or E6013 electrodes, and E7016 or E7018 electrodes.

EQUIPMENT AND MATERIALS NEEDED FOR THIS PRACTICE

1. Properly set-up and adjusted arc welding machine.

2. Proper safety protection (welding hood, safety glasses, wire brush, chipping hammer, leather gloves, pliers, long-sleeved shirt, long pants, and leather boots or shoes). Refer to Chapter 2 in the text for more specific safety information.

3. Arc welding electrodes with a 1/8-in. (3-mm) diameter.

4. Two or more pieces of mild steel plate, 6 in. (152 mm) long by 1/4 in. (6 mm) thick.

INSTRUCTIONS

1. Clamp the plates flat together and tack on each end.

2. Start the arc and weld the full length of the plate.

3. Make the weld bead as wide as the width of the edge joint.

4. Cool, chip, and inspect the weld for uniformity and defects.

5. Repeat the welds as necessary with all three (F) groups of electrodes until you can consistently make welds free of defects.

6. Turn off the welding machine and clean up your work area when you are finished welding.

INSTRUCTOR'S COMMENTS _____

_____

_____

■ PRACTICE 4-9

Name _____ Date _____ Electrode Used _____

Class _____ Instructor _____ Grade _____

OBJECTIVE: After completing this practice you should be able to weld an edge joint in the vertical down position using E6010 or E6011 electrodes, E6012 or E6013 electrodes, and E7016 or E7018 electrodes.

EQUIPMENT AND MATERIALS NEEDED FOR THIS PRACTICE

1. Properly set-up and adjusted arc welding machine.

2. Proper safety protection (welding hood, safety glasses, wire brush, chipping hammer, leather gloves, pliers, long-sleeved shirt, long pants and leather boots or shoes). Refer to Chapter 2 in the text for more specific safety information.

3. Arc welding electrodes with a 1/8-inch (3-mm) diameter.

4. Two or more pieces of mild steel plate, 6 in. (152 mm) long by 1/4 in. (6 mm) thick.

INSTRUCTIONS

1. Clamp the pieces flat together and tack on each end.

2. Start with the plates at a 45° angle and progressively increase this angle as skill develops until a vertical position is reached.

3. Start at the top and weld downward. Make the weld bead as wide as the width of the edge joint.

4. Cool, chip, and inspect the weld for uniformity and defects.

5. Repeat the welds as necessary with all three (F) groups of electrodes until you can consistently make welds free of defects.

6. Turn off the welding machine and clean up your work area when you are finished welding.

INSTRUCTOR'S COMMENTS _____

_____

_____

■ PRACTICE 4-10

Name _____   Date _____   Electrode Used _____

Class _____   Instructor _____   Grade _____

OBJECTIVE: After completing this practice you should be able to weld an edge joint in the vertical up position using E6010 or E6011 electrodes, E6012 or E6013 electrodes, and E7016 or E7018 electrodes.

EQUIPMENT AND MATERIALS NEEDED FOR THIS PRACTICE

1. Properly set-up and adjusted arc welding machine.

2. Proper safety protection (welding hood, safety glasses, wire brush, chipping hammer, leather gloves, pliers, long-sleeved shirt, long pants, and leather boots or shoes). Refer to Chapter 2 in the text for more specific safety information.

3. Arc welding electrodes with a 1/8-in. (3-mm) diameter.

4. Two or more pieces of mild steel plate, 6 in. (152 mm) long by 1/4 in. (6 mm) thick.

INSTRUCTIONS

1. Clamp the pieces flat together and tack on each end.

2. Start with the plates at a 45° angle and progressively increase this angle as skill develops until a vertical position is reached.

3. Start at the bottom and weld upward. Make the weld bead as wide as the width of the edge joint.

4. Cool, chip, and inspect the weld for uniformity and defects.

5. Repeat the welds as necessary with all three (F) groups of electrodes until you can consistently make welds free of defects.

6. Turn off the welding machine and clean up your work area when you are finished welding.

INSTRUCTOR'S COMMENTS _____

_____

_____

■ PRACTICE 4-11

Name _____ Date _____ Electrode Used _____

Class _____ Instructor _____ Grade _____

OBJECTIVE: After completing this practice you should be able to weld an edge joint in the horizontal position using E6010 or E6011 electrodes, E6012 or E6013 electrodes, and E7016 or E7018 electrodes.

EQUIPMENT AND MATERIALS NEEDED FOR THIS PRACTICE

1. Properly set-up and adjusted arc welding machine.

2. Proper safety protection (welding hood, safety glasses, wire brush, chipping hammer, leather gloves, pliers, long-sleeved shirt, long pants, and leather boots or shoes). Refer to Chapter 2 in the text for more specific safety information.

3. Arc welding electrodes with a 1/8-in. (3-mm) diameter.

4. Two or more pieces of mild steel plate, 6 in. (152 mm) long by 1/4 in. (6 mm) thick.

INSTRUCTIONS

1. Clamp the pieces flat together and tack on each end.

2. Start with the plates reclined at a slight angle and progressively increase this angle as skill develops until the plates are vertical and the weld bead is horizontal.

3. The "J" weave or stepped pattern work well for this joint.

4. Angle the electrode up and back toward the weld. This will cause more metal to be deposited along the top edge of the bead.

5. Cool, chip, and inspect the weld for uniformity and defects.

6. Repeat the welds as necessary with all three (F) groups of electrodes until you can consistently make welds free of defects.

7. Turn off the welding machine and clean up your work area when you are finished welding.

INSTRUCTOR'S COMMENTS _____

_____

_____

■ PRACTICE 4-12

Name _____   Date _____   Electrode Used _____

Class _____   Instructor _____   Grade _____

OBJECTIVE: After completing this practice you should be able to weld an edge joint in the overhead position using E6010 or E6011 electrodes, E6012 or E6013 electrodes, and E7016 or E7018 electrodes.

EQUIPMENT AND MATERIALS NEEDED FOR THIS PRACTICE

1. Properly set-up and adjusted arc welding machine.

2. Proper safety protection (welding hood, safety glasses, wire brush, chipping hammer, leather gloves, pliers, long-sleeved shirt, long pants, and leather boots or shoes). Refer to Chapter 2 in the text for more specific safety information.

3. Arc welding electrodes with a 1/8-in. (3-mm) diameter.

4. Two or more pieces of mild steel plate, 6 in. (152 mm) long by 1/4 in. (6 mm) thick.

INSTRUCTIONS

1. Clamp the pieces flat together, tack on each end and position the parts in the overhead position.

2. Strike the arc and keep the electrode in a slightly trailing angle. Keep a very short arc length.

3. Use the stepped pattern and move the electrode forward slightly when the molten weld pool grows to the correct size.

4. When the weld pool cools and begins to shrink, move the arc back near the center of the weld.

5. Hold the arc in this new position until the weld pool grows to the correct size.

6. Step the electrode forward again and keep repeating this pattern as the weld bead progresses along the entire length of the joint.

7. Cool, chip, and inspect the weld for uniformity and defects.

8. Repeat the welds as necessary with all three (F) groups of electrodes until you can consistently make welds free of defects.

9. Turn off the welding machine and clean up your work area when you are finished welding.

INSTRUCTOR'S COMMENTS _____

_____

_____

■ PRACTICE 4-13

Name _____ Date _____ Electrode Used _____

Class _____ Instructor _____ Grade _____

OBJECTIVE: After completing this practice you should be able to weld an outside corner joint in the flat position using E6010 or E6011 electrodes, E6012 or E6013 electrodes, and E7016 or E7018 electrodes.

EQUIPMENT AND MATERIALS NEEDED FOR THIS PRACTICE

1. Properly set-up and adjusted arc welding machine.

2. Proper safety protection (welding hood, safety glasses, wire brush, chipping hammer, leather gloves, pliers, long-sleeved shirt, long pants, and leather boots or shoes). Refer to Chapter 2 in the text for more specific safety information.

3. Arc welding electrodes with a 1/8-in. (3-mm) diameter.

4. Two or more pieces of mild steel plate, 6 in. (152 mm) long by 1/4 in. (6 mm) thick.

INSTRUCTIONS

1. Clamp the pieces in a 90° "V" configuration and tack on each end.

2. Working in the flat position, start at one end and make a straight bead the full length of the outside corner of the plates.

3. Remember to watch the molten weld pool as it is being made and not the arc itself.

4. Cool, chip, and inspect the weld for uniformity and defects.

5. Repeat the welds as necessary with all three (F) groups of electrodes until you can consistently make welds free of defects.

6. Turn off the welding machine and clean up your work area when you are finished welding.

INSTRUCTOR'S COMMENTS _____

_____

_____

■ PRACTICE 4-14

Name _____   Date _____   Electrode Used _____

Class _____   Instructor _____   Grade _____

OBJECTIVE: After completing this practice you should be able to weld an outside corner joint in the vertical down position using E6010 or E6011 electrodes, E6012 or E6013 electrodes, and E7016 or E7018 electrodes.

EQUIPMENT AND MATERIALS NEEDED FOR THIS PRACTICE

1. Properly set-up and adjusted arc welding machine.

2. Proper safety protection (welding hood, safety glasses, wire brush, chipping hammer, leather gloves, pliers, long-sleeved shirt, long pants, and leather boots or shoes). Refer to Chapter 2 in the text for more specific safety information.

3. Arc welding electrodes with a 1/8-in. (3-mm) diameter.

4. Two or more pieces of mild steel plate, 6 in. (152 mm) long by 1/4 in. (6 mm) thick.

INSTRUCTIONS

1. Clamp the pieces in a 90° "V" configuration and tack on each end.

2. Begin by positioning the plates at a 45° angle and progressively increase this angle as skill develops until the vertical position is reached.

3. Start at the top and make a straight bead downward along the full length of the outside corner of the plates.

4. Remember to watch the molten weld pool as it is being made and not the arc itself.

5. Cool, chip, and inspect the weld for uniformity and defects.

6. Repeat the welds as necessary with all three (F) groups of electrodes until you can consistently make welds free of defects.

7. Turn off the welding machine and clean up your work area when you are finished welding.

INSTRUCTOR'S COMMENTS _____

_____

_____

■ PRACTICE 4-15

Name _____ Date _____ Electrode Used _____

Class _____ Instructor _____ Grade _____

OBJECTIVE: After completing this practice you should be able to weld an outside corner joint in the vertical up position using E6010 or E6011 electrodes, E6012 or E6013 electrodes, and E7016 or E7018 electrodes.

EQUIPMENT AND MATERIALS NEEDED FOR THIS PRACTICE

1. Properly set-up and adjusted arc welding machine.

2. Proper safety protection (welding hood, safety glasses, wire brush, chipping hammer, leather gloves, pliers, long-sleeved shirt, long pants, and leather boots or shoes). Refer to Chapter 2 in the text for more specific safety information.

3. Arc welding electrodes with a 1/8-in. (3-mm) diameter.

4. Two or more pieces of mild steel plate, 6 in. (152 mm) long by 1/4 in. (6 mm) thick.

INSTRUCTIONS

1. Clamp the pieces in a 90° "V" configuration and tack on each end.

2. Begin by positioning the plates at a 45° angle and progressively increase this angle as skill develops until the vertical position is reached.

3. Start at the bottom and make a straight bead upward along the full length of the outside corner of the plates.

4. Remember to watch the molten weld pool as it is being made and not the arc itself.

5. Cool, chip, and inspect the weld for uniformity and defects.

6. Repeat the welds as necessary with all three (F) groups of electrodes until you can consistently make welds free of defects.

7. Turn off the welding machine and clean up your work area when you are finished welding.

INSTRUCTOR'S COMMENTS _____

_____

_____

■ PRACTICE 4-16

Name _____ Date _____ Electrode Used _____

Class _____ Instructor _____ Grade _____

OBJECTIVE: After completing this practice you should be able to weld an outside corner joint in the horizontal position using E6010 or E6011 electrodes, E6012 or E6013 electrodes, and E7016 or E7018 electrodes.

EQUIPMENT AND MATERIALS NEEDED FOR THIS PRACTICE

1. Properly set-up and adjusted arc welding machine.

2. Proper safety protection (welding hood, safety glasses, wire brush, chipping hammer, leather gloves, pliers, long-sleeved shirt, long pants, and leather boots or shoes). Refer to Chapter 2 in the text for more specific safety information.

3. Arc welding electrodes with a 1/8-in. (3-mm) diameter.

4. Two or more pieces of mild steel plate, 6 in. (152 mm) long by 1/4 in. (6 mm) thick.

INSTRUCTIONS

1. Clamp the pieces in a 90° "V" configuration and tack on each end.

2. Position the plates so that the joint is in a horizontal position.

3. Begin by reclining the plates at a slight angle and progressively increase this angle as skill develops until a full horizontal position is reached.

4. Start at one end and make a straight bead horizontally along the full length of the outside corner of the plates.

5. The "J" weave or stepped pattern work well for this position.

6. Angling the electrode up and back toward the weld will cause more metal to be deposited along the top edge of the weld.

7. Cool, chip, and inspect the weld for uniformity and defects.

8. Repeat the welds as necessary with all three (F) groups of electrodes until you can consistently make welds free of defects.

9. Turn off the welding machine and clean up your work area when you are finished welding.

INSTRUCTOR'S COMMENTS _____

_____

_____

■ PRACTICE 4-17

Name _____   Date _____   Electrode Used _____

Class _____   Instructor _____   Grade _____

OBJECTIVE: After completing this practice you should be able to weld an outside corner joint in the overhead position using E6010 or E6011 electrodes, E6012 or E6013 electrodes, and E7016 or E7018 electrodes.

EQUIPMENT AND MATERIALS NEEDED FOR THIS PRACTICE

1. Properly set-up and adjusted arc welding machine.

2. Proper safety protection (welding hood, safety glasses, wire brush, chipping hammer, leather gloves, pliers, long-sleeved shirt, long pants, and leather boots or shoes). Refer to Chapter 2 in the text for more specific safety information.

3. Arc welding electrodes with a 1/8-in. (3-mm) diameter.

4. Two or more pieces of mild steel plate, 6 in. (152 mm) long by 1/4 in. (6 mm) thick.

INSTRUCTIONS

1. Clamp the pieces in a 90° "V" configuration and tack on each end. Position the plates so that the joint is in the overhead position.

2. With the electrode pointed slightly into the joint, strike the arc in the joint. Keep a very short arc length.

3. Use the stepped pattern and move the electrode forward slightly when the molten weld pool grows to the correct size.

4. When the molten weld pool cools and begins to shrink, move the arc back near the center of the weld and hold it in this new position until the weld pool again grows to the correct size.

5. Step the electrode forward again and keep repeating this pattern until the weld progresses along the entire length of the joint.

6. Cool, chip, and inspect the weld for uniformity and defects.

7. Repeat the welds as necessary with all three (F) groups of electrodes until you can consistently make welds free of defects.

8. Turn off the welding machine and clean up your work area when you are finished welding.

INSTRUCTOR'S COMMENTS _____

_____

_____

■ PRACTICE 4-18

Name _____   Date _____   Electrode Used _____

Class _____   Instructor _____   Grade _____

OBJECTIVE: After completing this practice you should be able to weld a lap joint in the flat (1F) position using E6010 or E6011 electrodes, E6012 or E6013 electrodes, and E7016 or E7018 electrodes.

EQUIPMENT AND MATERIALS NEEDED FOR THIS PRACTICE

1. Properly set-up and adjusted arc welding machine.

2. Proper safety protection (welding hood, safety glasses, wire brush, chipping hammer, leather gloves, pliers, long-sleeved shirt, long pants, and leather boots or shoes). Refer to Chapter 2 in the text for more specific safety information.

3. Arc welding electrodes with a 1/8-in. (3-mm) diameter.

4. Two or more pieces of mild steel plate, 6 in. (152 mm) long by 1/4 in. (6 mm) thick.

INSTRUCTIONS

1. Clamp the pieces tightly together with no more than a 1/4 in. (6 mm) overlap and tack on each end. Position the plates so that the joint is in the flat position.

2. A small tack weld may be added at the center to prevent distortion during welding.

3. The "J," "C," or zigzag pattern work well for this joint.

4. Strike the arc at one end of the joint and establish a weld pool directly in the joint.

5. Move the electrode out on the bottom plate and then back to move the pool to the top edge of the top plate. Proceed in this manner along the entire length of the plate.

6. Follow the surface of the plates and not the trailing edge of the weld bead so that slag will not collect in the root.

7. Cool, chip, and inspect the weld for uniformity and defects.

8. Repeat the welds as necessary with all three (F) groups of electrodes until you can consistently make welds free of defects.

9. Turn off the welding machine and clean up your work area when you are finished welding.

INSTRUCTOR'S COMMENTS _____

_____

_____

■ PRACTICE 4-19

Name _____ Date _____ Electrode Used _____

Class _____ Instructor _____ Grade _____

OBJECTIVE: After completing this practice you should be able to weld a lap joint in the horizontal (2F) position using E6010 or E6011 electrodes, E6012 or E6013 electrodes, and E7016 or E7018 electrodes.

EQUIPMENT AND MATERIALS NEEDED FOR THIS PRACTICE

1. Properly set-up and adjusted arc welding machine.

2. Proper safety protection (welding hood, safety glasses, wire brush, chipping hammer, leather gloves, pliers, long-sleeved shirt, long pants, and leather boots or shoes). Refer to Chapter 2 in the text for more specific safety information.

3. Arc welding electrodes with a 1/8-in. (3-mm) diameter.

4. Two or more pieces of mild steel plate, 6 in. (152 mm) long by 1/4 in. (6 mm) thick.

INSTRUCTIONS

1. Clamp the pieces tightly together with no more than a 1/4 in. (6 mm) overlap and tack on each end. Position the plates so that the joint is in the horizontal position.

2. A small tack weld may be added at the center to prevent distortion during welding.

3. The "J," "C," or zigzag pattern work well for this joint.

4. Strike the arc at one end of the joint and establish a weld pool directly in the joint.

5. Make sure that the fillet is placed so that it is equally divided between both plates and continue along the entire length of the plate.

6. Follow the surface of the plates and not the trailing edge of the weld bead so that slag will not collect in the root.

7. Cool, chip, and inspect the weld for uniformity and defects.

8. Repeat the welds as necessary with all three (F) groups of electrodes until you can consistently make welds free of defects.

9. Turn off the welding machine and clean up your work area when you are finished welding.

INSTRUCTOR'S COMMENTS _____

_____

_____

■ PRACTICE 4-20

Name _____ Date _____ Electrode Used _____

Class _____ Instructor _____ Grade _____

OBJECTIVE: After completing this practice you should be able to weld a lap joint in the vertical up (3F) position using E6010 or E6011 electrodes, E6012 or E6013 electrodes, and E7016 or E7018 electrodes.

EQUIPMENT AND MATERIALS NEEDED FOR THIS PRACTICE

1. Properly set-up and adjusted arc welding machine.

2. Proper safety protection (welding hood, safety glasses, wire brush, chipping hammer, leather gloves, pliers, long-sleeved shirt, long pants, and leather boots or shoes). Refer to Chapter 2 in the text for more specific safety information.

3. Arc welding electrodes with a 1/8-in. (3-mm) diameter.

4. Two or more pieces of mild steel plate, 6 in. (152 mm) long by 1/4 in. (6 mm) thick.

INSTRUCTIONS

1. Clamp the pieces tightly together with no more than a 1/4 in. (6 mm) overlap and tack on each end.

2. Start practicing this joint with the plates inclined at a 45° angle and progressively increase this angle as skill develops until a full vertical position is reached.

3. Strike the arc at the bottom end of the plates and establish a weld pool in the root of the joint. The "J" or "T" patterns work well for this joint.

4. Use the "T" pattern to step ahead of the molten weld pool, allowing it to cool slightly. Do not deposit weld metal ahead of the molten weld pool.

5. As the weld pool begins to cool and shrink, move the electrode back down into the molten weld pool and quickly move the electrode from side to side to fill up the joint.

6. Cool, chip, and inspect the weld for uniformity and defects.

7. Repeat the welds as necessary with all three (F) groups of electrodes until you can consistently make welds free of defects.

8. Turn off the welding machine and clean up your work area when you are finished welding.

INSTRUCTOR'S COMMENTS _____

_____

_____

■ PRACTICE 4-21

Name _____   Date _____   Electrode Used _____

Class _____   Instructor _____   Grade _____

OBJECTIVE: After completing this practice you should be able to weld a lap joint in the overhead (4F) position using E6010 or E6011 electrodes, E6012 or E6013 electrodes, and E7016 or E7018 electrodes.

EQUIPMENT AND MATERIALS NEEDED FOR THIS PRACTICE

1. Properly set-up and adjusted arc welding machine.

2. Proper safety protection (welding hood, safety glasses, wire brush, chipping hammer, leather gloves, pliers, long-sleeved shirt, long pants, and leather boots or shoes). Refer to Chapter 2 in the text for more specific safety information.

3. Arc welding electrodes with a 1/8-in. (3-mm) diameter.

4. Two or more pieces of mild steel plate, 6 in. (152 mm) long by 1/4 in. (6 mm) thick.

INSTRUCTIONS

1. Clamp the pieces tightly together with no more than a 1/4 in. (6 mm) overlap and tack on each end. Position the plates in the overhead position.

2. With the electrode pointed slightly into the joint, strike the arc in the inside corner (root) of the lap joint. Keep a very short arc length.

3. Use the stepped pattern and move the electrode forward slightly when the molten weld pool has grown to the correct size.

4. When the molten weld pool cools and begins to shrink, move the electrode back near the center of the weld and hold it in this new position until the weld pool grows again to the correct size.

5. Step the electrode forward again keep repeating this pattern until the weld progresses along the entire length of the joint.

6. Cool, chip, and inspect the weld for uniformity and defects.

7. Repeat the welds as necessary with all three (F) groups of electrodes until you can consistently make welds free of defects.

8. Turn off the welding machine and clean up your work area when you are finished welding.

INSTRUCTOR'S COMMENTS _____

_____

_____

■ PRACTICE 4-22

Name _____ Date _____ Electrode Used _____

Class _____ Instructor _____ Grade _____

OBJECTIVE: After completing this practice you should be able to weld a tee joint in the flat (1F) position using E6010 or E6011 electrodes, E6012 or E6013 electrodes, and E7016 or E7018 electrodes.

EQUIPMENT AND MATERIALS NEEDED FOR THIS PRACTICE

1. Properly set-up and adjusted arc welding machine.

2. Proper safety protection (welding hood, safety glasses, wire brush, chipping hammer, leather gloves, pliers, long-sleeved shirt, long pants, and leather boots or shoes). Refer to Chapter 2 in the text for more specific safety information.

3. Arc welding electrodes with a 1/8-in. (3-mm) diameter.

4. Two or more pieces of mild steel plate, 6 in. (152 mm) long by 1/4 in. (6 mm) thick.

INSTRUCTIONS

1. Clamp the pieces together in the 90° "tee" configuration and tack on each end. Position the plates so that the joint is in the true flat position with each plate at 45° to the worktable.

2. Start at one end and establish a weld pool on both plates. Allow the weld pools to flow together before starting to move the bead along the plates.

3. Any of the weave patterns will work well on this joint.

4. To prevent slag inclusions, use a slightly higher than normal current setting.

5. Progress along the joint until the entire joint is welded being careful that there are no slag inclusions.

6. Cool, chip, and inspect the weld for uniformity and defects.

7. Repeat the welds as necessary with all three (F) groups of electrodes until you can consistently make welds free of defects.

8. Turn off the welding machine and clean up your work area when you are finished welding.

INSTRUCTOR'S COMMENTS _____

_____

_____

■ PRACTICE 4-23

Name _____   Date _____   Electrode Used _____

Class _____   Instructor _____   Grade _____

OBJECTIVE: After completing this practice you should be able to weld a tee joint in the horizontal (2F) position using E6010 or E6011 electrodes, E6012 or E6013 electrodes, and E7016 or E7018 electrodes.

EQUIPMENT AND MATERIALS NEEDED FOR THIS PRACTICE

1. Properly set-up and adjusted arc welding machine.

2. Proper safety protection (welding hood, safety glasses, wire brush, chipping hammer, leather gloves, pliers, long-sleeved shirt, long pants, and leather boots or shoes). Refer to Chapter 2 in the text for more specific safety information.

3. Arc welding electrodes with a 1/8-in. (3-mm) diameter.

4. Two or more pieces of mild steel plate, 6 in. (152 mm) long by 1/4 in. (6 mm) thick.

INSTRUCTIONS

1. Clamp the pieces together in the 90° "tee" configuration and tack on each end. Position the plates so that one plate is flat on the table and the other plate is vertical. The joint will be horizontal.

2. Start at one end and establish a weld pool on the flat plate. Allow the weld pool to flow onto the vertical as well before starting to move the bead along the plates.

3. The "J" or "C" weave pattern will work well on this joint.

4. Push the arc into the root and slightly up the vertical plate and keep the root of the joint fusing as the weld progresses.

5. Beware of undercutting on the vertical plate. If this occurs, lower the current slightly and direct the arc a little more toward the lower plate.

6. Cool, chip, and inspect the weld for uniformity and defects.

7. Repeat the welds as necessary with all three (F) groups of electrodes until you can consistently make welds free of defects.

8. Turn off the welding machine and clean up your work area when you are finished welding.

INSTRUCTOR'S COMMENTS _____

_____

_____

■ PRACTICE 4-24

Name _____ Date _____ Electrode Used _____

Class _____ Instructor _____ Grade _____

OBJECTIVE: After completing this practice you should be able to weld a tee joint in the vertical (3F) position using E6010 or E6011 electrodes, E6012 or E6013 electrodes, and E7016 or E7018 electrodes.

EQUIPMENT AND MATERIALS NEEDED FOR THIS PRACTICE

1. Properly set-up and adjusted arc welding machine.

2. Proper safety protection (welding hood, safety glasses, wire brush, chipping hammer, leather gloves, pliers, long-sleeved shirt, long pants, and leather boots or shoes). Refer to Chapter 2 in the text for more specific safety information.

3. Arc welding electrodes with a 1/8-in. (3-mm) diameter.

4. Two or more pieces of mild steel plate, 6 in. (152 mm) long by 1/4 in. (6 mm) thick.

INSTRUCTIONS

1. Clamp the pieces together in the 90° "tee" configuration and tack on each end.

2. Start practicing this joint with the plates inclined at a 45° angle and progressively increase this angle as skill develops until a full vertical position is reached.

3. The square, "J," "C," or "T" weave patterns will work well on this joint. The "T" or stepped pattern will give the best root penetration.

4. For this weld, undercutting may be a problem on both plates. Control this by holding the arc on a side only long enough for the filler metal to run down and fill it.

5. Proceed from the bottom to the top until the entire length of the joint is welded.

6. Cool, chip, and inspect the weld for uniformity and defects.

7. Repeat the welds as necessary with all three (F) groups of electrodes until you can consistently make welds free of defects.

8. Turn off the welding machine and clean up your work area when you are finished welding.

INSTRUCTOR'S COMMENTS _____

_____

_____

■ PRACTICE 4-25

Name _____ Date _____ Electrode Used _____

Class _____ Instructor _____ Grade _____

OBJECTIVE: After completing this practice you should be able to weld a tee joint in the overhead (4F) position using E6010 or E6011 electrodes, E6012 or E6013 electrodes, and E7016 or E7018 electrodes.

EQUIPMENT AND MATERIALS NEEDED FOR THIS PRACTICE

1. Properly set-up and adjusted arc welding machine.

2. Proper safety protection (welding hood, safety glasses, wire brush, chipping hammer, leather gloves, pliers, long-sleeved shirt, long pants, and leather boots or shoes). Refer to Chapter 2 in the text for more specific safety information.

3. Arc welding electrodes with a 1/8-in. (3-mm) diameter.

4. Two or more pieces of mild steel plate, 6 in. (152 mm) long by 1/4 in. (6 mm) thick.

INSTRUCTIONS

1. Clamp the pieces together in the 90° "tee" configuration and tack on each end. Position the plates in the overhead position.

2. Start the arc deep in the root of the joint. Keep a very short arc length.

3. The stepped pattern will work well for this joint and will give good root penetration.

4. For this weld, undercutting may be a problem on both plates. Control this by holding the arc on a side only long enough for the filler metal to run in and fill it.

5. Proceed from one end of the plates to the other until the entire length of the joint is welded.

6. Cool, chip, and inspect the weld for uniformity and defects.

7. Repeat the welds as necessary with all three (F) groups of electrodes until you can consistently make welds free of defects.

8. Turn off the welding machine and clean up your work area when you are finished welding.

INSTRUCTOR'S COMMENTS _____

_____

_____

CHAPTER 4: SHIELDED METAL ARC WELDING OF PLATE QUIZ

Name _____ Date _____

Class _____ Instructor _____ Grade _____

Instructions: Carefully read Chapter 4 in the text and answer the following questions.

A. FILL IN THE BLANK

Fill in the blank with the correct word. Answers may be more than one word.

1. SMAW can be used to make _____ on almost any type of metal, any shape, and in any position.

2. When striking the arc, scratch the electrode across the plate (like striking a _____).

3. Restarting a partially used electrode is more difficult than a new electrode and restarting one with the _____ can be very difficult.

4. Breaking the arc back over the weld prevents arc marks in _____ of the weld and results in the arc ending back away from the _____ of the weld bead.

5. Welding with the current set too low results in _____ and _____.

6. Higher amperage settings can also result in an increase in the amount of _____ and it is mostly _____.

7. Using smaller diameter electrodes requires _____ skill than using large diameter electrodes because the molten weld pool is _____ and easier to control.

8. Welders often use the terms heat and amperage inter-changeably when they are speaking about making changes to the _____.

9. The term work angle refers to the relationship between the _____ of the electrode and the _____ of the work.

10. A _____ electrode angle pushes molten metal and slag ahead of the weld.

11. A(n) _____ angle directs all of the electrodes jetting force directly into the joint.

12. A(n) _____ electrode angle pushes the molten metal away from the leading edge of the molten weld pool toward the back where it solidifies.

13. E6010 and E6011 electrodes both of these electrodes have _____-based fluxes.

14. _____-based flux electrodes have a softer arc smoother arc and produce less welding fumes and sparks than cellulose-based electrodes.

15. The tack weld must be _____, _____, and _____ so it does not adversely affect the finished weld.

16. An outside corner joint is made by placing the plates at a 90° angle to each other, with the edges forming a _____.

B. SHORT ANSWER

Write a brief answer in the space provided that will answer the question or complete the statement.

17. Name four factors that can prevent cold lap and slag inclusions.

a. _____

b. _____

c. _____

d. _____

18. List seven characteristics of a weld that can be controlled by the movement or weaving of the welding electrode.

a. _____

b. _____

c. _____

d. _____

e. _____

f. _____

g. _____

C. MATCHING

In the space provided to the left, write the letter from Column B that best answers the question or completes the statement in Column A.

Column A   Column B

_____ 19. What is the rate that the weld metal is added to the weld?   a. root opening

_____ 20. What is a large piece of metal that is used to absorb excessive heat?   b. stringer bead

_____ 21. What is a straight weld bead on the surface of a plate, with little or no side-to-side electrode movement called?   c. deposition rate

d. chill plate

_____ 22. What is the space between two plates to be welded called?

D. ESSAY

Provide complete answers for all of the following questions.

23. How much of the electrode end should be discolored if the current is set correctly?

_____

_____

24. What can happen to an electrode if it is overheated?

_____

_____

25. What three steps can a welder take to prevent a weld from being too hot?

_____

_____

26. Approximately how long is a correct arc length?

_____

_____

27. Which weave pattern works best?

_____

_____

28. How can an arc be started at the exact point where the welder wants it to start?

_____

_____

29. How is deep penetration obtained on a flat butt joint?

_____

_____

30. What effect does a short arc length have on a weld?

_____

_____

31. For what is the figure-8 weave pattern used?

_____

_____

32. Why should welders hold on to or lean against a stable object when they are welding?

_____

_____

33. Why should a welder start practicing vertical welds at a 45° angle?

_____

_____

34. How does a welder prevent slag from being trapped in the root of a flat lap welded joint?

_____

_____

INSTRUCTOR'S COMMENTS _____

_____

_____

Chapter 5

Shielded Metal Arc Welding of Pipe

JOINT PREPARATION FOR PIPE WELDING

It is important that the bevel be at the correct angle, which is 37.5° for vertical up welds and 30° for vertical down welds, and that each end of the mating pipes meet squarely, Figure 5-1(A) and (B).

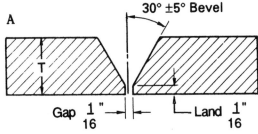

Figure 5-1A Standard joint design for vertical down welding.

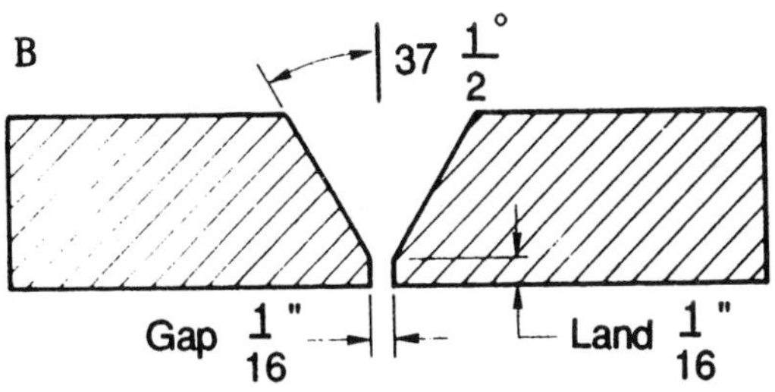

Figure 5-1B Standard joint design for vertical up and horizontal welding.

Weld bead sequencing for vertical down welding is shown in Figure 5-2.

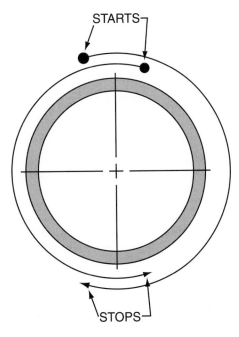

Figure 5-2

Figure 5-2 is for EXX10-type electrodes, designed for welding circumferential pipe joints in the vertical down position. The number of passes required for welding pipe will vary with the diameter of the electrode, the thickness of the pipe, the welding position, the type of current being used, and welding operators. The typical number of passes required to weld pipe using the vertical down technique is shown in Figure 5-3.

Pipe Wall Thickness	Number of Passes
1/4"	3
5/16"	4
3/8"	5
1/2"	7

Figure 5-3. Passes required for vertical down welding on pipe.

Weld bead sequencing for vertical up welding is shown in Figure 5-4.

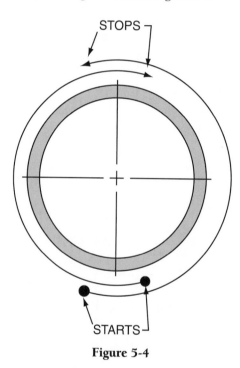

Figure 5-4

When welding carbon steel pipe, E6010 or E7010 electrodes are used for the first pass. All other passes normally are welded using an EXX18 (low-hydrogen) -type electrode. The number of passes required for welding pipe will vary with the diameter of the electrode, the thickness of the pipe, the welding position, the type of current being used, and welding operators. The typical number of passes required to weld pipe using the vertical up technique is shown in Figure 5-5.

Pipe Wall Thickness	Number of Passes
1/4"	2
5/16"	2
3/8"	3
1/2"	3
5/8"	4
3/4"	6
1"	7

Figure 5-5. Passes required for vertical down welding on pipe.

Weld bead sequencing for a horizontal butt weld on pipe is shown in Figure 5-6.

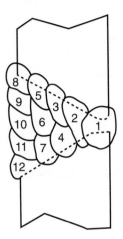

Figure 5-6

■ PRACTICE 5-1

Name _____   Date _____   Electrode Used _____

Class _____   Instructor _____   Grade _____

OBJECTIVE: After completing this practice, you should be able to weld stringer beads on carbon steel pipe in the 1G position using E6010, E6011, and E7018 electrodes.

EQUIPMENT AND MATERIALS NEEDED FOR THIS PRACTICE

1. A properly set-up and adjusted arc welding machine.

2. Proper safety protection (welding hood, safety glasses, wire brush, chipping hammer, leather gloves, long-sleeved shirt, long pants, leather boots or shoes, and a pair of pliers). Refer to Chapter 2 in the text for more specific safety information.

3. E6010 or E6011 and E7018 arc welding electrodes with a 1/8-in. (3-mm) diameter.

4. Schedule 40 mild steel pipe, 3 in. (76 mm) or larger in diameter.

5. A piece of soapstone.

6. A flexible straight edge.

INSTRUCTIONS

1. Be sure the pipe is clean and free of rust, grease, paint, or other contaminants that will affect the weld.

2. Use your flexible straightedge and soapstone and draw straight lines completely around the pipe with a spacing of 1/2 in.

3. Place the pipe horizontally on the welding table in a vee block made of angle iron. The vee block will hold the pipe steady and allow it to be moved easily between each weld bead.

4. Strike an arc on the pipe at the 11 o'clock position using an E6010 or E6011 electrode.

5. Make a stringer bead over the 12 o'clock position, stopping at the 1 o'clock position.

6. Roll the pipe until the end of the weld is at the 11 o'clock position.

7. Clean the weld crater by chipping and wire brushing.

8. Strike the arc again and establish a molten weld pool at the leading edge of the previous weld crater.

9. With the molten weld pool reestablished, move the electrode back on the weld bead just short of the last full ripple. This action will both reestablish good fusion and keep the weld bead size uniform.

10. Now that the new weld bead is tied or welded into the old weld, continue welding to the 1 o'clock position again.

11. Stop welding, roll the pipe, clean the crater, and resume welding.

12. Keep repeating this procedure until the weld is completely around the pipe.

13. Before the last weld is started, clean the beginning end of the first weld so that the end and beginning beads can be tied or welded together smoothly.

14. When you reach the beginning bead, swing your electrode around on both sides of the weld bead. The beginning of a weld bead is always high, narrow, and has little penetration. By swinging the weave pattern (the "C" pattern is best) on both sides of the bead, you can make the bead width uniform. The added heat will give deeper penetration at the starting point.

15. Hold the arc in the crater for a moment until the weld pool is built up.

16. Cool, chip, and inspect the bead for defects.

17. Repeat the beads as needed until they are defect free.

18. Repeat steps 1–17 using E7018 electrodes.

19. Turn off the welding machine and clean up your work area when you are finished welding.

INSTRUCTOR'S COMMENTS _____

_____

_____

■ PRACTICE 5-2

Name _____   Date _____   Electrode Used _____

Class _____   Instructor _____   Grade _____

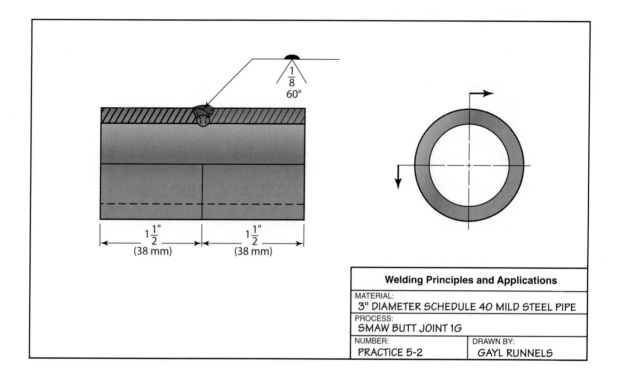

Welding Principles and Applications

MATERIAL: 3" DIAMETER SCHEDULE 40 MILD STEEL PIPE	
PROCESS: SMAW BUTT JOINT 1G	
NUMBER: PRACTICE 5-2	DRAWN BY: GAYL RUNNELS

OBJECTIVE: After completing this practice, you should be able to butt weld carbon steel pipe in the 1G position for low-pressure, light service applications.

EQUIPMENT AND MATERIALS NEEDED FOR THIS PRACTICE

1. A properly set-up and adjusted arc welding machine.

2. Proper safety protection (welding hood, safety glasses, wire brush, chipping hammer, leather gloves, long-sleeved shirt, long pants, leather boots or shoes, and a pair of pliers). Refer to Chapter 2 in the text for more specific safety information.

3. E6010 or E6011 arc welding electrodes with a 1/8-in. (3-mm) diameter.

4. Two or more pieces of Schedule 40 mild steel pipe, 3 in. (76 mm) or larger in diameter.

INSTRUCTIONS

1. Tack weld two pieces of pipe together.

2. Place the pipe horizontally in a vee block on the welding table.

3. Start the root pass at the 11 o'clock position.

4. Using a very short arc and high current setting, weld toward the 1 o'clock position.

5. Stop and roll the pipe, chip the slag, and repeat the weld until you have completed the root pass.

6. Clean the root pass by chipping and wire brushing. The root pass should not be ground this time.

7. Replace the pipe in the vee block on the table so that the hot pass can be done.

8. Turn up the machine amperage enough to remelt the root weld surface for the hot pass.

9. Use a step or whip electrode pattern, moving forward each time the molten weld pool washes out the slag, and returning each time the molten weld pool is nearly all solid.

10. Weld from the 11 o'clock position to the 1 o'clock position before stopping, rolling, and chipping the weld.

11. Repeat this procedure until the hot pass is complete. The filler pass and cover pass may be the same pass on this joint.

12. Turn down the machine amperage.

13. Use a "T," "J," "C," or zigzag pattern for the filler pass.

14. Start the weld at the 10 o'clock position and stop at the 12 o'clock position.

15. Sweep the electrode so that the molten weld pool melts out any slag trapped by the hot pass.

16. Watch the back edge of the bead to see that the molten weld pool is filling the groove completely.

17. Turn, chip, and continue the bead until the weld is complete.

18. Repeat this weld until you can consistently make welds free of defects.

19. Turn off the welding machine and clean up your work area when you are finished welding.

INSTRUCTOR'S COMMENTS _____

_____

_____

■ PRACTICE 5-3

Name _____ Date _____ Electrode Used _____

Class _____ Instructor _____ Grade _____

OBJECTIVE: After completing this practice, you should be able to butt weld carbon steel pipe in the 1G position using E6010 or E6011 electrodes for the root pass with E7018 electrodes for the filler and cover passes.

EQUIPMENT AND MATERIALS NEEDED FOR THIS PRACTICE

1. A properly set-up and adjusted arc welding machine.

2. Proper safety protection (welding hood, safety glasses, wire brush, chipping hammer, leather gloves, long-sleeved shirt, long pants, leather boots or shoes, and a pair of pliers). Refer to Chapter 2 in the text for more specific safety information.

3. E6010 or E6011 and E7018 arc welding electrodes with a 1/8-in. (3-mm) diameter.

4. Two or more pieces of Schedule 40 mild steel pipe, 3 in. (76 mm) or larger in diameter.

INSTRUCTIONS

1. Place the pipe on the arc welding table and tack weld two pieces together.

2. Strike an arc using, an E6010 or E6011 electrode and make a root weld that is as long as possible. If the root opening or gap is close and uniform, a straight, forward movement can be used. For wider gaps, a step or whip pattern must be used.

3. After completing and cleaning the root pass, make a hot pass. The hot pass need only burn out the slag to make the root pass clean. Undercut on the top of the pipe is acceptable. Use the E7018 electrode for these passes.

4. The filler and cover passes should be stringer beads or small weave beads. By keeping the molten weld pool size small, control is easier.

5. Cool, chip, and inspect the completed weld for uniformity and defects.

6. Repeat this weld until you can consistently make welds free of defects.

7. Turn off the welding machine and clean up your work area when you are finished welding.

INSTRUCTOR'S COMMENTS _____

_____

■ PRACTICE 5-4

Name _____   Date _____   Electrode Used _____

Class _____   Instructor _____   Grade _____

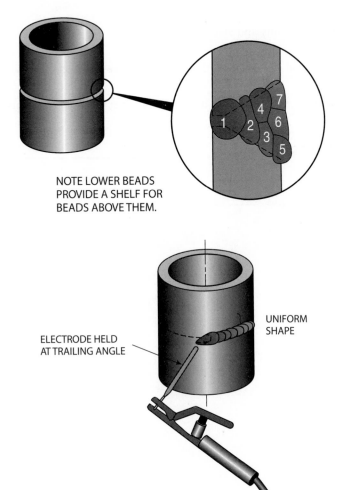

NOTE LOWER BEADS
PROVIDE A SHELF FOR
BEADS ABOVE THEM.

ELECTRODE HELD
AT TRAILING ANGLE

UNIFORM
SHAPE

OBJECTIVE: After completing this practice, you should be able to weld stringer beads on carbon steel pipe in the 2G position using E6010 or E6011 and E7018 electrodes.

EQUIPMENT AND MATERIALS NEEDED FOR THIS PRACTICE

1. A properly set-up and adjusted arc welding machine.

2. Proper safety protection (welding hood, safety glasses, wire brush, chipping hammer, leather gloves, long-sleeved shirt, long pants, leather boots or shoes, and a pair of pliers). Refer to Chapter 2 in the text for more specific safety information.

3. E6010 or E6011 and E7018 arc welding electrodes with a 1/8-in. (3-mm) diameter.

4. Schedule 40 mild steel pipe, 3 in. (76 mm) or larger in diameter.

INSTRUCTIONS

1. Place the pipe vertically on the welding table.

2. Hold the electrode at a 90° angle to the pipe and with a slight upward and trailing angle.

3. Use the "J" weave pattern for this practice.

4. Check the weld for uniformity and visual defects.

5. Repeat the weld until you can consistently make welds free of defects.

6. Turn off the welding machine and clean up your work area when you are finished welding.

INSTRUCTOR'S COMMENTS _____

_____

_____

■ PRACTICE 5-5

Name _____ Date _____ Electrode Used _____

Class _____ Instructor _____ Grade _____

$\frac{1}{8}$ IN. (3 MM) DI. E6010 OR E6011 ROOT PASS
E7018 FILLER AND COVER PASS

$\frac{1}{8}$
60°
G

Welding Principles and Applications

MATERIAL:
3" DIAMETER SCHEDULE 40 MILD STEEL PIPE

PROCESS:
SMAW BUTT JOINT 2G

NUMBER
PRACTICE 5-5

DRAWN BY:
LUCILLE GREENHAW

OBJECTIVE: After completing this practice you should be able to butt weld mild steel pipe in the 2G position using E6010 or E6011 electrodes.

EQUIPMENT AND MATERIALS NEEDED FOR THIS PRACTICE

1. Properly set-up and adjusted arc welding machine.

2. Proper safety protection (welding hood, safety glasses, wire brush, chipping hammer, leather gloves, pliers, long-sleeved shirt, long pants, and leather boots or shoes). Refer to Chapter 2 in the text for more specific safety information.

3. E6010 or E6011 electrodes with a 1/8-in. (3-mm) diameter.

4. Two or more pieces of mild steel pipe, 3 in. (76 mm) or larger in diameter.

INSTRUCTIONS

1. Align the pieces of pipe and tack weld them together in at least four places.

2. Place the pipe vertically on the welding table.

3. Hold the electrode at a 90° angle to the pipe and with a slight trailing angle. The electrode should be held tightly into the joint. If a burnthrough occurs, quickly push the electrode back over the burnthrough while increasing the trailing angle. This action forces the weld metal back into the opening.

4. When the root pass is complete, chip the surface slag and then clean out the trapped slag by grinding or chipping, or use a hot pass technique on the next pass.

5. After the weld is completed, visually inspect it for 100% penetration around 80% of the root bead length.

6. Cool, chip, and inspect the weld for uniformity and visual defects on the cover pass.

7. Repeat the weld until you can consistently make welds free of defects.

8. Turn off the welding machine and clean up your work area when you are finished welding.

INSTRUCTOR'S COMMENTS _____

_____

_____

■ PRACTICE 5-6

Name _____ Date _____ Electrode Used _____

Class _____ Instructor _____ Grade _____

HORIZONTAL

VERTICAL

WELD

OVERHEAD

CHANGING WELDING POSITIONS

$45° \pm 5°$

Welding Principles and Applications

MATERIAL:
3" DIAMETER SCHEDULE 40 MILD STEEL PIPE

PROCESS:
SMAW STRINGER BEADS 6G

NUMBER:
PRACTICE 5-6

DRAWN BY:

OBJECTIVE: After completing this practice you should be able to weld stringer beads on mild steel pipe in the 45° fixed 6G position using E6010 or E6011 and E7018 electrodes.

EQUIPMENT AND MATERIALS NEEDED FOR THIS PRACTICE

1. Properly set-up and adjusted arc welding machine.

2. Proper safety protection (welding hood, safety glasses, wire brush, chipping hammer, leather gloves, pliers, long-sleeved shirt, long pants, and leather boots or shoes). Refer to Chapter 2 in the text for more specific safety information.

3. E6010 or E6011 and E7018 electrodes with a 1/8-in. (3-mm) diameter.

4. One or more pieces of mild steel pipe, 3 in. (76 mm) or larger in diameter.

INSTRUCTIONS

1. Clamp the pipe in a fixed 45° position and at a height that is between waist-high and chest level.

2. Begin by using the E6010 or E6011 electrodes.

3. Using the straight stepped "T" or whipping pattern, start at the bottom of the pipe and establish a molten weld pool in the overhead position. Keep the molten weld pool small and narrow for easier control.

4. Move the molten weld pool up the side of the pipe, keeping the electrode on the downhill side with a trailing or pulling angle.

5. When the weld passes the side, decrease the downhill and trailing angles. This is done so that when the weld reaches the top, the electrode is perpendicular to the top of the pipe.

6. Repeat steps 3 through 5 using the E7018 electrodes.

7. Chip and wire brush the welds and check for appearance and defects.

8. Repeat this practice as needed until you can consistently make welds free of defects.

9. Turn off the welding machine and clean up your work area when you are finished welding.

INSTRUCTOR'S COMMENTS _____

_____

_____

■ PRACTICE 5-7

Name _____ Date _____ Electrode Used _____

Class _____ Instructor _____ Grade _____

HORIZONTAL

VERTICAL

BUTT JOINT

OVERHEAD

WELDING POSITIONS CHANGE AS THE BEAD IS MADE

45° ± 5°

Welding Principles and Applications

MATERIAL:
3" DIAMETER SCHEDULE 40 MILD STEEL PIPE

PROCESS:
SMAW BUTT WELD 6G

NUMBER:
PRACTICE 5-7

DRAWN BY:

OBJECTIVE: After completing this practice you should be able to make a butt weld joint on mild steel pipe in the 45° fixed 6G position using E6010 or E6011 electrodes.

EQUIPMENT AND MATERIALS NEEDED FOR THIS PRACTICE

1. Properly set-up and adjusted arc welding machine.

2. Proper safety protection (welding hood, safety glasses, wire brush, chipping hammer, leather gloves, pliers, long-sleeved shirt, long pants, and leather boots or shoes). Refer to Chapter 2 in the text for more specific safety information.

3. E6010 or E6011 electrodes with a 1/8-in. (3-mm) diameter.

4. Two or more pieces of mild steel pipe, 3 in. (76 mm) or larger in diameter.

INSTRUCTIONS

1. Tack weld two pieces of pipe securely together in four or more places.

2. Clamp the pipe in a fixed 45° position and at a height that is between waist-high and chest level.

3. Use the E6010 or E6011 electrodes.

4. Start at the top of the pipe, make a vertical down root pass that ends just beyond the bottom.

5. Repeat this weld on the other side.

6. Chip and wire brush the slag so that an uphill hot pass can be made.

7. Make the uphill hot pass. It must be kept small and concave so that more slag will not be trapped along the downhill side.

8. Clean the bead, turn down the amperage and complete the joint with stringer beads.

9. Chip and wire brush the welds and check for appearance and defects.

10. Repeat this practice as needed until you can consistently make welds free of defects.

11. Turn off the welding machine and clean up your work area when you are finished welding.

INSTRUCTOR'S COMMENTS _____

_____

_____

CHAPTER 5: SHIELDED METAL ARC WELDING OF PIPE QUIZ

Name _____ Date _____

Class _____ Instructor _____ Grade _____

Instructions: Carefully read Chapter 5 in the text and answer the following questions.

A. SHORT ANSWER

Write a brief answer in the space provided that will answer the question or complete the statement.

1. How does the measurement of tubing differ from that of pipe?

   _____
   _____
   _____

2. The wall thickness of pipe is determined by its _____

3. What is the purpose of the hot pass?

   _____
   _____
   _____

4. List three ways that pipe joints are beveled in preparation for welding.

   a. _____

   b. _____

   c. _____

5. What are consumable inserts used for?

   _____
   _____

B. IDENTIFICATION

In the space provided, identify the items shown in the illustration.

6. Correctly match the following from the illustration.

   _____ a. filler pass

   _____ b. cover pass

   _____ c. root pass

C. ESSAY

Provide complete answers for all of the following questions.

7. What is the difference between pipe and tubing?

_____

_____

8. What is meant by the phrase *schedule of a pipe*?

_____

_____

9. List four advantages of welded pipe joints as compared to threaded pipe joints.

   a. _____

   b. _____

   c. _____

   d. _____

10. When should a root pass be welded uphill and not downhill?

_____

_____

11. Why is the 6G position the most difficult pipe welding position?

_____

_____

D. DEFINITIONS

12. Define the following terms:

   root suck back _____

   icicles _____

   land _____

   root gap _____

   concave root surface _____

   root face _____

INSTRUCTOR'S COMMENTS _____

_____

_____

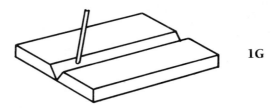

1G

1G Flat position—The welding position used to weld from the upper side of the joint; the face of the weld is approximately horizontal.

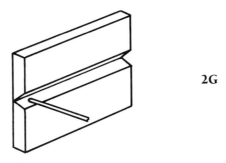

2G

2G Horizontal position—The position of welding in which the axis of the weld lies in an approximately horizontal plane and the face of the weld lies in an approximately vertical plane.

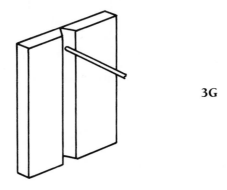

3G

3G Vertical position—The position of welding in which the axis of the weld is approximately vertical.

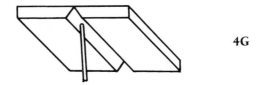

4G

4G Overhead position—The position in which welding is performed from the underside of the joint.

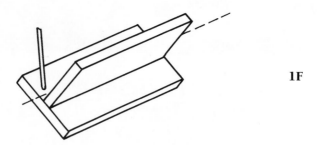

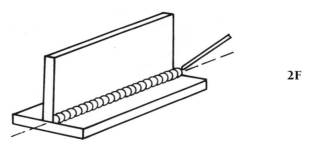

1F

1F Flat position—The welding position used to weld from the upper side of the joint; the face of the weld is approximately horizontal.

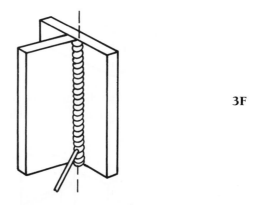

2F

2F Horizontal position—The position in which welding is performed on the upper side of an approximately horizontal surface and against an approximately vertical surface.

3F

3F Vertical position—The position of welding in which the axis of the weld is approximately vertical.

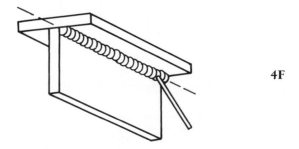

4F

4F Overhead position—The position in which welding is performed from the underside of the joint.

■ PRACTICE 6-1

Name _____   Date _____   Electrode Used _____

Class _____   Instructor _____   Grade _____

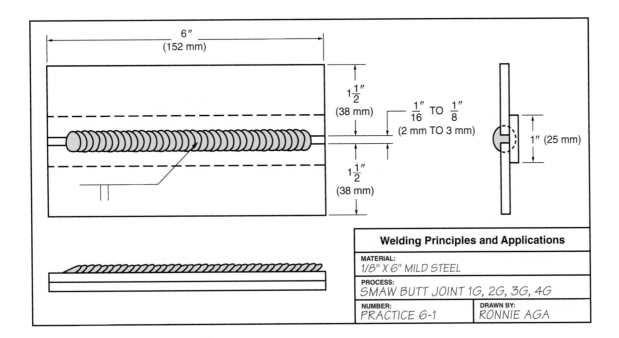

OBJECTIVE: After completing this practice, you should be able to weld a butt joint in all positions with a backing strip using E6010 or E6011 electrodes.

EQUIPMENT AND MATERIALS NEEDED FOR THIS PRACTICE

1. A properly set-up and adjusted welding machine.

2. Proper safety protection (welding hood, safety glasses, wire brush, chipping hammer, leather gloves, pliers, long-sleeved shirt, long pants, and leather boots or shoes). Refer to Chapter 2 in the text for more specific safety information.

3. E6010 or E6011 arc welding electrodes with a 1/8-in. (3-mm) diameter.

4. Three pieces of mild steel plate: (2) 1/8 in. (3 mm) thick by 1 1/2 in. (38 mm) wide by 6 in. (152 mm) long; and one strip of mild steel, (1) 1/8 in. (3 mm) thick by 1 in. (25 mm) wide by 6 in. (152 mm) long.

INSTRUCTIONS

1. Tack weld the plates together with a 1/16-in. (2-mm) to 1/8-in. (3-mm) root opening; see the above figure. Be sure there are no gaps between the backing strip and plates when the pieces are tacked together. If there is a small gap between the backing strip and the plates, it can be removed by placing the assembled test plates on an anvil and striking the tack weld with a hammer. This will close up the gap by compressing the tack welds.

2. Chip and wire brush tack welds.

3. Adjust the welding current by welding on a scrap plate.

4. Begin with plates in the flat position.

5. Use a straight step or "T" pattern for this root weld. Push the electrode into the root opening so that there is complete fusion with the backing strip and bottom edge of the plates. Failure to push the penetration deep into the joint will result in a cold lap at the root.

6. Watch the molten weld pool and keep its size, height, and width as uniform as possible. As the molten weld pool increases in size, move the electrode out of the weld pool. When the weld pool begins to cool, bring the electrode back into the molten weld pool. Use these weld pool indications to determine how far to move the electrode and when to return to the molten weld pool. After completing the weld, cut the plate and inspect the cross section of the weld for complete fusion at the edges.

7. Chip and wire brush the weld and examine to see if you have complete fusion. Repeat the welds as necessary until you can consistently make welds free of defects.

8. Repeat above steps in the 2G, 3G, and 4G positions.

9. Turn off the welding machine and clean up your work area when you are finished welding.

INSTRUCTOR'S COMMENTS _____

_____

_____

■ PRACTICE 6-2

Name _____ Date _____ Electrode Used _____

Class _____ Instructor _____ Grade _____

OBJECTIVE: After completing this practice, you should be able to weld a root pass with an open root using E6010 or E6011 electrodes in the 1G, 2G, 3G, and 4G positions.

EQUIPMENT AND MATERIALS NEEDED FOR THIS PRACTICE

1. A properly set-up and adjusted welding machine.

2. Proper safety protection (welding hood, safety glasses, wire brush, chipping hammer, leather gloves, pliers, long-sleeved shirt, long pants, and leather boots or shoes). Refer to Chapter 2 in the text for more specific safety information.

3. E6010 or E6011 arc welding electrodes with a 1/8-in. (3-mm) diameter.

4. Two pieces of mild steel plate, 1 1/2 in. (38 mm) wide by 6 in. (152 mm) long by 1/8 in. (3 mm) thick.

INSTRUCTIONS

1. Tack weld the plates together with a root opening of 0 in. (0 mm) to 1/16 in. (2 mm).

2. Chip and wire brush all tack welds.

3. Use a short arc length and high amperage to weld the joint.

4. Control penetration by the electrode angle used.

5. If burnthrough occurs, move back and lower the electrode angle.

6. After the weld is completed, it can be visually inspected for uniformity and complete penetration. Repeat this weld until you can consistently make it defect free.

7. Turn off the welding machine and clean up your work area when you are finished welding.

INSTRUCTOR'S COMMENTS _____

_____

_____

■ PRACTICE 6-3

Name _____ Date _____ Electrode Used _____

Class _____ Instructor _____ Grade _____

OBJECTIVE: After completing this practice, you should be able to weld an open root joint using the step technique with E6010 or E6011 electrodes in the 1G, 2G, 3G, and 4G positions.

EQUIPMENT AND MATERIALS NEEDED FOR THIS PRACTICE

1. A properly set-up and adjusted welding machine.

2. Proper safety protection (welding hood, safety glasses, wire brush, chipping hammer, leather gloves, pliers, long-sleeved shirt, long pants, and leather boots or shoes). Refer to Chapter 2 in the text for more specific safety information.

3. E6011 or E6011 arc welding electrodes with a 1/8-in. (3-mm) diameter.

4. Two pieces of mild steel plate 1 1/2 in. (38 mm) wide by 6 in. (152 mm) long by 1/8 in. (3 mm) thick.

INSTRUCTIONS

1. Tack weld the plates together with a root opening of 0 in. (0 mm) to 1/16 in. (2 mm).

2. Chip and wire brush all tack welds.

3. Begin with the plate in the flat position.

4. Starting at one end, make a stringer bead along the entire length using the straight step or "T"-weave pattern. Use a medium amperage setting.

5. As the bead is being made, push the electrode deeply into the root to establish a keyhole.

6. Once the keyhole is established, the electrode is moved out and back in the molten weld pool at a steady, rhythmic rate.

7. After the weld is completed, visually inspect it for uniformity and penetration. Repeat this weld until you can consistently make it defect free.

8. Repeat the above steps in the 2G, 3G, and 4G positions.

9. Turn off the welding machine and clean up your work area when you are finished welding.

INSTRUCTOR'S COMMENTS _____

_____

_____

■ PRACTICE 6-4

AWS SENSE LEVEL I

Name _____   Date _____

Class _____   Instructor _____   Grade _____

WELDING PROCEDURE SPECIFICATION (WPS)

Welding Procedure Specification No: <u>Practice 6-4</u>

TITLE:

Welding <u>SMAW</u> of <u>plate</u> to <u>plate</u>.

SCOPE:

This procedure is applicable for <u>V-groove plate with a backing strip</u> within the range of 1/8 in. (3 mm) through 1-1/2 in. (38 mm).

Welding may be performed in the following positions: <u>3G</u>.

BASE METAL:

The base metal shall conform to <u>Carbon Steel M-1 or P-1, Group 1 or 2</u>.

Backing Material Specification <u>Carbon Steel M-1 or P-1, Group 1, 2, or 3</u>.

FILLER METAL:

The filler metal shall conform to AWS specification no. <u>E7018</u> from AWS specification <u>A5.1</u>. This filler metal falls into F-number <u>F-4</u> and A-number <u>A-1</u>.

SHIELDING GAS:

The shielding gas, or gases, shall conform to the following compositions and purity:

<u>N/A</u>.

JOINT DESIGN AND TOLERANCES:

Joint Details:

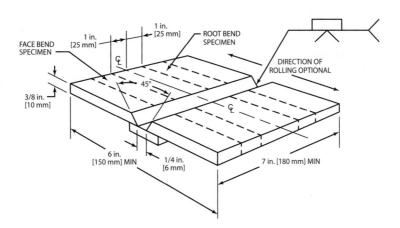

PREPARATION OF BASE METAL:

The bevel is to be flame or plasma cut on the edge of the plate before the parts are assembled. The beveled surface must be smooth and free of notches. Any roughness or notches that are deeper than 1/64 in. (0.4 mm) must be ground smooth.

All hydrocarbons and other contaminants, such as cutting fluids, grease, oil, and primers, must be cleaned off of all parts and filler metals before welding. This cleaning can be done with any suitable solvents or detergents. The backing strip, groove face, and inside and outside plate surface within 1 in. (25 mm) of the joint must be mechanically cleaned of slag, rust, and mill scale. Cleaning must be done with a wire brush or grinder down to bright metal.

ELECTRICAL CHARACTERISTICS:

The current shall be <u>direct-current electrode positive (DCEP)</u>. The base metal shall be on the <u>negative</u> side of the line.

WELDS	FILLER METAL DIA.	CURRENT	AMPERAGE RANGE
Tack	3/32 in. (2.4 mm)	DCEP	70 to 115
Root	1/8 in. (3.2 mm)	DCEP	115 to 165
Filler	5/32 in. (4 mm)	DCEP	150 to 220

PREHEAT:

The parts must be heated to a temperature higher than 50°F (10°C) before any welding is started.

BACKING GAS:

N/A.

SAFETY:

Proper protective clothing and equipment must be used. The area must be free of all hazards that may affect the welder or others in the area. The welding machine, welding leads, work clamp, electrode holder, and other equipment must be in safe working order.

WELDING TECHNIQUE:

Tack weld the plates together with the backing strip. There should be about a 1/4-in. (6-mm) root gap between the plates. Use the E7018 arc welding electrodes to make a root pass to fuse the plates and backing strip together. Clean the slag from the root pass, being sure to remove any trapped slag along the sides of the weld. Using the E7018 arc welding electrodes, make a series of stringer or weave filler welds, no thicker than 1/4 in. (6.4 mm), in the groove until the joint is filled.

INTERPASS TEMPERATURE:

The plate should not be heated to a temperature higher than 500°F (260°C) during the welding process. After each weld pass is completed, allow it to cool but never to a temperature below 50°F (10°C). The weldment must not be quenched in water.

CLEANING:

The slag must cleaned off between passes. The weld beads may be cleaned by a hand wire brush, a chipping hammer, a punch and hammer, or a needle scaler. All weld cleaning must be performed with the test plate in the welding position. A grinder may not be used to remove weld control problems such as undercut, overlap, trapped slag, and so forth.

INSPECTION:

Visual Inspection Criteria for Entry Welders: There shall be no cracks, no incomplete fusion. There shall be no incomplete joint penetration in groove welds except as permitted for partial joint penetration welds.

The Test Supervisor shall examine the weld for acceptable appearance and shall be satisfied that the welder is skilled in using the process and procedure specified for the test.

Undercut shall not exceed the lesser of 10% of the base metal thickness or 1/32 in. (0.8 mm).

Where visual examination is the only criterion for acceptance, all weld passes are subject to visual examination at the discretion of the Test Supervisor.

The frequency of porosity shall not exceed one in each 4 in. (100 mm) of weld length, and the maximum diameter shall not exceed 3/32 in. (2.4 mm).

Welds shall be free from overlap.

REPAIR:

No repairs of defects are allowed.

BEND TEST:

Transverse face bend. The weld is perpendicular to the longitudinal axis of the specimen and is bent so that the weld face becomes the tension surface of the specimen.

Transverse root bend. The weld is perpendicular to the longitudinal axis of the specimen and is bent so that the weld root becomes the tension surface of the specimen.

ACCEPTANCE CRITERIA FOR BEND TEST:

For acceptance, the convex surface of the face- and root-bend specimens shall meet both of the following requirements:

1. No single indication shall exceed 1/8 in. (3.2 mm) measured in any direction on the surface.

2. The sum of the greatest dimensions of all indications on the surface that exceed 1/32 in. (0.8 mm) but are less than or equal to 1/8 in. (3.2 mm) shall not exceed 3/8 in. (9.5 mm).

Cracks occurring at the corner of the specimens shall not be considered unless there is definite evidence that they result from slag or inclusions or other internal discontinuities.

INSTRUCTOR'S COMMENTS _____

_____

_____

■ PRACTICE 6-5

Name _____ Date _____

Class _____ Instructor _____ Grade _____

WELDING PROCEDURE SPECIFICATION (WPS)

Welding Procedure Specification No: <u>Practice 6-5</u>

TITLE:

Welding <u>SMAW</u> of <u>plate</u> to <u>plate</u>.

SCOPE:

This procedure is applicable for bevel and V-groove plates with a backing strip within the range of 3/8 in. (9.5 mm) through 3/4 in. (19 mm).

Welding may be performed in the following positions: <u>1G, 2G, 3G, and 4G</u>.

BASE METAL:

The base metal shall conform to <u>Carbon Steel M-1 or P-1, Group 1 or 2</u>.

Backing Material Specification <u>Carbon Steel M-1 or P-1, Group 1, 2, or 3</u>.

FILLER METAL:

The filler metal shall conform to AWS specification no. <u>E7018</u> from AWS specification <u>A5.1</u>. This filler metal falls into F-number <u>F-4</u> and A-number <u>A-1</u>.

SHIELDING GAS:

The shielding gas, or gases, shall conform to the following compositions and purity: <u>N/A</u>.

JOINT DESIGN AND TOLERANCES:

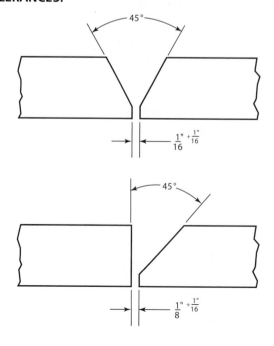

PREPARATION OF BASE METAL:

The bevel is to be flame or plasma cut on the edge of the plate before the parts are assembled. The beveled surface must be smooth and free of notches. Any roughness or notches that are deeper than 1/64 in. (0.4 mm) must be ground smooth.

All hydrocarbons and other contaminants, such as cutting fluids, grease, oil, and primers, must be cleaned off of all parts and filler metals before welding. This cleaning can be done with any suitable solvents or detergents. The backing strip, groove face, and inside and outside plate surface within 1 in. (25 mm) of the joint must be mechanically cleaned of slag, rust, and mill scale. Cleaning must be done with a wire brush or grinder down to bright metal.

ELECTRICAL CHARACTERISTICS:

The current shall be <u>direct-current electrode positive (DCEP)</u>. The base metal shall be on the <u>negative</u> side of the line.

AWS CLASSIFICATION AND POLARITY	ELECTRODE DIAMETER AND AMPERAGE RANGE		
	$\frac{3"}{32}$	$\frac{1"}{8}$	$\frac{5"}{32}$
E6010 - DCEP	40-80	70-130	110-165
E6011 - AC, DCEP	50-70	85-125	130-160
E7018 - AC, DCEP	70-110	90-165	125-220

PREHEAT:

The parts must be heated to a temperature higher than 50°F (10°C) before any welding is started.

BACKING GAS:

N/A.

SAFETY:

Proper protective clothing and equipment must be used. The area must be free of all hazards that may affect the welder or others in the area. The welding machine, welding leads, work clamp, electrode holder, and other equipment must be in safe working order.

WELDING TECHNIQUE:

Tack weld the plates together with the backing strip. There should be about a 1/8-in. (3.2-mm) root gap between the plates. Use the E7018 arc welding electrodes to make a root pass to fuse the plates and backing strip together. Clean the slag from the root pass, being sure to remove any trapped slag along the sides of the weld.

Using the E7018 arc welding electrodes, make a series of filler welds in the groove until the joint is filled.

INTERPASS TEMPERATURE:

The plate, outside of the heat-affected zone, should not be heated to a temperature higher than 500°F (260°C) during the welding process. After each weld pass is completed, allow it to cool; the weldment must not be quenched in water.

CLEANING:

The slag must cleaned off between passes. The weld beads may be cleaned by a hand wire brush, a chipping hammer, a punch and hammer, or a needle scaler. All weld cleaning must be performed with the test plate in the welding position. A grinder may not be used to remove weld control problems such as undercut, overlap, trapped slag, and so forth.

INSPECTION:

Visual Inspection Criteria for Entry Welders: There shall be no cracks, no incomplete fusion. There shall be no incomplete joint penetration in groove welds except as permitted for partial joint penetration welds.

The Test Supervisor shall examine the weld for acceptable appearance and shall be satisfied that the welder is skilled in using the process and procedure specified for the test.

Undercut shall not exceed the lesser of 10% of the base metal thickness or 1/32 in. (0.8 mm).

Where visual examination is the only criterion for acceptance, all weld passes are subject to visual examination at the discretion of the Test Supervisor.

The frequency of porosity shall not exceed one in each 4 in. (100 mm) of weld length, and the maximum diameter shall not exceed 3/32 in. (2.4 mm).

Welds shall be free from overlap.

REPAIR:

No repairs of defects are allowed.

SKETCHES:

1G, 2G, 3G, and 4G test plate drawing.

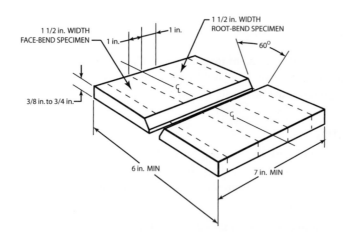

BEND TEST:

Transverse face bend. The weld is perpendicular to the longitudinal axis of the specimen and is bent so that the weld face becomes the tension surface of the specimen.

Transverse root bend. The weld is perpendicular to the longitudinal axis of the specimen and is bent so that the weld root becomes the tension surface of the specimen.

ACCEPTANCE CRITERIA FOR BEND TEST:

For acceptance, the convex surface of the face- and root-bend specimens shall meet both of the following requirements:

1. No single indication shall exceed 1/8 in. (3.2 mm) measured in any direction on the surface.

2. The sum of the greatest dimensions of all indications on the surface that exceed 1/32 in. (0.8 mm) but are less than or equal to 1/8 in. (3.2 mm) shall not exceed 3/8 in. (9.5 mm).

Cracks occurring at the corner of the specimens shall not be considered unless there is definite evidence that they result from slag or inclusions or other internal discontinuities.

INSTRUCTOR'S COMMENTS _____

_____

_____

■ PRACTICE 6-6

Name _____ Date _____

Class _____ Instructor _____ Grade _____

WELDING PROCEDURE SPECIFICATION (WPS)

Welding Procedures Specifications No.: <u>Practice 6-6</u>

TITLE:

Welding SMAW of pipe to pipe.

SCOPE:

This procedure is applicable for V-groove pipe within the range of 3 in. (76 mm) schedule 40 through 8 in. (203 mm) schedule 40.

Welding may be performed in the following positions: 1G, 2G, 5G, and 6G.

BASE METAL:

The base metal shall conform to Carbon Steel M-1/P/1S-1, Group 1 or 2. Backing material specification: N/A.

FILLER METAL:

The filler metal shall conform to AWS specification No. E6010 or E6011 root pass and E7018 for the cover pass from AWS specification A5.1. This filler metal falls into F-numbers F3 and F4 and A-number A-1.

SHIELDING GAS:

The shielding gas, or gases, shall conform to the following compositions and purity: N/A.

JOINT DESIGN AND TOLERANCES:

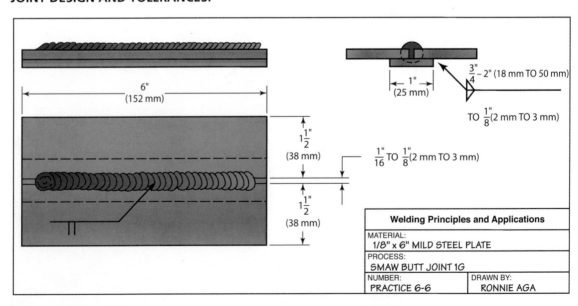

PREPARATION OF BASE METAL:

All workmanship part must prepared according to the AWS Workmanship Standard for Preparation of Base Metal.

ELECTRICAL CHARACTERISTICS:

The current shall be AC or DCEP. The base metal shall be on the work lead or negative side of the line.

PREHEAT:

The parts must be heated to a temperature higher than 70°F (21°C) before any welding is started.

BACKING GAS:

N/A

WELDING TECHNIQUE:

Tack weld the pipes together; there should be about a 1/8-in. (3-mm) root gap between the pipe ends. Use the E6010 or E6011 arc welding electrodes to make a root pass to fuse the pipe ends together. Clean the slag from the root pass and use either a hot pass or grinder to remove any trapped slag.

Using the E7018 arc welding electrodes, make a series of filler welds in the groove until the joint is filled. Figure 6-36 shows the recommended location sequence for the weld beads for the 1G and 5G positions, Figure 6-39 for the 2G position, and Figure 6-40 for the 6G position.

INTERPASS TEMPERATURE:

The plate should not be heated to a temperature higher than 350°F (175°C) during the welding process. After each weld pass is completed, allow it to cool; the weldment must not be quenched in water.

CLEANING:

The slag can be chipped and/or ground off between passes but can only be chipped off of the cover pass.

VISUAL INSPECTION CRITERIA FOR ENTRY-LEVEL WELDERS:

The weld must pass a visual inspection by the instructor or test supervisor based on the AWS Visual Inspection Criteria.

BEND TEST:

The weld is to be mechanically tested only after it has passed the visual inspection. Be sure that the test specimens are properly marked to identify the welder, the position, and the process.

SPECIMEN PREPARATION:

The weld must prepared for bend testing in accordance to the AWS Specimen Preparation Criteria.

ACCEPTANCE CRITERIA FOR BEND TEST:

The bent specimen must not exceed any of the acceptable limits of discontinuities as listed in the AWS Acceptance Criteria for Bend Test.

REPAIR:

No repairs of defects are allowed.

SKETCHES:

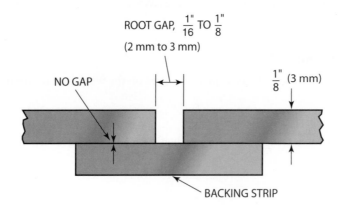

ROOT GAP, $\frac{1"}{16}$ TO $\frac{1"}{8}$
(2 mm to 3 mm)

NO GAP

$\frac{1"}{8}$ (3 mm)

BACKING STRIP

INSTRUCTOR'S COMMENTS _____

_____

_____

■ PRACTICE 6-7

AWS SENSE

Name _____   Date _____

Class _____   Instructor _____   Grade _____

WELDING PROCEDURE SPECIFICATION (WPS)

Welding Procedures Specifications No.: <u>Practice 6-7</u>

TITLE:

Welding SMAW of pipe to pipe.

SCOPE:

This procedure is applicable for V-groove pipe with or without a back ring within the range of 6 in. (150 mm) schedule 80 through 8 in. (203 mm) schedule 80.

Welding may be performed in the following positions: 6G (AWS SENSE Level II).

BASE METAL:

The base metal shall conform to Carbon Steel M-1/P/1S-1, Group 1 or 2.

BACKING RING:

Backing ring to suit diameter and nominal wall thickness of pipe.

FILLER METAL:

The filler metal shall conform to AWS specification No. E7018 for root, fill, and cover passes from AWS specification A5.1. This filler metal falls into F-numbers F4 and A-number A-1.

SHIELDING GAS:

The shielding gas, or gases, shall conform to the following compositions and purity: N/A.

JOINT DESIGN AND TOLERANCES:

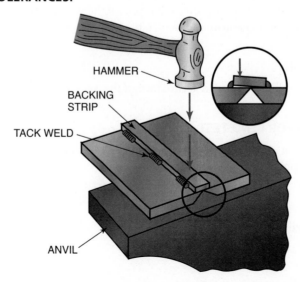

PREPARATION OF BASE METAL:

All workmanship part must prepared according to the AWS Workmanship Standard for Preparation of Base Metal

ELECTRICAL CHARACTERISTICS:

The current shall be AC or DCEP. The base metal shall be on the work lead or negative side of the line.

PREHEAT:

The parts must be heated to a temperature higher than 70°F (21°C) before any welding is started.

BACKING GAS:

N/A

WELDING TECHNIQUE:

Tack weld the pipes together; there should be about a 1/8-in. (3-mm) root gap between the pipe ends. Use E7018 arc welding electrodes to make a root pass to fuse the pipe ends together. Clean the slag from the root pass and use either a hot pass or grinder to remove any trapped slag.

Using the E7018 arc welding electrodes, make a series of filler welds in the groove until the joint is filled. Figure 6-40 shows the recommended location sequence for the weld beads for the 6G position.

INTERPASS TEMPERATURE:

The plate should not be heated to a temperature higher than 350°F (175°C) during the welding process. After each weld pass is completed, allow it to cool; the weldment must not be quenched in water.

CLEANING:

The slag can be chipped and/or ground off between passes but can only be chipped off of the cover pass.

VISUAL INSPECTION CRITERIA FOR ENTRY-LEVEL WELDERS:

The weld must pass a visual inspection by the instructor or test supervisor based on the AWS Visual Inspection Criteria.

BEND TEST:

The weld is to be mechanically tested only after it has passed the visual inspection. Be sure that the test specimens are properly marked to identify the welder, the position, and the process.

SPECIMEN PREPARATION:

The weld must prepared for bend testing in accordance to the AWS Specimen Preparation Criteria.

ACCEPTANCE CRITERIA FOR BEND TEST:

The bent specimen must not exceed any of the acceptable limits of discontinuities as listed in the AWS Acceptance Criteria for Bend Test.

REPAIR:

No repairs of defects are allowed.

SKETCHES:

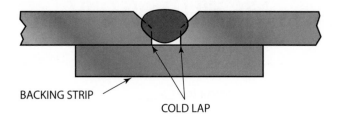

BACKING STRIP

COLD LAP

INSTRUCTOR'S COMMENTS _____

_____

_____

CHAPTER 6: ADVANCED SHIELDED METAL ARC WELDING QUIZ

Name _____ Date _____

Class _____ Instructor _____ Grade _____

Instructions: Carefully read Chapter 6 of the text and answer the following questions.

A. IDENTIFICATION

In the spaces provided, identify the items shown in the illustration.

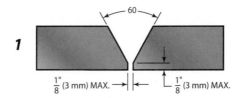

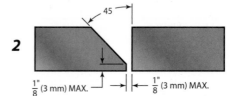

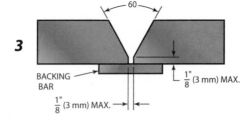

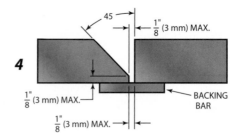

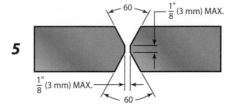

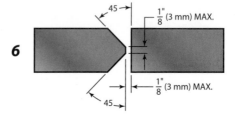

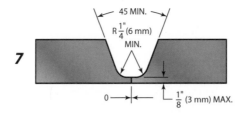

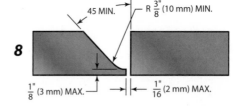

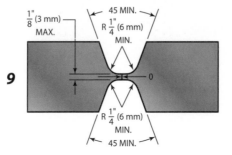

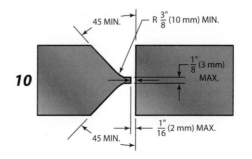

1. _____

2. _____

3. _____

4. _____

5. _____

6. _____

7. _____

8. _____

9. _____

10. _____

B. MATCHING

In the space provided to the left, write the letter from Column B that best answers the question or completes the statement in Column A.

Column A

_____ 11. What is the first weld bead of a multiple pass weld called?

_____ 12. What is the last weld bead on a multipass weld called?

_____ 13. What is the process of cutting a groove in the back side of a joint that has been welded called?

_____ 14. What is used to slow the cooling rate and reduce hardening?

_____ 15. This process helps to reduce cracking, hardness, distortion, and stresses.

_____ 16. What is the temperature of the plate during welding called?

_____ 17. Failure to clean a convex root weld will result in this type of discontinuity.

_____ 18. After the root pass is completed and it has been cleaned by grinding or with a hot pass, the groove is filled with weld metal. What are these weld beads called?

_____ 19. Which weld surface shape is more easily cleaned by chipping, wire brushing, or grinding?

_____ 20. What must be maintained to ensure 100% joint penetration?

Column B

a. filler passes

b. keyhole

c. wagon tracks

d. interpass temperature

e. root pass

f. concave

g. back gouging

h. cover pass/cap

i. preheating

j. postheating

C. ESSAY

Provide complete answers for all of the following questions.

21. When should a root weld pass be ground?

_____

_____

22. Describe how to prepare a guided-bend test specimen.

_____

_____

23. What are the maximum defects allowable on a guided-bend specimen in order for the specimen to pass?

_____

_____

24. For what reason is a hot pass used?

_____

_____

25. On an open root weld, how can burnthrough be controlled?

_____

_____

26. Why is thick plate beveled for welding?

_____

_____

27. What is the reason for back gouging some welds?

_____

_____

28. List four agencies that issue welding codes or standards.

_____

_____

29. List the steps required to prepare specimens for the guided-bend test.

_____

_____

30. How is a weld bead restarted without causing excessive buildup and loss of penetration?

_____

_____

31. Why should some plates be preheated before welding?

_____

_____

32. Why are some plates postheated after welding?

_____

_____

INSTRUCTOR'S COMMENTS _____

_____

_____

Chapter 7

Flame Cutting

EYE PROTECTION FOR FLAME CUTTING

A general guide for the selection of eye and face protection equipment is listed below.

TYPE OF CUTTING OPERATION	HAZARD	SUGGESTED SHADE NUMBER
Light cutting up to 1 in.	Sparks, harmful rays, molten metal, flying particles	3 or 4
Medium cutting, 1–6 in.		4 or 5
Heavy cutting, over 6 in.		5 or 6

Metal Thickness in. (mm)	Center Orifice Size No. Drill Size	Tip Cleaner No.*	Oxygen Pressure lb/in² (kPa)	Acetylene lb/in² (kPa)
1/8 (3)	60	7	10 (70)	3 (20)
1/4 (6)	60	7	15 (100)	3 (20)
3/8 (10)	55	11	20 (140)	3 (20)
1/2 (13)	55	11	25 (170)	4 (30)
3/4 (19)	55	11	30 (200)	4 (30)
1 (25)	53	12	35 (240)	4 (30)
2 (51)	49	13	45 (310)	5 (35)
3 (76)	49	13	50 (340)	5 (35)
4 (102)	49	13	55 (380)	5 (35)
5 (127)	45	**	60 (410)	5 (35)

*The tip cleaner number when counted from the small end toward the large end in a standard tip cleaner set.

**Larger than normally included in a standard tip cleaner set.

■ PRACTICE 7-1

Name _____   Date _____

Class _____   Instructor _____   Grade _____

OBJECTIVE: After completing this practice, you should be able to safely set-up a cutting torch.

EQUIPMENT AND MATERIALS NEEDED FOR THIS PRACTICE

1. Oxygen and fuel gas cylinders, security chains or straps, a place or object to secure cylinders, fuel and oxygen regulators, reverse flow valves or flashback arrestors, torch and tip, proper fitting wrench, leak-detecting solution, and tip cleaner.

2. Proper safety protection (proper shade gas welding/cutting face shield or goggles, safety glasses, wire brush, chipping hammer, leather gloves, pliers, long-sleeved shirt, long pants, and leather boots or shoes). Refer to Chapter 2 in the text for more specific safety information.

INSTRUCTIONS

1. The oxygen and acetylene cylinders must be securely chained to a cart or wall before the safety caps are removed.

2. After removing the safety caps, stand to one side and crack (open and quickly close) the cylinder valves, being sure there are no sources of possible ignition that may start a fire. Cracking the cylinder valves is done to blow out any dirt that may be in the valves.

3. Visually inspect all of the parts for any damage, needed repair, or cleaning.

4. Attach the regulators to the cylinder valves and tighten them securely with a wrench.

5. Attach a reverse flow valve or flashback arrestor, if the torch does not have them built in, to the hose connection on the regulator or to the hose connection on the torch body, depending on the type of reverse flow valve in the set. Do this on both the oxygen and acetylene lines. Occasionally, test each reverse flow valve by blowing through it to make sure it works properly.

6. If the torch you will be using is a combination-type torch, attach the cutting head at this time.

7. Last, install a cutting tip on the torch.

8. Before the cylinder valves are opened, back out the pressure regulating screws so that when the valves are opened the gauges will show zero pounds working pressure.

9. Stand to one side of the regulators as the cylinder valves are opened slowly.

10. The oxygen valve is opened all the way until it becomes tight, but do not overtighten, and the acetylene valve is opened no more than one-half turn.

11. Open one torch valve and then turn the regulating screw in slowly until 2 psig to 4 psig (14 kPag to 30 kPag) shows on the working pressure gauge. Allow the gas to escape so that the line is completely purged.

12. If you are using a combination welding and cutting torch, the oxygen valve nearest the hose connection must be opened before the flame adjusting valve or cutting lever will work.

13. Close the torch valve and repeat the purging process with the other gas.

14. Be sure there are no sources of possible ignition that may result in a fire.

15. With both torch valves closed, spray a leak-detecting solution on all connections including the cylinder valves. Tighten any connection that shows bubbles.

INSTRUCTOR'S COMMENTS _____

_____

_____

■ PRACTICE 7-2

Name _____   Date _____

Class _____   Instructor _____   Grade _____

OBJECTIVE: After completing this practice, you should have the ability and knowledge to clean a cutting tip properly.

EQUIPMENT AND MATERIALS NEEDED FOR THIS PRACTICE

1. Oxygen and fuel gas cylinders, security chains or straps, a place or object to secure cylinders, fuel and oxygen regulators, reverse flow valves or flashback arrestors, torch and tip, proper fitting wrench, leak-detecting solution, and tip cleaner.

2. Proper safety protection (proper shade gas welding/cutting face shield or goggles, safety glasses, wire brush, chipping hammer, leather gloves, pliers, long-sleeved shirt, long pants, and leather boots or shoes). Refer to Chapter 2 in the text for more specific safety information.

INSTRUCTIONS

1. Turn on a small amount of oxygen. This procedure is done to blow out any dirt loosened during the cleaning.

2. The end of the tip is first filed flat, using the file provided in the tip cleaning set.

3. Try several sizes of tip cleaners in a preheat hole until the correct size cleaner is determined. It should easily go all the way into the tip.

4. Push the cleaner in and out of each preheat hole several times. Tip cleaners are small, round files. Excessive use of them will greatly increase the orifice (hole) size. Avoid this.

5. Next, depress the cutting lever and, by trial and error, select the correct size tip cleaner for the center cutting orifice.

   NOTE: A tip cleaner should never be forced. If the tip needs additional care, refer back to the section on tip care in Chapter 4.

INSTRUCTOR'S COMMENTS _____

_____

_____

■ PRACTICE 7-3

Name _____ Date _____

Class _____ Instructor _____ Grade _____

OBJECTIVE: After completing this practice, you should be able to light and extinguish a torch safely.

EQUIPMENT AND MATERIALS NEEDED FOR THIS PRACTICE

1. Oxygen and fuel gas cylinders, security chains or straps, a place or object to secure cylinders, fuel and oxygen regulators, reverse flow valves or flashback arrestors, torch and tip, proper fitting wrench, leak-detecting solution, and tip cleaner.

2. Proper safety protection (proper shade gas welding/cutting face shield or goggles, safety glasses, wire brush, chipping hammer, leather gloves, pliers, long-sleeved shirt, long pants, and leather boots or shoes). Refer to Chapter 2 in the text for more specific safety information.

INSTRUCTIONS

1. Set the regulator working pressure for the tip size. If you do not know the correct pressure setting for the tip, start with the fuel set at 5 psig (35 kPag) and the oxygen set at 25 psig (170 kPag).

2. Point the torch tip upward and away from any equipment or other students.

3. Turn on just the acetylene valve and use only a spark lighter to ignite the acetylene. The torch may not stay lit. If this happens, close the valve slightly and try to relight the torch.

4. If the flame is small, it will produce heavy black soot and smoke. In this case, turn the flame up to stop the soot and smoke. The welder need not be concerned if the flame jumps slightly away from the torch tip.

5. With the acetylene flame burning almost smoke free, slowly open the oxygen valve and by using only the oxygen valve, adjust the flame to a neutral setting.

6. When the cutting oxygen lever is depressed, the flame may become slightly carbonizing. This may occur because of a drop in line pressure due to the high flow of oxygen through the cutting orifice.

7. With the cutting lever depressed, readjust the preheat flame to a neutral setting. The flame will become slightly oxidizing when the cutting lever is released. Since an oxidizing flame is hotter than a neutral flame, the metal being cut will be preheated faster. When the cut is started by depressing the lever, the flame automatically returns to the neutral setting and does not oxidize the top of the plate.

8. Extinguish the flame by turning off the oxygen and then the acetylene.

   CAUTION **Turning off the acetylene first will often cause a loud POP. This can often cause soot to clog the tip and torch.**

INSTRUCTOR'S COMMENTS _____

_____

_____

■ PRACTICE 7-4

Name _____ Date _____

Class _____ Instructor _____ Grade _____

OBJECTIVE: After completing this practice, you should be able to safely set the working pressure on a regulator using oxygen and acetylene.

EQUIPMENT AND MATERIALS NEEDED FOR THIS PRACTICE

1. Oxygen and fuel gas cylinders, security chains or straps, a place or object to secure cylinders, fuel and oxygen regulators, reverse flow valves or flashback arrestors, torch and tip, proper fitting wrench, leak-detecting solution, and tip cleaner.

2. Proper safety protection (oxyfuel goggles or tinted face shield, safety glasses, wire brush, chipping hammer, leather gloves, pliers, long-sleeved shirt, long pants, and leather boots or shoes). Refer to Chapter 2 in the text for more specific safety information.

INSTRUCTIONS

Setting the working pressure of the regulators can be done by following a table, or it can be set by watching the flame.

1. To set the regulator by watching the flame, first set the acetylene pressure at 2 psig to 4 psig (14 kPag to 30 kPag) and then light the acetylene flame.

2. Open the acetylene torch valve one to two turns and reduce the regulator pressure by backing out the setscrew until the flame starts to smoke.

3. Increase the pressure until the smoke stops and then increase it just a little more. This is the maximum fuel gas pressure the tip needs. With a larger tip and a longer hose, the pressure must be set higher. This is the best setting, and it is the safest one to use. With this lowest possible setting, there is less chance of a leak. If the hoses are damaged, the resulting fire will be much smaller than a fire burning from a hose with a higher pressure. There is also less chance of a leak with the lower pressure.

4. With the acetylene adjusted so that the flame almost stops smoking, slowly open the torch oxygen valve.

5. Adjust the torch to a neutral flame. When the cutting lever is depressed, the flame will become carbonizing, not having enough oxygen pressure.

6. While holding the cutting lever down, increase the oxygen regulator pressure slightly. Readjust the flame, as needed, to a neutral setting by using the oxygen valve on the torch.

7. Increase the pressure slowly and readjust the flame as you watch the length of the clear cutting stream in the center of the flame. The center stream will stay fairly long until a pressure is reached that causes turbulence disrupting the cutting stream. This turbulence will cause the flame to shorten in length considerably.

8. With the cutting lever still depressed, reduce the oxygen pressure until the flame lengthens once again. This is the maximum oxygen pressure that this tip can use without causing turbulence in the cutting stream. This turbulence will cause a very inferior cut. The lower pressure also will keep the sparks from being blown a longer distance from the work.

INSTRUCTOR'S COMMENTS _____

_____

_____

■ PRACTICE 7-5

Name _____ Date _____

Class _____ Instructor _____ Grade _____

OBJECTIVE: After completing this practice, you should be able to make a quality straight cut on a 1/4-in. (6-mm) steel plate.

EQUIPMENT AND MATERIALS NEEDED FOR THIS PRACTICE

1. Properly set-up and adjusted oxyfuel cutting equipment.

2. Proper safety protection (oxyfuel goggles or tinted face shield, safety glasses, wire brush, chipping hammer, leather gloves, pliers, long-sleeved shirt, long pants, and leather boots or shoes). Refer to Chapter 2 in the text for more specific safety information.

3. A straightedge.

4. A piece of soapstone.

5. A piece of mild steel plate, 6 in. (152 mm) square by 1/4 in. (6 mm) thick.

INSTRUCTIONS

1. Using a straightedge and soapstone, make several straight lines 1/2 in. (13 mm) apart. Starting at one end, make a cut along the entire length of plate.

2. The cut strip must fall free, be slag-free with vertical drag lines, and be within ±3/32 in. (2 mm) of a straight line and ±5° of being square.

3. Repeat this procedure until the cut is consistently produced.

4. Turn off the cylinder valves, bleed off the trapped pressure in the hoses, back off the regulator adjusting screws, and close the torch valves.

5. Coil up the hoses to prevent them from becoming damaged.

6. Clean up your work area when finished with this practice.

INSTRUCTOR'S COMMENTS _____

_____

_____

■ PRACTICE 7-6

Name _____ Date _____

Class _____ Instructor _____ Grade _____

OBJECTIVE: After completing this practice, you should be able to make a quality straight cut on a 1/2-in. (13-mm) steel plate.

EQUIPMENT AND MATERIALS NEEDED FOR THIS PRACTICE

1. Properly set-up and adjusted oxyfuel cutting equipment.

2. Proper safety protection (oxyfuel goggles or tinted face shield, safety glasses, wire brush, chipping hammer, leather gloves, pliers, long-sleeved shirt, long pants, and leather boots or shoes). Refer to Chapter 2 in the text for more specific safety information.

3. A straightedge.

4. A piece of soapstone.

5. A piece of mild steel plate, 6 in. (152 mm) square by 1/2 in. (13 mm) thick.

INSTRUCTIONS

1. Using a straightedge and soapstone, make several straight lines 1/2 in. (13 mm) apart. Starting at one end, make a cut along the entire length of plate.

2. The strip must fall free, be slag-free with vertical drag lines, and be within ±3/32 in. (2 mm) of a straight line and ±5° of being square.

3. Repeat this procedure until the cut is consistently produced.

4. Turn off the cylinder valves, bleed off the trapped pressure in the hoses, back off the regulator adjusting screws, and close the torch valves.

5. Coil up the hoses to prevent them from becoming damaged.

6. Clean up your work area when finished with this practice.

   NOTE: Remember that starting a cut in thick plate will take longer and the cutting speed will be slower.

INSTRUCTOR'S COMMENTS _____

_____

_____

■ PRACTICE 7-7

Name _____   Date _____

Class _____   Instructor _____   Grade _____

OBJECTIVE: After completing this practice you should be able to make a flat, straight cut in sheet metal.

EQUIPMENT AND MATERIALS NEEDED FOR THIS PRACTICE

1. Properly set-up and adjusted oxyfuel cutting equipment.

2. Proper safety protection (oxyfuel goggles or tinted face shield, safety glasses, wire brush, chipping hammer, leather gloves, pliers, long-sleeved shirt, long pants, and leather boots or shoes). Refer to Chapter 2 in the text for more specific safety information.

3. A piece of soapstone.

4. A straightedge.

5. A piece of mild steel sheet 10 in. (254 mm) square by 18 gauge to 11 gauge thick.

INSTRUCTIONS

1. Using a straightedge and soapstone, mark the plate in strips 1/2 in. (13 mm) wide.

2. Hold the torch at a very sharp leading angle and cut along the line.

3. The cut must be smooth and straight with as little slag as possible.

4. Repeat this procedure until the cut can be made smooth, straight, and slag-free.

5. Turn off the cylinder valves, bleed off the trapped pressure in the hoses, back off the regulator adjusting screws, and close the torch valves.

6. Coil up the hoses to prevent them from becoming damaged.

7. Clean up your work area when finished with this practice.

INSTRUCTOR'S COMMENTS _____

_____

_____

■ PRACTICE 7-8

Name _____ Date _____

Class _____ Instructor _____ Grade _____

OBJECTIVE: After completing this practice you should be able to cut holes in flat plate.

EQUIPMENT AND MATERIALS NEEDED FOR THIS PRACTICE

1. Properly set-up and adjusted oxyfuel cutting equipment.

2. Proper safety protection (oxyfuel goggles or tinted face shield, safety glasses, wire brush, chipping hammer, leather gloves, pliers, long-sleeved shirt, long pants, and leather boots or shoes). Refer to Chapter 2 in the text for more specific safety information.

3. A piece of soapstone.

4. A circle template or circular pattern.

5. A piece of mild steel plate 6 in. (152 mm) square by 1/4 in. (6 mm) thick.

INSTRUCTIONS

1. Using a circle template or circular pattern and soapstone, mark several 1/2-in. (13-mm) diameter and 1-in. (25-mm) diameter circles.

2. Start in the center of the circle and make an outward spiraling cut until the hole is the desired size.

3. The hole must be within ±3/32 in. (2 mm) of being round and ±5° of being square to metal.

4. The hole may have slag on the bottom.

5. Repeat this procedure until both the small and large holes can be made with consistent accuracy.

6. Turn off the cylinder valves, bleed off the trapped pressure in the hoses, back off the regulator adjusting screws, and close the torch valves.

7. Coil up the hoses to prevent them from becoming damaged.

8. Clean up your work area when finished with this practice.

INSTRUCTOR'S COMMENTS _____

_____

_____

■ PRACTICE 7-9

Name _____ Date _____

Class _____ Instructor _____ Grade _____

OBJECTIVE: After completing this practice, you should be able to bevel 3/8-inch plate.

EQUIPMENT AND MATERIALS NEEDED FOR THIS PRACTICE

1. Properly set-up and adjusted oxyfuel cutting equipment.

2. Proper safety protection (oxyfuel goggles or tinted face shield, safety glasses, wire brush, chipping hammer, leather gloves, pliers, long-sleeved shirt, long pants, and leather boots or shoes). Refer to Chapter 2 in the text for more specific safety information.

3. A piece of soapstone.

4. A straightedge.

5. A piece of mild steel plate, 6 in. (152 mm) square by 3/8 in. (10 mm) thick.

INSTRUCTIONS

1. Using a straightedge and soapstone, mark the plate in strips 1/2 in. (13 mm) wide.

2. Set the tip for beveling and cut a bevel. The bevel should be within ±3/32 in. (2 mm) of a straight line and ±5° of a 45° angle. There may be some soft slag, but no hard slag, on the beveled plate.

3. Repeat this practice until the beveled angle cut is consistently produced.

4. Turn off the cylinder valves, bleed off the trapped pressure in the hoses, back off the regulator adjusting screws, and close the torch valves.

5. Coil up the hoses to prevent them from becoming damaged.

6. Clean up your work area when finished with this practice.

INSTRUCTOR'S COMMENTS _____

_____

_____

■ PRACTICE 7-10

Name _____   Date _____

Class _____   Instructor _____   Grade _____

OBJECTIVE: After completing this practice, you should be able to cut 1/4-in. to 3/8-in. (6-mm to 10-mm) carbon steel in both the vertical up and down positions.

EQUIPMENT AND MATERIALS NEEDED FOR THIS PRACTICE

1. Properly set-up and adjusted oxyfuel cutting equipment.

2. Proper safety protection (oxyfuel goggles or tinted face shield, safety glasses, wire brush, chipping hammer, leather gloves, pliers, long-sleeved shirt, long pants, and leather boots or shoes). Refer to Chapter 2 in the text for more specific safety information.

3. A piece of soapstone.

4. A straightedge.

5. A piece of mild steel plate, 6 in. (152 mm) square by 1/4 in. (6 mm) to 3/8 in. (10 mm) thick, marked in strips 1/2 in. (13 mm) wide all along the 1/4-in. and 3/8-in. (6-mm and 10-mm) thick carbon steel plate.

INSTRUCTIONS

1. Using a straightedge and soapstone, make several straight lines 1/2 in. (13 mm) apart on the steel plate.

2. Place the plate in a vertical position.

3. Start your first cut at the top of the plate and travel downward. Then, starting at the bottom, make the next cut upward. Make sure that the sparks do not cause a safety hazard and that the metal being cut off will not fall on any person or object.

4. The cut must be free of hard to remove dross and within ±3/32 in. (2 mm) of a straight line and ±5° of being square.

5. Repeat these cuts until they are consistently produced.

6. Turn off the cylinder valves, bleed off the trapped pressure in the hoses, back off the regulator adjusting screws, and close the torch valves.

7. Coil up the hoses to prevent them from becoming damaged.

8. Clean up your work area when finished with this practice.

INSTRUCTOR'S COMMENTS _____

_____

_____

■ PRACTICE 7-11

Name _____ Date _____

Class _____ Instructor _____ Grade _____

OBJECTIVE: After completing this practice you should be able to make an overhead straight cut.

EQUIPMENT AND MATERIALS NEEDED FOR THIS PRACTICE

1. Properly set-up and adjusted oxyfuel cutting equipment.

2. Proper safety protection (oxyfuel goggles or tinted face shield, safety glasses, wire brush, chipping hammer, leather gloves, pliers, long-sleeved shirt, long pants, and leather boots or shoes). Refer to Chapter 2 in the text for more specific safety information.

3. A piece of soapstone.

4. A straightedge.

5. A piece of mild steel sheet 10 in. (254 mm) square by 1/4 in. (6 mm) to 3/8 in. (10 mm) thick.

INSTRUCTIONS

1. Using a straightedge and soapstone, mark the plate in strips 1/2 in. (13 mm) wide.

2. Be sure that you are completely protected from the hot sparks which will fall from the cut.

3. Angle the torch so that most of the sparks and slag will be blown away from you.

4. Starting at one edge of the plate, make a cut in the overhead position. The plate should fall free when the cut is completed.

5. The cut must be smooth and straight with as little slag as possible and within 1/8 in. (3 mm) of being a straight line and within ±5° of being square to metal.

6. Repeat this procedure until the cut can be made smooth, straight, and slag-free.

7. Turn off the cylinder valves, bleed off the trapped pressure in the hoses, back off the regulator adjusting screws, and close the torch valves.

8. Coil up the hoses to prevent them from becoming damaged.

9. Clean up your work area when finished with this practice.

INSTRUCTOR'S COMMENTS _____

_____

_____

■ PRACTICE 7-16

Name _____ Date _____

Class _____ Instructor _____ Grade _____

OBJECTIVE: After completing this practice you should be able to make a square cut on pipe in the 2G (vertical) position.

EQUIPMENT AND MATERIALS NEEDED FOR THIS PRACTICE

1. Properly set-up and adjusted oxyfuel cutting equipment.

2. Proper safety protection (oxyfuel goggles or tinted face shield, safety glasses, wire brush, chipping hammer, leather gloves, pliers, long-sleeved shirt, long pants, and leather boots or shoes). Refer to Chapter 2 in the text for more specific safety information.

3. A piece of soapstone and a flexible straightedge which can be wrapped around the pipe.

4. A piece of schedule 40 mild steel pipe 3 in. (76 mm) in diameter by 6 in. (152 mm) long.

INSTRUCTIONS

1. Using the straightedge and soapstone, wrap the straightedge around the pipe and mark rings every 1/2 in. (13 mm) along the pipe.

2. Place the pipe vertically on the cutting table and place a plate over the open end of the pipe to contain the sparks.

3. Start at one side and make the cut horizontally around the pipe.

4. When the cut is completed the ring may have to be tapped free.

5. When the pipe is placed upright on a flat surface, it must stand within 5° of vertical and have no gaps greater than 1/8 in. (3 mm) between the cut edge and the flat surface.

6. Repeat this procedure until the cut can consistently be made correctly.

7. Turn off the cylinder valves, bleed off the trapped pressure in the hoses, back off the regulator adjusting screws, and close the torch valves.

8. Coil up the hoses to prevent them from becoming damaged.

9. Clean up your work area when finished with this practice.

INSTRUCTOR'S COMMENTS _____

_____

_____

CHAPTER 7: FLAME CUTTING QUIZ

Name _____   Date _____

Class _____   Instructor _____   Grade _____

Instructions: Carefully read Chapter 7 in the text and answer the following questions.

A. SHORT ANSWER

Write a brief answer in the space provided that will answer the question or complete the statement.

1. List the parts of the oxyfuel cutting torch.

a. _____

b. _____

c. _____

d. _____

e. _____

2. List five fuel gases that are used for flame cutting.

a. _____

b. _____

c. _____

d. _____

e. _____

3. What fuel gas is most frequently used?

_____

4. Explain the differences between a combination torch set and a cutting torch only.

   _____

5. What could happen if acetylene was used in a tip with more than eight preheat holes, or in a two-piece tip?

   _____

   _____

   _____

   _____

   _____

6. From the illustrations shown below, which torch is classified as an injector mixing torch?

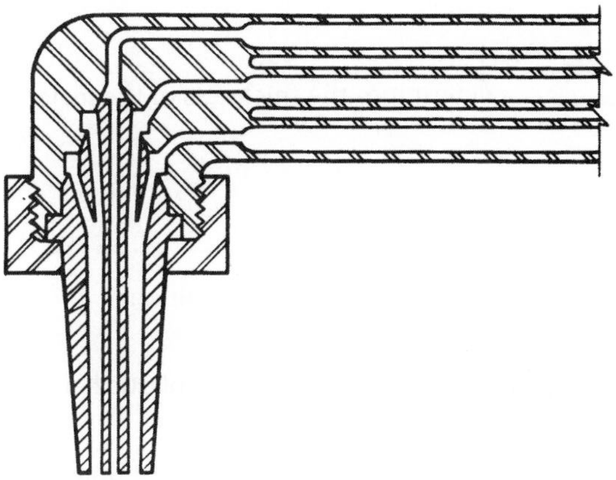

   A. 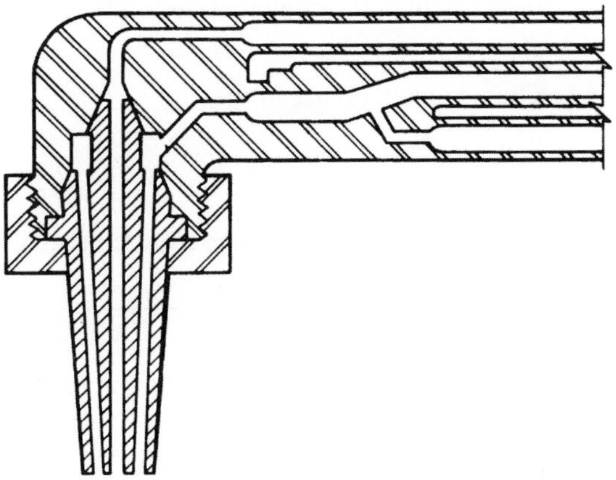   **B.**

   _____

   _____

7. Define *coupling* distance.

   _____

   _____

8. What are the two major methods for controlling distortion while making a cut?

   _____

   _____

9. Why is a slight forward torch angle used when cutting a straight line? Why not on patterns?

   _____

   _____

B. FILL IN THE BLANK

Fill in the blank with the correct word. Answers may be more than one word.

10. _____ cutting is done in short-run production and one-of-a-kind fabrication, as well as in demolition and scrapping operations.

11. The OFC process will work easily on any _____ that will rapidly oxidize at an elevated temperature.

12. _____ and _____ are two terms that have the same meaning they refer to the lowest temperature at which a material will combust or start burning.

13. The dedicated cutting torches are usually _____ and have larger gas flow passages than the combination torches.

14. The diameter, or size of the _____, determines the thickness of the metal that can be cut.

15. The amount of preheat flame required to make a perfect cut is determined by the _____ and by the material _____, _____, and surface condition.

16. If the cutting tip is stuck in the torch head, _____ of the head with a plastic hammer.

17. It is important that the regulator keep the lower pressure _____ over a range of flow rates.

18. This pressure difference results from the resistance to the gas flow, which is referred to as _____.

19. Fittings should screw together _____ and require only _____ wrench pressure to be leak tight.

20. High-pressure valve seats that leak result in a _____ or rising pressure on the working side of the regulator.

21. A flashback that reaches the cylinder may cause a _____ or a(n) _____.

22. Fuel-gas hoses must be red and have _____ threaded fittings.

23. Dirty tips can be cleaned using a set of _____.

24. When making a cut with a hand torch, it is important for the welder to be _____ in order to make the cut as smooth as possible.

25. _____ slag is very porous, brittle, and easily removed from a cut.

C. IDENTIFICATION

In the space provided, identify the items shown on the illustration.

_____ 26. Travel speed too fast

_____ 27. Preheat flames too high above the surface

_____ 28. Travel speed too slow

_____ 29. Cutting oxygen pressure too high

_____ 30. Correct cut

_____ 31. Preheat flames too close to the surface

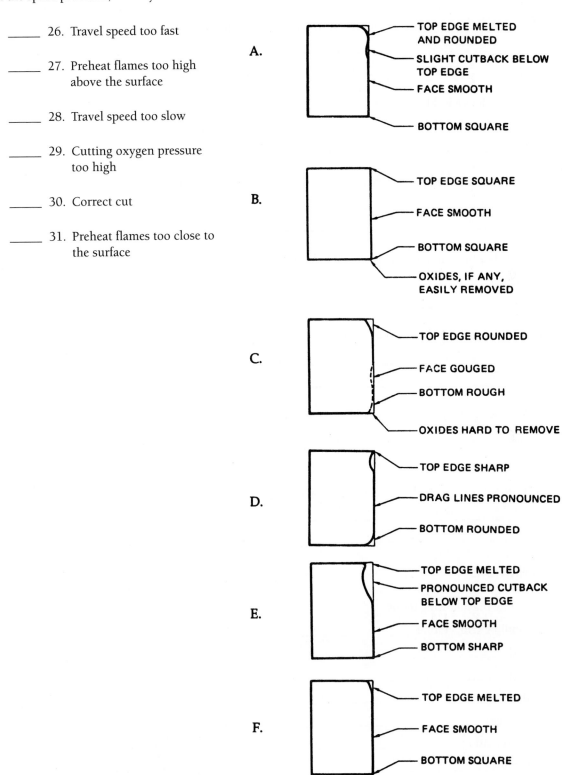

A.
- TOP EDGE MELTED AND ROUNDED
- SLIGHT CUTBACK BELOW TOP EDGE
- FACE SMOOTH
- BOTTOM SQUARE

B.
- TOP EDGE SQUARE
- FACE SMOOTH
- BOTTOM SQUARE
- OXIDES, IF ANY, EASILY REMOVED

C.
- TOP EDGE ROUNDED
- FACE GOUGED
- BOTTOM ROUGH
- OXIDES HARD TO REMOVE

D.
- TOP EDGE SHARP
- DRAG LINES PRONOUNCED
- BOTTOM ROUNDED

E.
- TOP EDGE MELTED
- PRONOUNCED CUTBACK BELOW TOP EDGE
- FACE SMOOTH
- BOTTOM SHARP

F.
- TOP EDGE MELTED
- FACE SMOOTH
- BOTTOM SQUARE

D. ESSAY

Provide complete answers for all of the following questions.

32. List the advantages of a combination welding and cutting torch.

_____

_____

33. When does a welder need to use more preheating time?

_____

_____

34. How are MPS and propane two-piece tips different from each other? Why?

_____

_____

35. How is a tip removed from the torch head if it is stuck?

_____

_____

36. How is a tip seal checked for a leak?

_____

_____

37. Why is the oxygen turned on while a welder is cleaning a tip?

_____

_____

38. Without using a chart, how can a welder correctly set the gas pressures?

_____

_____

E. DEFINITIONS

39. Define the following terms:

    kindling temperature _____

    kerf _____

    slag _____

    flashback arrestor _____

    MPS gas _____

    hand cutting _____

    venturi _____

INSTRUCTOR'S COMMENTS _____

_____

_____

Chapter 8

Plasma Arc Cutting

■ PRACTICE 8-1

Name _____ Date _____

Base Metal _____ Thickness _____

Class _____ Instructor _____ Grade _____

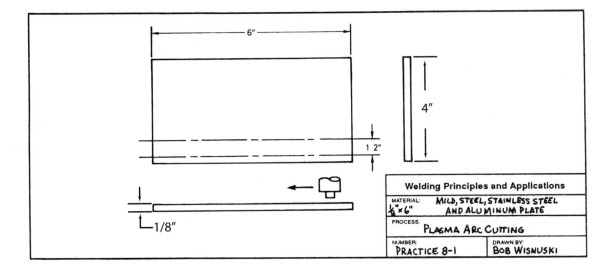

OBJECTIVE: After completing this practice, you should be able to safely set up and operate a PAC machine to make straight cuts on thin gauge mild steel, stainless steel, and aluminum in the flat position.

EQUIPMENT AND MATERIALS NEEDED FOR THIS PRACTICE

1. A properly set-up and adjusted PAC machine.

2. Proper safety protection (welding hood, safety glasses, leather gloves, pliers, long-sleeved shirt, long pants, and leather boots or shoes). Refer to Chapter 2 in the text for more specific safety information.

3. One or more pieces of mild steel, stainless steel, and aluminum, 6 in. (152 mm) long by 4 in. (100 mm) wide and 16 gauge and 1/8 in. (3 mm) thick.

INSTRUCTIONS

1. Starting at one end of the piece of metal that is 1/8 in. (3 mm) thick, hold the torch as close as possible to a 90° angle.

2. Lower your hood and establish a plasma cutting stream.

3. Move the torch in a straight line down the plate toward the other end.

4. If the width of the kerf changes, speed up or slow down the travel rate to keep the kerf the same size for the entire length of the plate.

5. Repeat the cut using both thicknesses of all three types of metals until you can make consistently smooth cuts that are within ±3/32 in. (2.3 mm) of a straight line and ±5° of being square to metal.

6. Turn off the PAC equipment and clean up your work area when you have finished cutting.

INSTRUCTOR'S COMMENTS _____
_____
_____

■ PRACTICE 8-2

Name _____ Date _____

Base Metal _____ Thickness _____

Class _____ Instructor _____ Grade _____

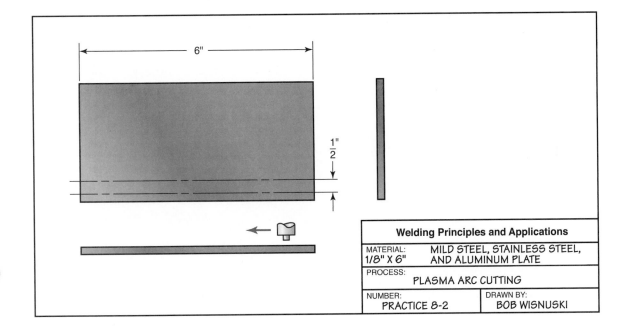

Welding Principles and Applications

MATERIAL: 1/8" X 6"	MILD STEEL, STAINLESS STEEL, AND ALUMINUM PLATE	
PROCESS: 	PLASMA ARC CUTTING	
NUMBER: PRACTICE 8-2		DRAWN BY: BOB WISNUSKI

OBJECTIVE: After completing this practice, you should be able to safely set up and operate a PAC machine to make straight cuts on thicker mild steel, stainless steel, and aluminum plates ranging in thickness from 1/4 to 1/2 in. (6 to 13 mm) in the flat position.

EQUIPMENT AND MATERIALS NEEDED FOR THIS PRACTICE

1. A properly set-up and adjusted PAC machine.

2. Proper safety protection (welding hood, safety glasses, leather gloves, pliers, long-sleeved shirt, long pants, and leather boots or shoes). Refer to Chapter 2 in the text for more specific safety information.

3. One or more pieces of mild steel, stainless steel, and aluminum, 6 in. (152 mm) long by 4 in. (100 mm) wide and 1/4 in. (6 mm) and 1/2 in. (13 mm) thick.

INSTRUCTIONS

1. Starting at one end of the piece of metal that is 1/4 in. (6 mm) thick, hold the torch as close as possible to a 90° angle.

2. Lower your hood and establish a plasma cutting stream.

3. Move the torch in a straight line down the plate toward the other end.

4. If the width of the kerf changes, speed up or slow down the travel rate to keep the kerf the same size for the entire length of the plate.

5. Repeat the cut using both thicknesses of all three types of metals until you can make consistently smooth cuts that are within ±3/32 in. (2.3 mm) of a straight line and ±5° of being square to metal.

INSTRUCTOR'S COMMENTS _____

_____

_____

■ PRACTICE 8-3

Name _____ Date _____

Base Metal _____ Thickness _____

Class _____ Instructor _____ Grade _____

OBJECTIVE: After completing this practice, you should be able to safely set up and operate a PAC machine to cut 1/2-in. (13-mm) and 1-in. (25-mm) diameter holes in mild steel, stainless steel, and aluminum plates ranging in thickness from 16 gauge sheet metal to 1/2 in. (6 to 13 mm) plate in the flat position.

EQUIPMENT AND MATERIALS NEEDED FOR THIS PRACTICE

1. A properly set-up and adjusted PAC machine.

2. Proper safety protection (welding hood, safety glasses, leather gloves, pliers, long-sleeved shirt, long pants, and leather boots or shoes). Refer to Chapter 2 in the text for more specific safety information.

3. One or more pieces of mild steel, stainless steel, and aluminum, 16 gauge, 1/8 in. (3 mm), 1/4 in. (6 mm), and 1/2 in. (13 mm) thick.

INSTRUCTIONS

1. Starting with the piece of metal that is 1/8 in. (3 mm) thick, hold the torch as close as possible to a 90° angle.

2. Lower your hood and establish a plasma cutting stream.

3. Move the torch in an outward spiral until the hole is the desired size.

4. Repeat the whole cutting process until both sizes of holes are made using all the thicknesses of all three types of metals until you can make consistently smooth cuts that are within ±3/32 in. (2.3 mm) of being round and ±5° of being square to metal.

INSTRUCTOR'S COMMENTS _____

_____

_____

■ PRACTICE 8-4

Name _____   Date _____

Base Metal _____   Thickness _____

Class _____   Instructor _____   Grade _____

Welding Principles and Applications

MATERIAL: ¼"AND ½" MILD STEEL, STAINLESS STEEL, ALUMINUM PLATE

PROCESS: PLASMA ARC CUTTING

NUMBER: PRACTICE 8-4   DRAWN BY: LEN HEBERT

OBJECTIVE: After completing this practice, you should be able to safely set up and operate a PAC machine to bevel mild steel, stainless steel, and aluminum plates ranging from 1/4 to 1/2 in. (6 to 13 mm) thick plate.

EQUIPMENT AND MATERIALS NEEDED FOR THIS PRACTICE

1. A properly set-up and adjusted PAC machine.

2. Proper safety protection (welding hood, safety glasses, leather gloves, pliers, long-sleeved shirt, long pants, and leather boots or shoes). Refer to Chapter 2 in the text for more specific safety information.

3. One or more pieces of mild steel, stainless steel, and aluminum, 6 in. (152 mm) long by 4 in. (100 mm) wide and 1/4 in. (6 mm) and 1/2 in. (13 mm) thick. You will cut a 45° bevel down the length of the plate.

INSTRUCTIONS

1. Starting at one end of the piece of metal that is 1/4 in. (6 mm) thick, hold the torch as close as possible to a 45° angle.

2. Lower your hood and establish a plasma cutting stream.

3. Move the torch in a straight line down the plate toward the other end.

4. Repeat the cut using both thicknesses of all three types of metals until you can make consistently smooth cuts that are within ±3/32 in. (2.3 mm) of a straight line and ±5° of a 45° angle.

INSTRUCTOR'S COMMENTS _____

_____

_____

■ PRACTICE 8-5

Name _____   Date _____

Class _____   Instructor _____   Grade _____

Welding Principles and Applications

MATERIAL: 1/4" AND 1/2" MILD STEEL, STAINLESS STEEL, AND ALUMINUM PLATE

PROCESS: PLASMA ARC GOUGING

NUMBER: PRACTICE 8-5   DRAWN BY: MARY HEBERT

OBJECTIVE: After completing this practice, you should be able to safely set up and operate a PAC machine to make a U-groove in steel, stainless steel, and aluminum.

EQUIPMENT AND MATERIALS NEEDED FOR THIS PRACTICE

1. A properly set-up and adjusted PAC machine.

2. Proper safety protection (welding hood, safety glasses, leather gloves, pliers, long-sleeved shirt, long pants, and leather boots or shoes). Refer to Chapter 2 in the text for more specific safety information.

3. One or more pieces of mild steel, stainless steel, and aluminum, 6 in. (152 mm) long by 4 in. (100 mm) wide and 1/4 in. (6 mm) or 1/2 in. (13 mm) thick.

INSTRUCTIONS

1. Starting at one end of the piece of metal, hold the torch as close as possible to a 30° angle.

2. Lower your hood and establish a plasma cutting stream.

3. Move the torch in a straight line down the plate toward the other end.

4. If the width of the U-groove changes, speed up or slow down the travel rate to keep the groove the same width and depth for the entire length of the plate.

5. Repeat the gouging of the U-groove using all three types of metals until you can make consistently smooth grooves that are within ±3/32 in. (2.3 mm) of a straight line and uniform in width and depth.

INSTRUCTOR'S COMMENTS _____
_____
_____

■ PRACTICE 8-6

Name _____ Date _____

Class _____ Instructor _____ Grade _____

OBJECTIVE: After completing this practice, you should be able to safely set up and operate a PAC machine to cut a 1/2-in. (13 mm) wide cut on pipe or round stock.

EQUIPMENT AND MATERIALS NEEDED FOR THIS PRACTICE

1. A properly set-up and adjusted PAC machine.

2. Proper safety protection (welding hood, safety glasses, leather gloves, pliers, long-sleeved shirt, long pants, and leather boots or shoes). Refer to Chapter 2 in the text for more specific safety information.

3. A piece of soapstone and a flexible straightedge, which can be wrapped around the pipe or round stock.

4. One or more pieces of pipe or round stock 6 in. (152 mm) long.

INSTRUCTIONS

1. Using the straightedge and soapstone, wrap the straightedge around the pipe or round stock and mark rings every 1/2 inch (13 mm).

2. Hold the torch so it is pointed parallel to the round piece and in line with the line to be cut.

3. Lower your hood, and pull the trigger to start the pilot arc and the cutting plasma stream.

4. Use one of the piercing techniques to begin the cut.

5. If the round stock is 1/2 in. (13 mm) or smaller, keep the torch pointed in the same direction, and move the torch across the piece.

6. If the round stock is 1/2 in. (13 mm) or thicker than the PAC torch can cut through easily, then you must move the torch back and forth to make the kerf wider to allow the cut to go all the way through the round stock.

7. Repeat the cut using both thicknesses of all three types of metals until you can make consistently smooth cuts that are within ± 3/32 in. (2.3 mm) of a straight line and ± 5° of being square. Turn off the PAC equipment and clean up your work area when you are finished cutting.

INSTRUCTOR'S COMMENTS _____

_____

_____

■ PRACTICE 8-7

Name _____   Date _____

Class _____   Instructor _____   Grade _____

OBJECTIVE: After completing this practice, you should be able to safely set up and operate a PAC pipe beveling machine to cut a 45° angle bevel on the pipe.

EQUIPMENT AND MATERIALS NEEDED FOR THIS PRACTICE

1. A properly set-up and adjusted PAC machine.

2. Proper safety protection (welding hood, safety glasses, leather gloves, pliers, long-sleeved shirt, long pants, and leather boots or shoes). Refer to Chapter 2 in the text for more specific safety information.

3. A piece of soapstone and a flexible straightedge, which can be wrapped around the pipe or round stock.

4. One or more pieces of 6 to 8 in. (150 to 400 mm) diameter pipe.

INSTRUCTIONS

1. Attach the pipe beveling machine to the pipe.

2. Attach a machine plasma torch to the beveling machine carriage.

3. Set the torch to a 45o angle.

4. Run the carriage all the way around the pipe to check that it has freedom of movement and that the torch nozzle tip is traveling at the same height all the way around the pipe.

5. Clear any obstacles to the torch moving freely.

6. Change the beveling machine spacers if the tip height changes as the torch is moved all the way around the pipe.

7. Using the running start technique of piercing the pipe, start the carriage moving at the same time the plasma stream is started.

8. Watch the spark stream to determine that the cut is progressing at the correct speed.

9. Make any adjustments in speed to keep the spark stream flowing outward in a steady stream.

10. If the beveled piece is to be used as a welding test coupon, you should hold it with a pair of pliers so that it does not drop to the floor. The hot pipe piece can easily be bent out of round if it hits the floor.

11. Repeat the cut until you can make consistently smooth beveled cuts. Turn off the PAC equipment and clean up your work area when you are finished cutting.

INSTRUCTOR'S COMMENTS _____

_____

_____

CHAPTER 8: PLASMA ARC CUTTING QUIZ

Name _____ Date _____

Class _____ Instructor _____ Grade _____

Instructions: Carefully read Chapter 8 in the text and answer the following questions.

A. MATCHING

In the space provided to the left, write the letter from Column B that best answers the question or completes the statement in Column A.

Column A

Column B

_____ 1. What is the state of matter that is found in the region of an electrical discharge?

a. cup

_____ 2. What do we call the spacing between the electrode tip and the nozzle tip?

b. plasma

_____ 3. The nozzle is sometimes called the ___.

c. dross

_____ 4. What is the space left in the metal as the metal is removed during a cut?

d. electrode setback

_____ 5. What is the metal compound that resolidifies and attaches itself to the bottom of a cut?

e. kerf

B. IDENTIFICATION

In the space provided, identify the items shown in the illustration.

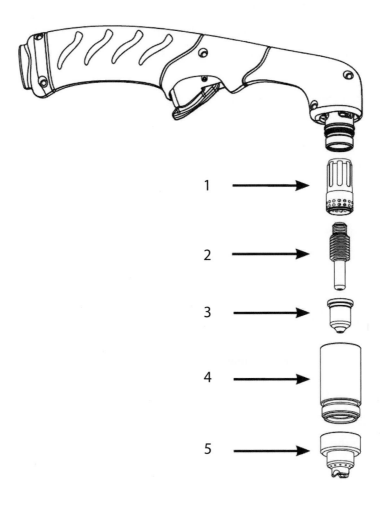

6. Identify the replaceable torch components.

1. _____

2. _____

3. _____

4. _____

5. _____

C. FILL IN THE BLANK

Fill in the blank with the correct word. Answers may be more than one word.

7. The plasma arc can be used for _____, _____, _____,

and _____.

8. Plasma welding torches produce a _____ velocity plasma and cutting torches

produce a _____ velocity plasma stream.

9. Manual-type torch bodies are made of a special plastic that is resistant to _____, _____, and _____.

10. _____ major benefit to plasma electrodes is its resistance to oxidation at elevated temperatures, so air can be used as the plasma gas.

11. Medium- and high-amperage torches may be _____ cooled.

12. The voltage for a plasma arc process ranges from _____ to _____ volts closed circuit and _____ to _____ volts open circuit.

13. _____ — this process uses a concentrated column of water surrounding the plasma column.

14. Any material that is _____ can be cut using the PAC process.

15. The torch _____ is the distance from the nozzle tip to the work.

16. There are three commonly used techniques—_____, _____, and _____.

D. SHORT ANSWER

Write a brief answer in the space provided that will answer the question or complete the statement.

17. What is the function of a water shroud nozzle?

_____

_____

18. List eight factors that can affect the width of a PAC kerf.

a. _____

b. _____

c. _____

d. _____

e. _____

f. _____

g. _____

h. _____

19. Why do all plasma cuts have a slight bevel?

_____

_____

20. List seven factors for controlling dross when making a plasma arc cut.

    a. _____

    b. _____

    c. _____

    d. _____

    e. _____

    f. _____

    g. _____

21. Define the term *plasma arc*.

    _____

    _____

22. Why are some plasma arc torches water cooled?

    _____

    _____

23. List the purpose of four of the wires, hoses, and/or cables that attach the PAC torch to the power supply.

    _____

    _____

24. Why is there less distortion when parts are cut using a plasma torch as compared to a standard oxyfuel torch?

    _____

    _____

25. What effect do the higher cutting speeds have on the hardness zone formed along the cut edge?

    _____

    _____

26. Why must plates be preheated before they are cut using oxyfuel?

    _____

    _____

27. Why were aluminum, stainless steel, and copper the metals most often cut using plasma?

    _____

    _____

28. How can nonconductive materials be cut with the PAC process?

_____

_____

29. Explain how the plasma cut is started.

_____

_____

30. Define a nontransfer arc.

_____

_____

31. Why are some plasma torches designed to swirl the plasma gas?

_____

_____

32. List four of the effects on the cut that changing the PAC gas(es) will have.

a. _____

b. _____

c. _____

d. _____

33. List six things that can affect the formation of dross on the bottom of a PAC plate.

a. _____

b. _____

c. _____

d. _____

e. _____

f. _____

34. Why are dyes added to the water in a PAC water table?

_____

_____

35. Why are most manual cutting machines limited to 100 amperes or less?

_____

_____

36. Why must the power be off before working on any part of the plasma arc torch, machine, cables, and so forth?

_____

_____

37. Explain each of the following special safety concerns as they apply to plasma arc cutting:

    a. electrical shock _____

    b. moisture _____

    c. noise _____

    d. light _____

    e. fumes _____

    f. gases _____

    g. sparks _____

E. DEFINITIONS

38. Define the following terms:

    plasma arc gouging _____

    water tables _____

    stack cutting _____

    water injection _____

    pilot arc _____

    nontransfer _____

    high-frequency alternating current _____

    kerf _____

    dross _____

    heat-affected zone _____

    travel speed _____

    water shroud _____

    gas lens _____

    electrode tip _____

    nozzle insulator _____

    nozzle tip _____

    nozzle _____

    electrode setback _____

    cup _____

INSTRUCTOR'S COMMENTS _____

_____

_____

Chapter 9

Related Cutting Processes

Air Carbon Arc Electrode Current Recommendations									
	Electrode Diameter								
Current	5/32	3/16	1/4	5/16	3/8	1/2	5/8	3/4	FLAT
Min DC	90	200	300	350	450	800	1000	1250	300
Max DC	150	250	400	450	600	1000	1250	1600	500
Min AC	—	200	300	325	350	500	—	—	—
Max AC	—	250	400	425	450	600	—	—	—

Air Carbon Arc Cutting Electrode Recommendations			
Metal	Electrode	Current	Polarity
Steel	DC	DC	DCRP
Cast Iron:			
Gray, Ductile,	DC	DC	DCRP
Malleable	DC	DC	DCSP
Copper Alloys	DC	DC	DCSP
	DC	DC	—
Nickel Alloys	AC	AC or DC	DCSP

■ PRACTICE 9-1

Name_____   Date _____

Class _____   Instructor _____   Grade _____

OBJECTIVE: After completing this practice, you should be able to safely set up and operate an air carbon arc cutting torch to make a U-groove gouge on carbon steel plate.

EQUIPMENT AND MATERIALS NEEDED FOR THIS PRACTICE

1. An air carbon arc cutting torch and welding power supply that has been safely set up in accordance with the manufacturer's specific instructions in the owner's manual.

2. Proper safety protection equipment (safety glasses, welding helmet, leather gloves, pliers, long-sleeved shirt, long pants, and leather boots or shoes). Refer to Chapter 2 and Chapter 9 in the text for more specific safety information.

3. A mild steel plate, 3/8 in. (9 mm) by 4 in. (100 mm) by 6 in. (150 mm).

INSTRUCTIONS

1. Adjust the air pressure to approximately 80 psi.

2. Set the amperage within the range for the diameter electrode you are using by referring to the box that the electrodes came in.

3. Check to see that the stream of sparks will not start a fire or cause any damage to anyone or anything in the area.

4. Make sure the area is safe and turn on the welding machine.

5. Using a suitable dry leather glove to avoid electrical shock, insert the electrode in the torch jaws so that about 6 in. (152 mm) is extending outward. Be sure not to touch the electrode to any metal parts because it may short out.

6. Turn on the air at the torch head.

7. Lower your arc welding helmet.

8. Slowly bring the electrode down at about a 30° angle so it will make contact with the plate near the starting edge. Be prepared for a loud sharp sound when the arc starts.

9. Once the arc is struck, move the electrode in a straight line down the plate toward the other end. Keep the speed and angle of the torch constant. You should be attempting to skim the surface of the plate at a depth of 1/16 in. (2 mm) to 3/32 in. (2.3 mm) with each skimming pass made with the carbon arc electrode. Keep skimming the gouged surface until you have reached the depth required.

10. When you reach the other end, lift the torch so the arc will stop.

11. Raise your helmet and turn off the air.

12. Remove the remaining electrode from the torch so it will not accidently touch anything.

13. When the metal is cool, chip or brush any slag or dross off of the plate. This material should remove easily. The groove must be within ±1/8 in. (3 mm) of being straight and within ±3/32 in. (2.3 mm) of uniformity in width and depth.

14. Repeat this cut until it can be made within these tolerances.

15. Turn off the welding machine and clean up your work area when you are finished welding.

INSTRUCTOR'S COMMENTS _____

_____

_____

■ PRACTICE 9-2

Name_____   Date _____

Class _____   Instructor _____   Grade _____

OBJECTIVE: After completing this practice, you should be able to safely set up and operate an air carbon arc cutting torch to make a J-groove along the edge of a plate.

EQUIPMENT AND MATERIALS NEEDED FOR THIS PRACTICE

1. An air carbon arc cutting torch and welding power supply that has been safely set up in accordance with the manufacturer's specific instructions in the owner's manual.

2. Proper safety protection equipment (safety glasses, welding helmet, leather gloves, pliers, long-sleeved shirt, long pants, and leather boots or shoes). Refer to Chapter 2 and Chapter 9 in the text for more specific safety information.

3. A mild steel plate, 3/8 in. (9 mm) by 4 in. (100 mm) by 6 in. (150 mm).

INSTRUCTIONS

1. Adjust the air pressure to approximately 80 psi.

2. Set the amperage within the range for the diameter electrode you are using.

3. Check to see that the stream of sparks will not start a fire or cause any damage to anyone or anything in the area.

4. Make sure the area is safe and turn on the welding machine.

5. Using a suitable dry leather glove to avoid electrical shock, insert the electrode in the torch jaws so that about 6 in. (152 mm) is extending outward. Be sure not to touch the electrode to any metal parts because it may short out.

6. Turn on the air at the torch head.

7. Lower your arc welding helmet.

8. Slowly bring the electrode down at about a 30° angle so it will make contact with the plate near the starting edge.

9. Once the arc is struck, move the electrode in a straight line down the edge of the plate toward the other end. Keep the speed and angle of the torch constant. You should be trying to just skim the edge of the plate until you have gouged a J-groove along the plate edge.

10. When you reach the other end, lift the torch so the arc will stop.

11. Raise your helmet and turn off the air.

12. Remove the remaining electrode from the torch so it will not accidently touch anything.

13. When the metal is cool, chip or brush any slag or dross off of the plate. This material should remove easily. The groove must be within ±1/8 in. (3 mm) of being straight and within ±3/32 in. (2.3 mm) of uniformity in width and depth.

14. Repeat this cut until it can be made within these tolerances.

15. Turn off the welding machine and clean up your work area when you are finished welding.

INSTRUCTOR'S COMMENTS _____

_____

_____

■ PRACTICE 9-3

Name _____ Date _____ Electrode Used _____

Class _____ Instructor _____ Grade _____

OBJECTIVE: After completing this practice, you should be able to make a U-groove along the root face of a weld joint on a plate (back side).

EQUIPMENT AND MATERIALS NEEDED FOR THIS PRACTICE

1. An air carbon arc cutting torch and welding power supply that has been safely set up in accordance with the manufacturer's specific instructions in the owner's manual.

2. Proper safety protection equipment (safety glasses, welding helmet, leather gloves, pliers, long-sleeved shirt, long pants, and leather boots or shoes). Refer to Chapter 2 and Chapter 9 in the text for more specific safety information.

3. Two plates 6 in. (152) in length by 4 in. in width x 3/8 in. (10 mm) thick that have been joined with a suitable partial joint penetration groove-type weld.

INSTRUCTIONS

1. Adjust the air pressure to approximately 80 metric?

2. Set the amperage within the range for the diameter electrode you are using by referring to the box that the electrodes came in.

3. Check to see that the stream of sparks will not start a fire or cause any damage to anyone or anything in the area.

4. Make sure the area is safe and turn on the welding machine.

5. Using a suitable dry leather glove to avoid electrical shock, insert the electrode in the torch jaws so that about 6 in. (152 mm) is extending outward. Be sure not to touch the electrode to any metal parts because it may short out.

6. Take the groove welded plate and place it with the root face side up. (Back side of the plate up.)

7. Turn on the air at the torch head.

8. Lower your arc welding helmet.

9. Start the arc at the joint between the two plates on the side opposite where the weld bead is.

10. Once the arc is struck, move the electrode in a straight line down the edge of the plate toward the other end. Watch the bottom of the cut to see that it is deep enough. If there is a line along the bottom of the groove, it needs to be deeper. Once the groove depth is determined, keep the speed and angle of the torch constant. Try to gouge a depth of 3/32 in. (2.3 mm) with each skim pass of the carbon arc electrode.

11. When you reach the other end, break the arc off.

12. Raise your helmet and turn off the air.

13. Remove the remaining electrode from the torch so it will not accidently touch anything.

14. When the metal is cool, chip or brush any slag or dross off of the plate. This material should remove easily. The groove must be within ±1/8 in. (3 mm) of being straight, but it may vary in depth so long as all of the unfused root of the weld has been removed.

15. Repeat this cut until it can be made within these tolerances.

16. Turn off the welding machine and clean up your work area when you are finished welding.

INSTRUCTOR'S COMMENTS _____

_____

_____

■ PRACTICE 9-4

Name _____ Date _____ Electrode Used _____

Class _____ Instructor _____ Grade _____

OBJECTIVE: After completing this practice, you should be able to make a U-groove to remove a weld from a plate.

EQUIPMENT AND MATERIALS NEEDED FOR THIS PRACTICE

1. An air carbon arc cutting torch and welding power supply that has been safely set up in accordance with the manufacturer's specific instructions in the owner's manual.

2. Proper safety protection equipment (safety glasses, welding helmet, leather gloves, pliers, long-sleeved shirt, long pants, and leather boots or shoes). Refer to Chapter 2 and Chapter 9 in the text for more specific safety information.

3. Two steel plates that have been welded together, minimum length 6 in. (152 mm).

INSTRUCTIONS

1. Adjust the air pressure to approximately 80 psi.

2. Set the amperage within the range for the diameter electrode you are using by referring to the box that the electrodes came in.

3. Check to see that the stream of sparks will not start a fire or cause any damage to anyone or anything in the area.

4. Make sure the area is safe and turn on the welding machine.

5. Using a suitable dry leather glove to avoid electrical shock, insert the electrode in the torch jaws so that about 6 in. (152 mm) is extending outward. Be sure not to touch the electrode to any metal parts because it may short out.

6. Turn on the air at the torch head.

7. Lower your arc welding helmet.

8. Start the arc on one end of the weld bead.

9. Once the arc is struck, move the electrode in a straight line down the weld toward the other end. Watch the bottom of the cut to see that it is deep enough. If there is not a line along the bottom of the groove, it needs to be deeper. Once the groove depth is determined, keep the speed and angle of the torch constant.

10. When you reach the other end, break the arc off.

11. Raise your helmet and turn off the air.

12. Remove the remaining electrode from the torch so it will not accidently touch anything.

13. When the metal is cool, chip or brush any slag or dross off of the plate. This material should remove easily. The groove must be within ±1/8 in. (3 mm) of being straight, but it may vary in depth so that all of the weld metal has been removed. You should be able to completely separate the two plates.

14. Repeat this cut until it can be made within these tolerances.

15. Turn off the welding machine and clean up your work area when you are finished welding.

INSTRUCTOR'S COMMENTS _____

_____

_____

CHAPTER 9: RELATED CUTTING PROCESSES QUIZ

Name _____   Date _____

Class _____   Instructor _____   Grade _____

Instructions: Carefully read Chapter 9 in the text and answer the following questions.

A. SHORT ANSWER

Write a brief answer in the space provided that will answer the question or complete the statement.

1. List six advantages of laser cutting.

    a. _____

    b. _____

    c. _____

    d. _____

    e. _____

    f. _____

2. List three advantages of using a laser to drill a hole.

    a. _____

    b. _____

    c. _____

3. What are the four major differences between an air carbon arc torch and a shielded metal arc welding electrode holder?

    a. _____

    b. _____

    c. _____

    d. _____

4. What are the three shapes of air carbon arc electrodes?

    a. _____

    b. _____

    c. _____

5. List four metals that can be cut using the air carbon arc process.

    a. _____

    b. _____

    c. _____

    d. _____

B. FILL IN THE BLANK

Fill in the blank with the correct word. Answers may be more than one word.

6. Lasers are used for welding (LBW) materials that are too _____ or too _____ to be welded with other heat sources.

7. When the laser beam exits the _____, it can be treated as any other type of light.

8. Laser light beams can be focused into a very _____.

9. When a filler wire is automatically fed into the GTA welding arc it is referred to as _____ Gas Tungsten Arc Welding.

10. Unlike the oxyfuel process, the air carbon arc cutting process does not require that the base metal be _____ with the cutting stream.

11. If the CAC-A _____ is lower than the minimum, the arc will tend to sputter out, and it will be hard to make clean cuts.

12. CAC-A gouging is the removal of a quantity of metal to form a _____ or _____.

13. The CAC-A sound level is high enough to cause _____ if proper ear protection is not used.

14. The _____ can be used to cut reinforced concrete.

15. If an abrasive is used, the small water jet _____ will wear out faster.

C. ESSAY

Provide complete answers for all of the following questions.

16. List four things that manufacturers use lasers to do.

   a. _____

   b. _____

   c. _____

   d. _____

17. List four materials that can be used to produce a laser light.

a. _____

b. _____

c. _____

d. _____

18. What are the two major types or divisions of lasers?

a. _____

b. _____

19. List four gases or mixtures of gases that can be used to produce laser light.

a. _____

b. _____

c. _____

d. _____

20. What surface condition hinders the laser's ability to penetrate it?

_____

_____

21. What does the phrase "threshold of the surface temperature" mean in relation to lasers?

_____

_____

22. Explain how a laser is used to drill a hole through thick materials.

_____

_____

23. What is the power range of most lasers?

_____

_____

24. Why was early carbon arc cutting limited to vertical and overhead positions?

_____

_____

25. What is the purpose of the following air carbon arc electrodes?

a. round _____

b. flat _____

26. Why are some air carbon arc electrodes able to be joined together?

_____

_____

27. Why will a small diameter hose have an adverse effect on air flowing to the air carbon arc torch?

_____

_____

28. What might become a hazard to your health without proper safety precautions, including adequate ventilation, when air carbon arc cutting?

_____

_____

29. Why would air carbon arc "washing" be used?

_____

_____

30. List and explain six air carbon arc safety considerations.

a. _____

b. _____

c. _____

d. _____

e. _____

f. _____

31. How are most oxygen lances "started" for cutting?

_____

_____

32. When water jet cutting, what will cause the kerf to be wider?

_____

_____

D. DEFINITIONS

33. Define the following terms:

monochromatic _____

synchronized wave form _____

YAG _____

delamination _____

exothermic gases _____

graphite _____

gouging _____

washing _____

INSTRUCTOR'S COMMENTS _____

_____

_____

Chapter 10

Gas Metal Arc Welding Equipment, Setup, and Operation

CHAPTER 10: GAS METAL ARC WELDING EQUIPMENT, SETUP, AND OPERATION QUIZ

Name _____ Date _____

Class _____ Instructor _____ Grade _____

Instructions: Carefully read Chapter 10 of the text and answer the following questions.

A. MATCHING

In the space provided to the left, write the letter from Column B that best answers the question or completes the statement in Column A.

Column A

_____ 1. What is the short period at the end of the weld when the wire feed stops but the current does not?

_____ 2. What technique has good joint visibility and makes welds with less joint penetration?

_____ 3. What is the distance from the contact tube to the arc measured along the wire?

_____ 4. What is the measure of the amount of weld metal deposited?

_____ 5. What is the rate at which the wire is being consumed by the arc?

Column B

a. deposition rate

b. electrode extension

c. burn-back time

d. wire melting rate

e. forehand welding

Correctly match the following terms to the illustrations below.

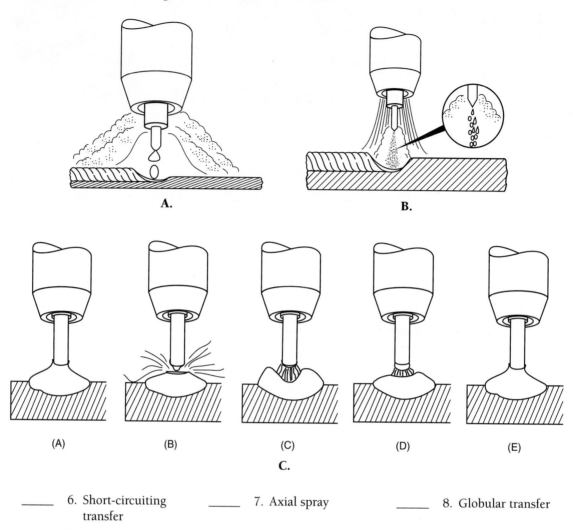

A.

B.

(A)   (B)   (C)   (D)   (E)

C.

_____ 6. Short-circuiting   _____ 7. Axial spray   _____ 8. Globular transfer
         transfer

B. FILL IN THE BLANK

Fill in the blank with the correct word. Answers may be more than one word.

9. Gas metal arc welding (GMAW) is also known as _____, _____, or

   _____.

10. The wire is in direct contact with the molten weld pool in the _____

    mode of transfer.

11. A glob of metal falls across the arc, landing in the molten weld pool in the _____

    transfer.

12. In the _____ metal transfer mode very small drops are projected across the

    arc gap to the molten weld pool.

13. The deposition rate is nearly always less that the _____ rate because not all of the wire is converted to weld metal.

14. The amount of weld metal deposited in ratio to the wire used is called the

_____.

15. The distance from the contact tube to the arc measured along the wire is called the

_____.

16. The forehand technique is sometimes referred to as _____ the weld bead, and backhand may be referred to as _____ the weld bead.

C. IDENTIFICATION

In the space provided, identify the items shown in the illustration, the major parts of the GMA welding equipment.

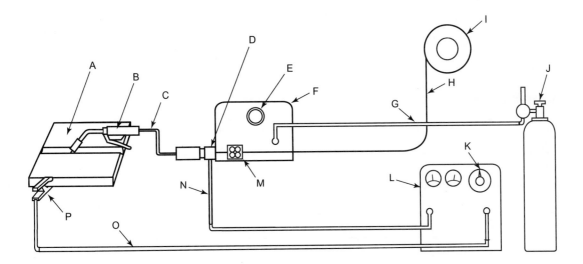

17. _____ welding machine

18. _____ wire-feed unit

19. _____ wire

20. _____ gun

21. _____ gas hose

22. _____ wire-speed feed control

23. _____ work cable

24. _____ shielding gas source

25. _____ voltage control

26. _____ wire reel

27. _____ welding gun cable

28. _____ power terminal

29. _____ work

30. _____ welding cable

31. _____ wire-feed drive rollers

32. _____ work clamp

D. SHORT ANSWER

Write a brief answer in the space provided that will answer the question or complete the statement.

33. List three methods of transferring filler metal with the GMAW process.

    a. _____

    b. _____

    c. _____

34. List three advantages of the GMA spot welding process.

    a. _____

    b. _____

    c. _____

E. ESSAY

Provide complete answers for all of the following questions.

35. What effect does changing the inductance have on a GMA weld?

    _____

    _____

36. What effect does changing the slope have on a GMA weld?

    _____

    _____

37. Why does changing the gun angle affect the weld?

    _____

    _____

38. Why does changing the electrode extension length affect the weld?

    _____

    _____

39. How does a linear electrode feed change the wire speed?

    _____

    _____

INSTRUCTOR'S COMMENTS _____

_____

_____

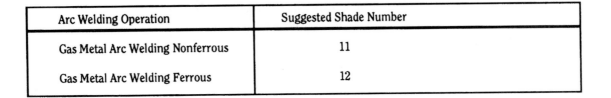

Chapter 11

Gas Metal Arc Welding

Arc Welding Operation	Suggested Shade Number
Gas Metal Arc Welding Nonferrous	11
Gas Metal Arc Welding Ferrous	12

Troubleshooting for GMAW	
Problem	**Correction**
Arc Blow	1. Change gun angle. 2. Move ground clamp. 3. Use backup bars, brass or copper. 4. Demagnetize part.
Cracked Welds	1. Check filler wire compatibility with base metal. 2. Use preheat and postheat on weldment. 3. Use a convex weld bead. 4. Check design of root opening. 5. Change welding speed. 6. Change shielding gas.
Dirty Welds	1. Decrease gun angle. 2. Hold gun nozzle closer to work. 3. Increase gas flow. 4. Clean weld joint area, gas flow. 5. Check for draft that may be blowing shielding gas away. 6. Check gun nozzle for damaged or worn parts. 7. Center contact tip in gun nozzle. 8. Clean filler wire before it enters wire drive. 9. Check cables and gun for air or water leaks. 10. Keep unused filler wire in shipping containers.
Wide Weld Bead	1. Increase welding speed. 2. Reduce current. 3. Use a different welding technique. 4. Shorten arc length.
Incomplete Penetration	1. Increase current. 2. Reduce welding speed. 3. Shorten arc length. 4. Increase root opening. 5. Change gun angle.

Problem	Correction
Irregular Arc Start	1. Use wire cutters to cut off the end of the filler wire before starting new weld. 2. Check ground. 3. Check contact tip. 4. Check polarity. 5. Check for drafts. 6. Increase gas flow.
Irregular Wire-Feed Burn-back	1. Check contact tip. 2. Check wire-feed speed. 3. Increase drive roll pressure. 4. Check voltage. 5. Check polarity. 6. Check wire spool for kinks or bends. 7. Clean or replace worn conduit liner.
Welding Cables Overheating	1. Check for loose cable connections. 2. Use larger cables. 3. Use shorter cables. 4. Decrease welding time.
Porosity	1. Check for drafts. 2. Check shielding gas. 3. Increase gas flow. 4. Decrease gun angle. 5. Hold nozzle close to work. 6. Do not weld if metal is wet. 7. Clean weld joint area. 8. Center contact tip with gun nozzle. 9. Check gun nozzle for damage. 10. Check gun and cables for air or water leaks.
Spatter	1. Change gun angle. 2. Shorten arc length. 3. Decrease wire speed. 4. Check for draft.
Undercutting	1. Reduce current. 2. Change gun angle. 3. Use different welding technique. 4. Reduce welding speed. 5. Shorten arc length.
Incomplete Fusion	1. Increase current. 2. Change welding technique. 3. Shorten arc length. 4. Check joint preparation. 5. Clean weld joint area.
Unstable Arc	1. Clean weld area. 2. Check contact tip. 3. Check for loose cable connections.

SHIELDING GASES AND GAS MIXTURES USED FOR GAS METAL ARC WELDING		
Shielding Gas	**Chemical Behavior**	**Uses and Usage Notes**
1. Argon	Inert	Welding virtually all metals except steel
2. Helium	Inert	Al and Cu alloys for greater heat and to minimize porosity
3. Ar & He (20–80 to 50–50%)	Inert	Al and Cu alloys for greater heat and to minimize porosity, but with quieter, more readily controlled arc action
4. N_2	Reducing	On Cu, very powerful arc
5. Ar + 25–30% N_2	Reducing	On Cu, powerful but smoother operating, more readily controlled arc than with N_2
6. Ar + 1–2% O_2	Oxidizing	Stainless and alloy steels, also for some deoxidized copper alloys
7. Ar + 3–5% O_2	Oxidizing	Plain carbon, alloy, and stainless steels (generally requires highly deoxidized wire)
8. Ar + 3–5% O_2	Oxidizing	Various steels using deoxidized wire
9. Ar + 20–30% O_2	Oxidizing	Various steels, chiefly with short-circuiting arc
10. Ar + 5% O_2 + 15% CO_2	Oxidizing	Various steels using deoxidized wire
11. CO_2	Oxidizing	Plain-carbon and low-alloy steels, deoxidized wire essential
12. CO_2 + 3–10% O_2	Oxidizing	Various steels using deoxidized wire
13. CO_2 + 20% O_2	Oxidizing	Steels

GENERALLY RECOMMENDED FILLER METALS AND SHIELDING GASES FOR GAS METAL ARC WELDING VARIOUS BASE METALS			
Base Metal Type	**Shielding Gas Composition**	**Specific Alloy to Be Welded**	**Filler Metal Type Electrode**
Aluminum and its alloys	Pure argon or helium-argon (75–25%)	1100 2219 3003, 3004 5050 5052 5154, 5254 5083, 5084, 5486 6061 7039	1100 or 4043 4145 or 2319 319 4043 4043 or 5554 5554 or 5154 5554 or 5154 5556 or 5356 4043 or 5556 5556, 5356, or 5183
Magnesium alloys	Pure argon	AZ31B, 61A, 81A ZE10XA ZK20XA AZ31B, 61A, 63A 80A, 81A, 91C, 92A AM80A, 100A ZE10XA XK20XA	AZ61A AZ92A
Copper	Helium-argon mixture 75–25%) pure argon on thin sections	Deoxidized copper	Deoxidized copper Silicon-0.25% Tin-0.75% Mn-0.15%
Copper-nickel alloy	Pure argon	Cu-Ni alloy: 70–30 90–10	Titanium deoxidized 70–30 Cu-Ni 70–30 or 90–10
Bronzes	Pure argon Argon + 5% O_2	Manganese bronze Aluminum bronze Nickel-aluminum bronze Tin bronze	Aluminum bronze Aluminum bronze Aluminum bronze Phosphor bronze
Nickel and nickel alloys	Helium-argon mixture (75–25) or pure argon	Nickel Nickel-copper (Monel) Nickel–chromium (Inconel)	Similar to base metal, titanium deoxidized (see supplier)
Plain low-carbon steel	CO_2; argon + 10 to 30% CO_2; or argon + 2 to 5% O_2	Hot or cold rolled sheet or plate ASTM A7, A36 A285, A373, or equivalent	Deoxidized plain carbon steel
Low-alloy carbon steel	Argon + 1–2% O_2 or argon + 10–30% CO_2	Hot or cold rolled sheet of various grades	Deoxidized low-alloy steel
Stainless steel	Argon + 1–5% O_2	302, 304, 321, 347, 309, 310 316, etc.	Electrode to match base alloy

TYPICAL SHIELDING GAS FLOW RATES* FOR GAS METAL ARC WELDING VARIOUS MATERIALS

Spray-Type Arc — 1/16 in. dia. filler wire

Materials	Argon	Helium	Argon-Helium (25–75%)	Argon-Oxygen (1–5 O_2)	CO_2
Aluminum	50	100	80	—	—
Magnesium	50	100	80	—	—
Plain C Steel	—	—	—	40	40
Low-alloy steel	—	—	—	40	40
Stainless steel	40	—	—	40	—
Nickel	50	100	80	—	—
Ni-Cu alloy	50	100	80	—	—
Ni-Cr Fe alloy	50	100	80	—	—
Copper	50	100	80	—	—
Cu-Ni alloy (70–30%)	50	100	80	—	—
Si bronze	40	80	60	—	—
Al bronze	50	100	80	—	—
Phos. bronze	40	80	80	40	—

Short-Circuiting Arc — 0.035 dia filler wire

Materials	Argon	Argon-Oxygen (1–5% O_2)	Argon-CO_2 (75–25%)	CO_2
Aluminum	35	—	—	—
Magnesium	35	—	—	—
Plain C Steel	25	25	25	35
Low-alloy steel	25	25	25	35
Stainless steel	25	25	—	35
Nickel	35	—	—	—
Ni-Cu alloy	35	—	—	—
Ni-Cr-Fe alloy	35	—	—	—
Copper	30	—	—	—
Cu-Ni alloy (70–30)	30	—	—	—
Si bronze	25	25	—	—
Al bronze	25	—	—	—
Phos. bronze	25	25	—	—

*All rates are in cubic feet per hour and are plus or minus 40%. The lower rates would be most suitable for indoor work and moderate amperage welding. The higher rates would be more suitable for high current, maximum speed, and outdoor welding.

OPERATING INSTRUCTIONS FOR GMAW

1. Set polarity.

   Making sure that the machine is turned off, connect the electrode cable to either the positive stud for reverse polarity or to the negative stud for straight polarity. Check with your instructor for the correct polarity. Some machines have a polarity switch instead of changing cables; this makes the changing of the polarity much easier.

2. Start the welder.

   Place the "power switch" to the "on" position. On some machines, you will also have to place the "power switch" on the wire feeder to the "on" position.

3. Open the shielding gas valve.

   Slowly open the flowmeter valve, simultaneously squeezing the gun trigger until the flowmeter reads approximately 15 to 20 cubic feet per hour.

4. Set voltage.

   Set the voltage by either increasing or decreasing the voltage control knob on the machine. The open circuit voltage should be approximately 2 volts higher than the desired welding voltage.

5. Set amperage (wire-speed control).

   Most amperage (wire-speed control) adjustment knobs are calibrated in inches per minute. Set the control knob to the desired wire speed. If you are not sure of what setting you need, place the adjustment knob in the 9 to 10 o'clock position and then fine-tune the amperage while making a sample weld.

6. The final setting for both voltage and amperage must be made while welding.

■ PRACTICE 11-1

Name _____ Date _____

Class _____ Instructor _____ Grade _____

OBJECTIVE: After completing this practice, you should be able to safely set up the GMAW equipment.

EQUIPMENT AND MATERIALS NEEDED FOR THIS PRACTICE

1. GMAW power source.

2. Welding gun.

3. Electrode feed unit.

4. Electrode supply.

5. Shielding gas supply.

6. Shielding gas regulator and flowmeter.

7. Electrode conduit.

8. Power and work leads.

9. Shielding gas hoses.

10. Assorted hand tools (see instructor).

INSTRUCTIONS

1. Be sure that the power to the machine is "off." If the shielding gas supply is a cylinder, it must be chained securely in place before the valve protection cap is removed. Standing to one side of the cylinder, quickly crack the valve to blow out any dirt in the valve before the flowmeter regulator is attached. Attach the correct hose from the regulator to the "gas-in" connection on the electrode feed unit or machine.

2. Install the reel of electrode (welding wire) on the holder and secure it.

3. Check the roller size to ensure that it matches the wire size.

4. The conduit liner size should be checked to be sure that it is compatible with the wire size.

5. Connect the conduit to the feed unit. The conduit or an extension should be aligned with the groove in the roller and set as close to the roller as possible without touching. Misalignment at this point can contribute to a bird's nest. Bird nesting of the electrode wire results when the feed roller pushes the wire into a tangled ball because the wire would not go through the outfeed side conduit. The welding wire appears to look like a bird's nest.

6. Be sure that the power is "off" before attaching the welding cables. The electrode and work leads should be attached to the proper terminals. The electrode lead should be attached to electrode or positive (+). If necessary, it is also attached to the power cable part of the gun lead. The work lead should be attached to work or negative (–).

7. The shielding "gas-out" side of the solenoid is then attached to the gun lead. If a separate splice is required from the gun switch circuit to the feed unit, it should be connected at this time.

8. Check to see that the welding contactor circuit is connected from the feed unit to the power source. The welding gun should be permanently attached to the main lead cable and conduit.

9. There should be a gas diffuser attached to the end of the conduit liner to ensure proper gas flow.

10. A contact tube (tip) of the correct size to match the electrode wire size being used should be installed.

11. A shielding gas nozzle is attached to complete the assembly.

12. Recheck all fittings and connections for tightness. Loose fittings can leak; loose connections can cause added resistance, reducing the welding efficiency.

13. Some manufacturers include detailed setup instructions with their equipment.

INSTRUCTOR'S COMMENTS _____

_____

_____

■ PRACTICE 11-2

Name _____ Date _____

Class _____ Instructor _____ Grade _____

OBJECTIVE: After completing this practice, you should be able to safely thread the GMAW wire into the equipment.

EQUIPMENT AND MATERIALS NEEDED FOR THIS PRACTICE

1. GMAW power source.

2. Welding gun.

3. Electrode feed unit.

4. Electrode supply.

5. Shielding gas supply.

6. Shielding gas regulator and flowmeter.

7. Electrode conduit.

8. Power and work leads.

9. Shielding gas hoses.

10. Assorted hand tools (see instructor).

INSTRUCTIONS

1. Check to see that the unit is assembled correctly according to the manufacturer's specifications.

2. Switch on the power and check the gun switch circuit by depressing the switch. The power source relays, feed relays, gas solenoid, and feed motor should all activate.

3. Cut the end of the electrode wire free. Hold it tightly so that it does not unwind. The wire has a natural curve that is known as its cast. The cast is measured by the diameter of the circle that the wire would make if it were loosely laid on a flat surface. The cast helps the wire make a good electrical contact as it passes through the contact tube. However, the cast can be a problem when threading the system.

4. To make threading easier, straighten about 12 in. (305 mm) of the end of the wire and cut any kinks off.

5. Separate the wire-feed rollers and push the wire first through the guides, then between the rollers, and finally into the conduit liner.

6. Reset the rollers so that there is a slight amount of compression on the wire.

7. Set the wire-feed speed control to a slow speed.

8. Hold the welding gun so that the electrode conduit and cable are as straight as possible.

9. Remove the nozzle and contact tip from the gun.

10. Press the gun switch. The wire should start feeding into the liner. Watch to make certain that the wire feeds smoothly and release the gun switch as soon as the end comes through the gun.

CAUTION **If the wire stops feeding before it reaches the end of the contact tube, stop and check the system. If no obvious problem can be found, mark the wire with tape and remove it from the gun. It then can be held next to the system to determine the location of the problem.**

11. With the wire feed running, adjust the feed roller compression so that the wire reel can be stopped easily by a slight pressure. Too light a roller pressure will cause the wire to feed erratically. Too high a pressure can turn a minor problem into a major disaster. If the wire jams at a high roller pressure, the feed rollers keep feeding the wire, causing it to bird nest and possibly short out. With a light pressure, the wire can stop, preventing bird nesting. This is very important with soft wires. The other advantage of a light pressure is that the feed will stop if something like clothing or gas hoses are caught in the reel.

12. With the feed running, adjust the spool drag so that the reel stops when the feed stops. The reel should not coast to a stop because the wire can be snagged easily. Also, when the feed restarts, a jolt occurs when the slack in the wire is taken up. This jolt can be enough to momentarily stop the wire, possibly causing a discontinuity in the weld or may even snap the wire.

13. When the test runs are completed, the wire can either be rewound or cut off. Some wire-feed units have a retract button. This allows the feed driver to reverse and retract the wire automatically. To rewind the wire on units without this retract feature, release the rollers and turn them backward by hand. If the machine will not allow the feed rollers to be released without upsetting the tension, you must cut the wire.

14. Replace the tip and nozzle on the gun.

CAUTION **Do not discard pieces of wire on the floor. They present a hazard to safe movement around the machine. In addition, a small piece of wire can work its way into a filter screen on the welding power source. If the piece of wire shorts out inside the machine, it could become charged with high voltage, which may cause injury or death. Always wind the wire tightly into a ball or cut it into short lengths before discarding it in the proper waste container.**

15. Use wire nippers and nip the electrode wire to the correct stick out length.

INSTRUCTOR'S COMMENTS _____

_____

_____

■ PRACTICE 11-3

Name _____ Date _____

Class _____ Instructor _____ Grade _____

Welding Principles and Applications

MATERIAL:
16 GA AND 1/8" MILD STEEL SHEET 12" X 3"

PROCESS:
GMAW STRINGER BEAD FLAT POSITION

NUMBER:	DRAWN BY:
PRACTICE 11–3	GAVIN DUBOIS

OBJECTIVE: After completing this practice, you should be able to make stringer beads on the 1/8-in. (3-mm) and 16-gauge carbon steel using the short-circuiting metal transfer method in the flat position.

EQUIPMENT AND MATERIALS NEEDED FOR THIS PRACTICE

1. A properly set-up and adjusted GMA welding machine.

2. Proper safety protection (welding hood, safety glasses, leather gloves, wire cutting pliers, long-sleeved shirt, long pants, leather shoes or boots). Refer to Chapter 2 in the text for more specific safety information.

3. 0.035-in. and/or 0.045-in. (0.9-mm and/or 1.2-mm) diameter wire.

4. Two or more pieces of mild steel sheet, 12-in. (305-mm) long by 3-in. (75-mm) wide and 16-gauge and 1/8-in. (3-mm) thick.

INSTRUCTIONS

1. Starting at one end of the plate and using either a pushing or dragging technique, make a weld bead along the entire 12 in. (305 mm) length of the metal.

2. After the weld is complete, check its appearance. Make any needed changes to correct the weld.

3. Repeat the weld and make additional adjustments.

4. After the machine is set, start to work on improving the straightness and uniformity of the weld. Keeping the bead straight and uniform can be hard because of the limited visibility due to the small amount of light and the size of the molten weld pool. The welder's view is further restricted by the shielding gas nozzle. Even with limited visibility, it is possible to make a satisfactory weld by watching the edge of the molten weld pool, the sparks, and the weld bead produced. Watching the leading edge of the molten weld pool (forehand welding, push technique) will show you the molten weld pool fusion and width. Watching the trailing edge of the molten weld pool (backhand welding, drag technique) will show you the amount of buildup and the relative heat input. The quantity and size of sparks produced can indicate the relative location of the filler wire in the molten weld pool. The number of sparks will increase as the wire strikes the solid metal ahead of the molten weld pool. The gun itself will begin to vibrate or bump as the wire momentarily pushes against the cooler, unmelted base metal before it melts. Changes in weld width, buildup, and proper joint tracking can be seen by watching the bead as it appears from behind the shielding gas nozzle.

5. Repeat each type of bead on both plate thicknesses as needed until consistently good beads are obtained.

6. Turn off the welding machine and shielding gas, and clean up your work area when you are finished welding.

INSTRUCTOR'S COMMENTS _____

_____

_____

■ PRACTICE 11-4

Name _____ Date _____

Class _____ Instructor _____ Grade _____

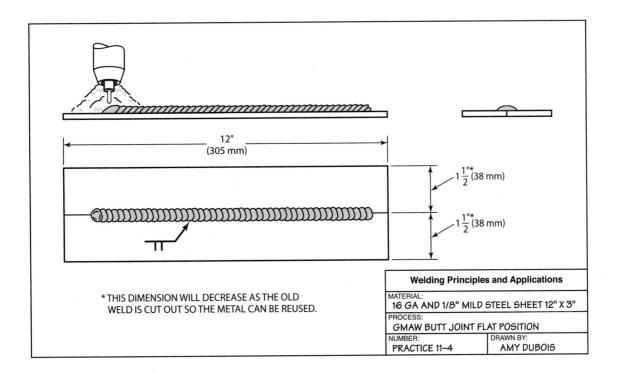

Welding Principles and Applications

MATERIAL:
16 GA AND 1/8" MILD STEEL SHEET 12" X 3"

PROCESS:
GMAW BUTT JOINT FLAT POSITION

NUMBER:	DRAWN BY:
PRACTICE 11–4	AMY DUBOIS

* THIS DIMENSION WILL DECREASE AS THE OLD
WELD IS CUT OUT SO THE METAL CAN BE REUSED.

OBJECTIVE: After completing this practice, you should be able to weld a butt, lap, and tee joint in the flat position using 1/8-in. (3-mm) and 16-gauge carbon steel.

EQUIPMENT AND MATERIALS NEEDED FOR THIS PRACTICE

1. A properly set-up and adjusted GMA welding machine.

2. Proper safety protection (welding hood, safety glasses, leather gloves, wire cutting pliers, long-sleeved shirt, long pants, leather shoes or boots). Refer to Chapter 2 in the text for more specific safety information.

3. 0.035-in. and/or 0.045-in. (0.9-mm and/or 1.2-mm) diameter wire.

4. Two or more pieces of mild steel sheet, 12-in. (305-mm) long by 1 1/2-in. (38-mm) wide and 16-gauge and 1/8-in. (3-mm) thick.

INSTRUCTIONS

1. Tack weld the sheets together in a butt joint and place them flat on the welding table.

2. Starting at one end, run a bead along the joint. Watch the molten weld pool and bead for signs that a change in technique may be required.

3. Make any needed changes as the weld progresses. By the time the weld is complete, you should be making a nearly perfect weld.

4. Using the same technique that was established in the last weld, make another weld. This time, the entire 12 in. (305 mm) of weld should be defect free.

5. Repeat each type of joint with both thicknesses of metal as needed until consistently good quality beads are obtained.

6. Repeat the above steps for the lap joint and tee joint.

7. Turn off the welding machine and shielding gas, and clean up your work area when you are finished welding.

INSTRUCTOR'S COMMENTS _____

_____

_____

■ PRACTICE 11-5

Name _____ Date _____

Class _____ Instructor _____ Grade _____

OBJECTIVE: After completing this practice, you should be able to make butt, lap, and tee joints. All with 100% penetration.

EQUIPMENT AND MATERIALS NEEDED FOR THIS PRACTICE

1. A properly set-up and adjusted GMA welding machine.

2. Proper safety protection (welding hood, safety glasses, leather gloves, wire cutting pliers, long-sleeved shirt, long pants, leather shoes or boots). Refer to Chapter 2 in the text for more specific safety information.

3. 0.035-in. and/or 0.045-in. (0.9 mm and/or 1.2-mm) diameter wire.

4. Two or more pieces of mild steel sheet, 12-in. (305-mm) long by 1 1/2-in. (38-mm) wide and 16-gauge and 1/8-in. (3-mm) thick.

INSTRUCTIONS

1. Tack weld the sheets together in a butt joint and place them in the flat position.

2. Start at one end of the plate and run a bead along the joint.

3. Use gun angle, current, and travel speed which will produce 100% penetration.

4. Make any changes as needed as the weld progresses.

5. Repeat until consistently good welds can be made with 100% penetration on both thicknesses of metal.

6. Repeat the above steps with the lap joint and tee joint.

7. Turn off the welding machine and shielding gas, and clean up your work area when you are finished welding.

INSTRUCTOR'S COMMENTS _____

_____

_____

■ PRACTICE 11-6

Name _____ Date _____

Class _____ Instructor _____ Grade _____

OBJECTIVE: After completing this practice, you should be able to make a stringer bead in the vertical up position at a 45° angle.

EQUIPMENT AND MATERIALS NEEDED FOR THIS PRACTICE

1. A properly set-up and adjusted GMA welding machine.

2. Proper safety protection (welding hood, safety glasses, leather gloves, wire cutting pliers, long-sleeved shirt, long pants, leather shoes or boots). Refer to Chapter 2 in the text for more specific safety information.

3. 0.035-in. and/or 0.045-in. (0.9-mm and/or 1.2-mm) diameter wire.

4. One or more pieces of mild steel sheet, 12-in. (305-mm) long by 1 1/2-in. (38-mm) wide and 16-gauge and 1/8-in. (3-mm) thick.

INSTRUCTIONS

1. Brace one of the plates on the welding table at a 45° inclined angle.

2. Start at the bottom of the plate and hold the welding gun at a slight angle to the plate.

3. Brace yourself, lower your hood, and begin to weld.

4. Depending on the machine settings and type of shielding gas used, you will make a weave pattern.

5. If the molten weld pool is large and fluid (hot), use a "C" or "J" weave pattern to allow a longer time for the molten weld pool to cool.

6. Do not make the weave so long or fast that the wire is allowed to strike the metal ahead of the molten weld pool. If this happens, spatter increases and a spot or zone of incomplete fusion may occur.

7. If the molten weld pool is small and controllable, use a small "C," zigzag, or "J" weave pattern to control the width and buildup of the weld. A slower travel speed can also be used.

8. Watch for complete fusion along the leading edge of the molten weld pool.

9. A weld that is high and has little or no fusion is too "cold." Changing the welding technique will not correct this problem. The welder must stop welding and make the needed adjustments to the welding machine.

10. As the weld progresses up the plate, the back or trailing edge of the molten weld pool will cool, forming a shelf to support the molten metal. Watch the shelf to be sure that molten metal does not run over, forming a drip. When it appears that the metal may flow over the shelf, either increase the weave lengths or stop and start the current for brief moments to allow the weld to cool. Stopping for brief moments will not allow the shielding gas to be lost.

11. Continue to weld along the entire 12 in. (305 mm) length of plate.

12. Repeat this weld as needed until a straight and uniform weld bead is produced on both thicknesses of metal.

13. Turn off the welding machine and shielding gas, and clean up your work area when you are finished welding.

INSTRUCTOR'S COMMENTS _____

_____

_____

■ PRACTICE 11-7

Name _____ Date _____

Class _____ Instructor _____ Grade _____

OBJECTIVE: After completing this practice you should be able to make stringer beads on sheet steel in the vertical up position.

EQUIPMENT AND MATERIALS NEEDED FOR THIS PRACTICE

1. A properly set-up and adjusted GMA welding machine.

2. Proper safety protection (welding hood, safety glasses, wire brush, chipping hammer, leather gloves, pliers, long-sleeved shirt, long pants, and leather boots or shoes). Refer to Chapter 2 in the text for more specific safety information.

3. 0.035-in. and/or 0.045-in. (0.9-mm and/or 1.2-mm) diameter welding wire.

4. Two or more pieces of mild steel sheet, 12-in. (305-mm) long by 1 1/2-in. (38-mm) wide and 16-gauge and 1/8-in. (3-mm) thick.

INSTRUCTIONS

1. Place the sheet steel on the welding table at a 90° angle, vertical up angle.

2. Start at the bottom of the plate and hold the gun at a slight up angle to the plate.

3. Brace yourself, lower your hood, and begin to weld.

4. Depending on the machine settings and type of shielding gas, you may make a weave pattern.

5. If the weld pool is large and fluid (hot), use a "C" or "J" weave pattern to allow a longer time for the molten weld pool to cool.

6. Do not make the weave so long or so fast that the wire is allowed to strike the metal ahead of the molten weld pool. If this happens, spatter increases and a spot or zone of incomplete fusion may occur.

7. If the molten weld pool is small and controllable, use a small "C," zigzag, or "J" weave pattern to control the width and buildup of the bead. A slower travel speed can also be used.

8. Watch for complete fusion along the leading edge of the molten weld pool.

9. A weld that is too high and has little or no fusion is too "cold." Changing the welding technique will not correct the problem. The welder must stop welding and make the needed adjustments to the welding machine.

10. As the weld progresses up the plate, the back or trailing edge of the molten weld pool will cool forming a shelf to support the molten metal. Watch the shelf to be sure that molten weld metal does not run over, forming a drip. When it appears that the metal may flow over the shelf, either increase the weave lengths or stop and start the current for brief moments to allow the weld to cool. Stopping for brief moments will not allow the shielding gas to be lost.

11. Continue to weld along the entire 12 in. (305 mm) length of the plate.

12. Repeat the weld as needed on both thicknesses of plate until a straight and uniform weld bead with proper penetration can be made.

13. Turn off the welding machine and shielding gas, and clean up your work area when finished with this practice.

INSTRUCTOR'S COMMENTS _____

_____

_____

■ PRACTICE 11-8

Name _____ Date _____

Class _____ Instructor _____ Grade _____

OBJECTIVE: After completing this practice you should be able to make butt, lap, and tee joints in the vertical up position at a 45° angle.

EQUIPMENT AND MATERIALS NEEDED FOR THIS PRACTICE

1. A properly set-up and adjusted GMA welding machine.

2. Proper safety protection (welding hood, safety glasses, wire brush, chipping hammer, leather gloves, pliers, long-sleeved shirt, long pants, and leather boots or shoes). Refer to Chapter 2 in the text for more specific safety information.

3. 0.035-in. and/or 0.045-in. (0.9-mm and/or 1.2-mm) diameter welding wire.

4. Two or more pieces of mild steel sheet, 12-in. (305-mm) long by 1 1/2-in. (38-mm) wide and 16-gauge and 1/8-in. (3-mm) thick.

INSTRUCTIONS

1. Tack weld the pieces together in the butt configuration, lap, configuration, and tee configuration and place them on the welding table at a 45° angle, vertical up angle.

2. Start at the bottom of the plate and hold the gun at a slight up angle to the plate.

3. Brace yourself, lower your hood, and begin to weld.

4. Depending on the machine settings and type of shielding gas, you may make a weave pattern.

5. If the weld pool is large and fluid (hot), use a "C" or "J" weave pattern to allow a longer time for the molten weld pool to cool.

6. Do not make the weave so long or so fast that the wire is allowed to strike the metal ahead of the molten weld pool. If this happens, spatter increases and a spot or zone of incomplete fusion may occur.

7. If the molten weld pool is small and controllable, use a small "C," zigzag, or "J" weave pattern to control the width and buildup of the bead. A slower travel speed can also be used.

8. Watch for complete fusion along the leading edge of the molten weld pool.

9. A weld that is too high and has little or no fusion is too "cold." Changing the welding technique will not correct the problem. The welder must stop welding and make the needed adjustments to the welding machine.

10. As the weld progresses up the plate, the back or trailing edge of the molten weld pool will cool forming a shelf to support the molten metal. Watch the shelf to be sure that molten weld metal does not run over, forming a drip. When it appears that the metal may flow over the shelf, either increase the weave lengths or stop and start the current for brief moments to allow the weld to cool. Stopping for brief moments will not allow the shielding gas to be lost.

11. Continue to weld along the entire 12 in. (305 mm) length of the plate.

12. Repeat the weld as needed on both thicknesses of plate until a straight and uniform weld bead with proper penetration can be made.

13. Turn off the welding machine and shielding gas, and clean up your work area when finished with this practice.

INSTRUCTOR'S COMMENTS _____

_____

_____

■ PRACTICE 11-9

Name _____ Date _____

Class _____ Instructor _____ Grade _____

OBJECTIVE: After completing this practice, you should be able to make stringer beads on sheet steel in the 45° vertical down position.

EQUIPMENT AND MATERIALS NEEDED FOR THIS PRACTICE

1. A properly set-up and adjusted GMA welding machine.

2. Proper safety protection (welding hood, safety glasses, leather gloves, wire cutting pliers, long-sleeved shirt, long pants, leather shoes or boots). Refer to Chapter 2 in the text for more specific safety information.

3. 0.035-in. and/or 0.045-in. (0.9-mm and/or 1.2-mm) diameter wire.

4. Two or more pieces of mild steel sheet, 12-in. (305-mm) long by 1 1/2-in. (38-mm) wide and 16-gauge and 1/8-in. (3-mm) thick.

INSTRUCTIONS

1. Place the sheet steel on the welding table at a 45° angle, vertical down position.

2. Start at the top of the plate and hold the welding gun at a slight dragging angle to the plate.

3. Brace yourself, lower your hood, and begin to weld.

4. Depending on the machine settings and type of shielding gas used, you may or may not make a weave pattern.

5. If the molten weld pool is large and fluid (hot), use a "C" weave pattern to allow a longer time for the molten weld pool to cool.

6. The leading edge and sides should flow into the base metal not curl over onto it. (Cold lap.)

7. Some changes in gun angle may increase penetration. Experiment with this.

8. Watch for complete fusion along the leading edge and sides of the molten weld pool.

9. A weld that is high and has little or no fusion is too "cold." Changing the welding technique will not correct this problem. The welder must stop welding and make the needed adjustments to the welding machine.

10. Continue to weld along the entire 12 in. (305 mm) length of plate.

11. Repeat this weld as needed on both thicknesses of plate until a straight weld bead is produced that is uniform in height and width.

12. Turn off the welding machine and shielding gas, and clean up your work area when you are finished welding.

INSTRUCTOR'S COMMENTS _____

_____

_____

■ PRACTICE 11-10

Name _____ Date _____

Class _____ Instructor _____ Grade _____

OBJECTIVE: After completing this practice, you should be able to make stringer beads on sheet steel in the vertical down position.

EQUIPMENT AND MATERIALS NEEDED FOR THIS PRACTICE

1. A properly set-up and adjusted GMA welding machine.

2. Proper safety protection (welding hood, safety glasses, leather gloves, wire cutting pliers, long-sleeved shirt, long pants, leather shoes or boots). Refer to Chapter 2 in the text for more specific safety information.

3. 0.035-in. and/or 0.045-in. (0.9-mm and/or 1.2-mm) diameter wire.

4. Two or more pieces of mild steel sheet, 12-in. (305-mm) long by 1 1/2-in. (38-mm) wide and 16-gauge and 1/8-in. (3-mm) thick.

INSTRUCTIONS

1. Place the sheet steel on the welding table at a 90° angle, vertical down position.

2. Start at the top of the plate and hold the welding gun at a slight dragging angle to the plate.

3. Brace yourself, lower your hood, and begin to weld.

4. Depending on the machine settings and type of shielding gas used, you may or may not make a weave pattern.

5. If the molten weld pool is large and fluid (hot), use a "C" weave pattern to allow a longer time for the molten weld pool to cool.

6. Watch the sides and leading edge of the molten weld pool for proper fusion.

7. Change the gun angle as needed to get proper penetration.

8. Continue to weld along the entire 12 in. (305 mm) length of plate.

9. Repeat this weld as needed on both thicknesses of plate until a straight weld bead is produced that is uniform in height and width and is defect free.

10. Turn off the welding machine and shielding gas, and clean up your work area when you are finished welding.

INSTRUCTOR'S COMMENTS _____

_____

_____

■ PRACTICE 11-11

Name _____ Date _____

Class _____ Instructor _____ Grade _____

OBJECTIVE: After completing this practice, you should be able to make butt, lap, and tee joints in the vertical down position.

EQUIPMENT AND MATERIALS NEEDED FOR THIS PRACTICE

1. A properly set-up and adjusted GMA welding machine.

2. Proper safety protection (welding hood, safety glasses, leather gloves, wire cutting pliers, long-sleeved shirt, long pants, leather shoes or boots). Refer to Chapter 2 in the text for more specific safety information.

3. 0.035-in. and/or 0.045-in. (0.9-mm and/or 1.2-mm) diameter wire.

4. Two or more pieces of mild steel sheet, 12-in. (305-mm) long by 1 1/2-in. (38-mm) wide and 16-gauge and 1/8-in. (3-mm) thick.

INSTRUCTIONS

1. Tack weld the metal pieces of the butt, lap, and tee joints together and brace them on the welding table at a 90° angle for vertical down welding.

2. Start at the top of the plate, holding the welding gun with a slight dragging angle.

3. Brace yourself, lower your hood, and begin to weld.

4. Depending on the machine settings and type of shielding gas used, you will make a weave pattern.

5. If the molten weld pool is large and fluid (hot), use a "C" weave pattern to allow a longer time for the molten weld pool to cool.

6. Do not make the weave so long or fast that the wire is allowed to strike the metal ahead of the molten weld pool. If this happens, spatter increases and a spot or zone of incomplete fusion may occur.

7. If the molten weld pool is small and controllable, use a small "C," zigzag, or "J" weave pattern to control the width and buildup of the weld. A slower travel speed can also be used.

8. Watch for complete fusion along the leading edge and sides of the molten weld pool.

9. A weld that is high and has little or no fusion is too "cold." Changing the welding technique will not correct this problem. The welder must stop welding and make the needed adjustments to the welding machine.

10. Continue to weld along the entire 12 in. (305 mm) length of plate.

11. Repeat this weld as needed on both thicknesses of plate until a straight weld bead is produced that is uniform in height and width.

12. Turn off the welding machine and shielding gas, and clean up your work area when you are finished welding.

INSTRUCTOR'S COMMENTS _____

_____

_____

■ PRACTICE 11-12

Name _____ Date _____

Class _____ Instructor _____ Grade _____

OBJECTIVE: After completing this practice you should be able to make butt and tee joints in the vertical down position with 100% penetration.

EQUIPMENT AND MATERIALS NEEDED FOR THIS PRACTICE

1. A properly set-up and adjusted GMA welding machine.

2. Proper safety protection (welding hood, safety glasses, wire brush, chipping hammer, leather gloves, pliers, long-sleeved shirt, long pants, and leather boots or shoes). Refer to Chapter 2 in the text for more specific safety information.

3. 0.035 in. and/or 0.045-in. (0.9-mm and/or 1.2-mm) diameter welding wire.

4. Two or more pieces of mild steel sheet, 12-in. (305-mm) long by 1 1/2-in. (38-mm) wide and 16-gauge and 1/8-in. (3-mm) thick.

INSTRUCTIONS

1. Tack weld the pieces together in the butt configuration and tee configuration and place them on the welding table at a 90° vertical down position.

2. It may be necessary to adjust the root openings and machine settings to get 100% penetration.

3. Brace yourself, lower your hood and, starting at the top, begin to weld downward.

4. Depending on the machine settings and type of shielding gas, you may make a weave pattern.

5. If the weld pool is large and fluid (hot), use a "C" weave pattern to allow a longer time for the molten weld pool to cool.

6. Do not make the weave so long or so fast that the wire is allowed to strike the metal ahead of the molten weld pool. If this happens, spatter increases and a spot or zone of incomplete fusion may occur.

7. If the molten weld pool is small and controllable, use a small "C," zigzag, or "J" weave pattern to control the width and buildup of the bead. A slower travel speed can also be used.

8. Watch for complete fusion along the sides and leading edge of the molten weld pool.

9. A weld that is too high and has little or no fusion is too "cold." Changing the welding technique will not correct the problem. The welder must stop welding and make the needed adjustments to the welding machine.

10. Continue to weld along the entire 12 in. (305 mm) length of the plate.

11. Repeat the weld as needed on both thicknesses of plate until a straight and uniform weld bead with 100% penetration can be made.

12. Turn off the welding machine and shielding gas, and clean up your work area when finished with this practice.

INSTRUCTOR'S COMMENTS _____

_____

_____

■ PRACTICE 11-13

Name _____ Date _____

Class _____ Instructor _____ Grade _____

OBJECTIVE: After completing this practice, you should be able to make horizontal stringer beads on a plate with a reclined angle.

EQUIPMENT AND MATERIALS NEEDED FOR THIS PRACTICE

1. A properly set-up and adjusted GMA welding machine.

2. Proper safety protection (welding hood, safety glasses, leather gloves, wire cutting pliers, long-sleeved shirt, long pants, leather shoes or boots). Refer to Chapter 2 in the text for more specific safety information.

3. 0.035-in. and/or 0.045-in. (0.9-mm and/or 1.2-mm) diameter wire.

4. One or more pieces of mild steel sheet, 12-in. (305-mm) long by 1 1/2-in. (38-mm) wide and 16-gauge and 1/8-in. (3-mm) thick.

INSTRUCTIONS

1. Place the mild steel plate on the welding table with a 45° reclined angle.

2. Start at one end of the plate with the gun pointed in a slightly upward direction.

3. You may use a pushing or a dragging gun angle, depending on the current setting and penetration desired.

4. Undercutting along the top edge and overlap along the bottom edge are problems with both gun angles. Careful attention must be paid to the manipulation of the "weave" technique used to overcome these problems. The most successful weave patterns are the "C" and "J" patterns. The "J" pattern is the most frequently used. The "J" pattern allows weld metal to be deposited along a shelf created by the previous weave. The length of the "J" can be changed to control the weld bead size.

5. Small weld beads are easier to control than large ones.

6. The weld must be straight and uniform in size and have complete fusion.

7. Repeat these welds on both thicknesses of plate until you have established the rhythm and technique that work well for you.

8. Turn off the welding machine and shielding gas, and clean up your work area when you are finished welding.

INSTRUCTOR'S COMMENTS _____

_____

_____

■ PRACTICE 11-14

Name _____ Date _____

Class _____ Instructor _____ Grade _____

OBJECTIVE: After completing this practice, you should be able to make stringer beads in the horizontal position on mild steel plate.

EQUIPMENT AND MATERIALS NEEDED FOR THIS PRACTICE

1. A properly set-up and adjusted GMA welding machine.

2. Proper safety protection (welding hood, safety glasses, leather gloves, wire cutting pliers, long-sleeved shirt, long pants, leather shoes or boots). Refer to Chapter 2 in the text for more specific safety information.

3. 0.035-in. and/or 0.045-in. (0.9-mm and/or 1.2-mm) diameter wire.

4. One or more pieces of mild steel sheet, 12-in. (305-mm) long by 1 1/2-in. (38-mm) wide and 16-gauge and 1/8-in. (3-mm) thick.

INSTRUCTIONS

1. Place the mild steel plate in a horizontal position on the welding table.

2. Start at one end of the plate with the gun pointed in a slightly upward direction.

3. You may use a pushing or a dragging gun angle, depending on the current setting and penetration desired.

4. Undercutting along the top edge and overlap along the bottom edge are problems with both gun angles. Careful attention must be paid to the manipulation of the "weave" technique used to overcome these problems. The most successful weave patterns are the "C" and "J" patterns. The "J" pattern is the most frequently used. The "J" pattern allows weld metal to be deposited along a shelf created by the previous weave. The length of the "J" can be changed to control the weld bead size.

5. Small weld beads are easier to control than large ones.

6. The weld must be straight and uniform in size and have complete fusion.

7. Repeat these welds with both thicknesses of plate until you have established the rhythm and technique that work well for you.

8. Turn off the welding machine and shielding gas, and clean up your work area when you are finished welding.

INSTRUCTOR'S COMMENTS _____

_____

_____

■ PRACTICE 11-15

Name _____ Date _____

Class _____ Instructor _____ Grade _____

OBJECTIVE: After completing this practice, you should be able to weld a butt and tee joint in the horizontal position on mild steel plate.

EQUIPMENT AND MATERIALS NEEDED FOR THIS PRACTICE

1. A properly set-up and adjusted GMA welding machine.

2. Proper safety protection (welding hood, safety glasses, leather gloves, wire cutting pliers, long-sleeved shirt, long pants, leather shoes or boots). Refer to Chapter 2 in the text for more specific safety information.

3. 0.035-in. and/or 0.045-in. (0.9-mm and/or 1.2-mm) diameter wire.

4. Two or more pieces of mild steel sheet, 12-in. (305-mm) long by 1 1/2-in. (38-mm) wide and 16-gauge and 1/8-in. (3-mm) thick.

INSTRUCTIONS

1. Tack weld the metal pieces of the butt and tee joints together and brace them on the welding table in the horizontal position.

2. Start at one end of the plate with the gun pointed in a slightly upward direction.

3. You may use a pushing or a dragging gun angle, depending on the current setting and penetration desired.

4. Undercutting along the top edge and overlap along the bottom edge are problems with both gun angles. Careful attention must be paid to the manipulation of the "weave" technique used to overcome these problems. The most successful weave patterns are the "C" and "J" patterns. The "J" pattern is the most frequently used. The "J" pattern allows weld metal to be deposited along a shelf created by the previous weave. The length of the "J" can be changed to control the weld bead size.

5. Small weld beads are easier to control than large ones.

6. The weld must be straight and uniform in size and have complete fusion.

7. Repeat these welds on both thicknesses of metal until you have established the rhythm and technique that work well for you.

8. Turn off the welding machine and shielding gas, and clean up your work area when you are finished welding.

INSTRUCTOR'S COMMENTS _____

_____

_____

■ PRACTICE 11-16

Name _____ Date _____

Class _____ Instructor _____ Grade _____

OBJECTIVE: After completing this practice, you should be able to weld a butt and tee joint in the horizontal position on mild steel plate with 100% penetration.

EQUIPMENT AND MATERIALS NEEDED FOR THIS PRACTICE

1. A properly set-up and adjusted GMA welding machine.

2. Proper safety protection (welding hood, safety glasses, leather gloves, wire cutting pliers, long-sleeved shirt, long pants, leather shoes or boots). Refer to Chapter 2 in the text for more specific safety information.

3. 0.035-in. and/or 0.045-in. (0.9-mm and/or 1.2-mm) diameter wire.

4. Two or more pieces of mild steel sheet, 12-in. (305-mm) long by 1 1/2-in. (38-mm) wide and 16-gauge and 1/8-in. (3-mm) thick.

INSTRUCTIONS

1. Tack weld the metal pieces of the butt and tee joints together and brace them on the welding table in the horizontal position.

2. Start at one end of the plate with the gun pointed in a slightly upward direction.

3. You may use a pushing or a dragging gun angle, depending on the current setting and penetration desired.

4. Undercutting along the top edge and overlap along the bottom edge are problems with both gun angles. Careful attention must be paid to the manipulation of the "weave" technique used to overcome these problems. The most successful weave patterns are the "C" and "J" patterns. The "J" pattern is the most frequently used. The "J" pattern allows weld metal to be deposited along a shelf created by the previous weave. The length of the "J" can be changed to control the weld bead size.

5. Small weld beads are easier to control than large ones.

6. The weld must be straight and uniform in size and have 100% penetration.

7. Repeat these welds until you have established the rhythm and technique that work well for you.

8. Turn off the welding machine and shielding gas, and clean up your work area when you are finished welding.

INSTRUCTOR'S COMMENTS _____

_____

_____

■ PRACTICE 11-17

Name _____ Date _____

Class _____ Instructor _____ Grade _____

OBJECTIVE: After completing this practice, you should be able to weld stringer beads in the overhead position on both 1/16-in. (2-mm) and 1/8-in. (3-mm) mild steel.

EQUIPMENT AND MATERIALS NEEDED FOR THIS PRACTICE

1. A properly set-up and adjusted GMA welding machine.

2. Proper safety protection (welding hood, safety glasses, leather gloves, wire cutting pliers, long-sleeved shirt, long pants, leather shoes or boots). Refer to Chapter 2 in the text for more specific safety information.

3. 0.035-in. and/or 0.045-in. (0.9-mm and/or 1.2-mm) diameter wire.

4. Two or more pieces of mild steel sheet, 12-in. (305-mm) long by 1 1/2-in. (38-mm) wide and 16-gauge and 1/8-in. (3-mm) thick.

INSTRUCTIONS

1. In the overhead position, the molten weld pool should be kept small. This can be achieved by using lower current settings, traveling faster, or by pushing the molten weld pool.

2. When welding overhead, extra personal protection is required to reduce the danger of burns. Leather sleeves or leather jackets should be worn.

3. Make several short weld beads using various techniques to establish the method that is most successful and most comfortable for you. After each weld, stop and evaluate it before making a change.

4. When you have decided on the technique to be used, make a welded stringer bead that is 12 in. (305 mm) long.

5. Repeat the weld until it can be made straight, uniform in size, and free from any visual defects.

6. Turn off the welding machine and shielding gas, and clean up your work area when you are finished welding.

INSTRUCTOR'S COMMENTS _____

_____

_____

■ PRACTICE 11-18

Name _____ Date _____

Class _____ Instructor _____ Grade _____

OBJECTIVE: After completing this practice, you should be able to weld a butt, lap, and tee joint in the overhead position.

EQUIPMENT AND MATERIALS NEEDED FOR THIS PRACTICE

1. A properly set-up and adjusted GMA welding machine.

2. Proper safety protection (welding hood, safety glasses, leather gloves, wire cutting pliers, long-sleeved shirt, long pants, leather shoes or boots). Refer to Chapter 2 in the text for more specific safety information.

3. 0.035-in. and/or 0.045-in. (0.9-mm and/or 1.2-mm) diameter wire.

4. Two or more pieces of mild steel sheet, 12-in. (305-mm) long by 1 1/2-in. (38-mm) wide and 16-gauge and 1/8-in. (3-mm) thick.

INSTRUCTIONS

1. Tack weld the pieces of metal together in the butt, lap, and tee configurations and secure them in the overhead position.

2. When welding overhead, extra personal protection is required to reduce the danger of burns. Leather sleeves or leather jackets should be worn.

3. Make several short weld beads using various techniques to establish the method that is most successful and most comfortable for you. After each weld, stop and evaluate it before making a change.

4. When you have decided on the technique to be used, make butt, lap, and tee welded joints.

5. Repeat the welds on both metal thicknesses until they can be made straight, uniform in size, and free from any visual defects.

6. Turn off the welding machine and shielding gas, and clean up your work area when you are finished welding.

INSTRUCTOR'S COMMENTS _____

_____

_____

■ PRACTICE 11-19

Name _____ Date _____

Class _____ Instructor _____ Grade _____

OBJECTIVE: After completing this practice, you should be able to weld a butt and tee joint in the overhead position with 100% penetration.

EQUIPMENT AND MATERIALS NEEDED FOR THIS PRACTICE

1. A properly set-up and adjusted GMA welding machine.

2. Proper safety protection (welding hood, safety glasses, leather gloves, wire cutting pliers, long-sleeved shirt, long pants, leather shoes or boots). Refer to Chapter 2 in the text for more specific safety information.

3. 0.035-in. and/or 0.045-in. (0.9-mm and/or 1.2-mm) diameter wire.

4. Two or more pieces of mild steel sheet, 12-in. (305-mm) long by 1 1/2-in. (38-mm) wide and 16-gauge and 1/8-in. (3-mm) thick.

INSTRUCTIONS

1. Tack weld the pieces of metal together in the butt and tee configurations and secure them in the overhead position.

2. When welding overhead, extra personal protection is required to reduce the danger of burns. Leather sleeves or leather jackets should be worn.

3. It may be necessary to adjust the root opening to allow 100% penetration.

4. Make several short weld beads using various techniques to establish the method that is most successful and most comfortable for you. After each weld, stop and evaluate it before making a change.

5. Use a dragging torch angle with a "C" or "J" weave pattern.

6. Repeat the welds on both metal thicknesses until they can be made straight, uniform in size, and free from any visual defects.

7. Turn off the welding machine and shielding gas, and clean up your work area when you are finished welding.

INSTRUCTOR'S COMMENTS _____

_____

_____

■ PRACTICE 11-20

Name _____ Date _____

Class _____ Instructor _____ Grade _____

OBJECTIVE: After completing this practice you should be able to make a stringer bead in the flat (1G) position using pulsed-arc globular metal transfer.

EQUIPMENT AND MATERIALS NEEDED FOR THIS PRACTICE

1. A properly set-up and adjusted GMA welding machine.

2. Proper safety protection (welding hood, safety glasses, wire brush, chipping hammer, leather gloves, pliers, long-sleeved shirt, long pants, and leather boots or shoes). Refer to Chapter 2 in the text for more specific safety information.

3. 0.035-in. and/or 0.045-in. (0.9-mm and/or 1.2-mm) diameter welding wire.

4. One or more pieces of mild steel plate, 12-in. (305-mm) long by 1 1/2-in. (38-mm) wide by 1/4-in. (6-mm) thick.

INSTRUCTIONS

1. Have your instructor show you how to correctly set the controls on the welding machine so that you will be welding with a pulsed-arc and globular transfer. The current and wire-feed speed as well as the frequency, amplitude, and pulse width must be correct for the metal being welded.

2. Place the metal flat on the weld table and start at one end using either a push or dragging technique to make a bead along the entire 12 in. (305 mm) of the plate.

3. Keep the arc near the center of the weld pool. If it is too far forward, the spatter will be increased and if it is too far rearward, leading edge fusion may be reduced.

4. Use a weave pattern that allows the arc to follow the leading edge of the molten weld pool.

5. Inspect the appearance of the bead and make any necessary changes in the weld parameters, gun angle, or weave pattern to correct for discrepancies.

6. Repeat the weld until the bead can be made straight, uniform, free of any visual defects, and with good penetration.

7. Turn off the welding machine and shielding gas, and clean up your work area when finished with this practice.

INSTRUCTOR'S COMMENTS _____

_____

_____

■ PRACTICE 11-21

Name _____ Date _____

Class _____ Instructor _____ Grade _____

OBJECTIVE: After completing this practice you should be able to make a butt joint in the flat (1G) position using pulsed-arc globular metal transfer.

EQUIPMENT AND MATERIALS NEEDED FOR THIS PRACTICE

1. A properly set-up and adjusted GMA welding machine.

2. Proper safety protection (welding hood, safety glasses, wire brush, chipping hammer, leather gloves, pliers, long-sleeved shirt, long pants, and leather boots or shoes). Refer to Chapter 2 in the text for more specific safety information.

3. 0.035-in. and/or 0.045-in. (0.9-mm and/or 1.2-mm) diameter welding wire.

4. Two or more pieces of mild steel plate, 12-in. (305-mm) long by 1 1/2-in. (38-mm) wide by 1/4-in. (6-mm) thick. Prepare the long edges by grinding or burning a 45° bevel on the plates.

INSTRUCTIONS

1. Have your instructor show you how to correctly set the controls on the welding machine so that you will be welding with a pulsed-arc and globular transfer. The current and wire-feed speed as well as the frequency, amplitude, and pulse width must be correct for the metal being welded.

2. Tack the plates together. Leave a 1/8 in. (3 mm) root opening between the plates.

3. Place the metal flat on the weld table. Start at one end using either a push or dragging technique to make a single pass weld along the entire 12 in. (305 mm) of the plate.

4. Use a weave pattern that follows the contour of the groove.

5. Inspect the appearance of the bead and make any necessary changes in the weld parameters, gun angle, or weave pattern to correct for discrepancies.

6. Repeat the weld until the bead can be made straight, uniform, free of any visual defects, and with good penetration.

7. Turn off the welding machine and shielding gas, and clean up your work area when finished with this practice.

INSTRUCTOR'S COMMENTS _____

_____

_____

■ PRACTICE 11-22

Name _____ Date _____

Class _____ Instructor _____ Grade _____

OBJECTIVE: After completing this practice you should be able to make a butt joint in the flat (1G) position using pulsed-arc globular metal transfer. The weld must have 100% penetration.

EQUIPMENT AND MATERIALS NEEDED FOR THIS PRACTICE

1. A properly set-up and adjusted GMA welding machine.

2. Proper safety protection (welding hood, safety glasses, wire brush, chipping hammer, leather gloves, pliers, long-sleeved shirt, long pants, and leather boots or shoes). Refer to Chapter 2 in the text for more specific safety information.

3. 0.035-in. and/or 0.045-in. (0.9-mm and/or 1.2-mm) diameter welding wire.

4. Two or more pieces of mild steel plate, 12-in. (305-mm) long by 1 1/2-in. (38-mm) wide by 1/4-in. (6-mm) thick. Prepare the long edges by grinding or burning a 45° bevel on the plates.

INSTRUCTIONS

1. Have your instructor show you how to correctly set the controls on the welding machine so that you will be welding with a pulsed-arc and globular transfer. The current and wire-feed speed as well as the frequency, amplitude, and pulse width must be correct for the metal being welded.

2. Tack the plates together. Leave a 1/8 in. (3 mm) root opening between the plates.

3. Place the metal flat on the weld table. Start at one end and regulate your weld parameters, travel speed, gun angle, and weave pattern so that 100% penetration is achieved.

4. If the weld burns through, appears to sink or will not fill up, increase the travel speed.

5. Inspect the appearance of the bead and make any necessary changes in the weld parameters, gun angle, or weave pattern to correct for discrepancies.

6. Repeat the weld until the bead can be made straight, uniform, free of any visual defects, and with 100% penetration.

7. Turn off the welding machine and shielding gas, and clean up your work area when finished with this practice.

INSTRUCTOR'S COMMENTS _____

_____

_____

■ PRACTICE 11-23

Name _____ Date _____

Class _____ Instructor _____ Grade _____

OBJECTIVE: After completing this practice you should be able to make a tee joint and a lap joint in the flat (1F) position using pulsed-arc globular metal transfer.

EQUIPMENT AND MATERIALS NEEDED FOR THIS PRACTICE

1. A properly set-up and adjusted GMA welding machine.

2. Proper safety protection (welding hood, safety glasses, wire brush, chipping hammer, leather gloves, pliers, long-sleeved shirt, long pants, and leather boots or shoes). Refer to Chapter 2 in the text for more specific safety information.

3. 0.035-in. and/or 0.045-in. (0.9-mm and/or 1.2-mm) diameter welding wire.

4. Two or more pieces of mild steel plate, 12-in. (305-mm) long by 1 1/2-in. (38-mm) wide by 1/4-in. (6-mm) thick. Edge preparation is not required.

INSTRUCTIONS

1. Have your instructor show you how to correctly set the controls on the welding machine so that you will be welding with a pulsed-arc and globular transfer. The current and wire-feed speed as well as the frequency, amplitude, and pulse width must be correct for the metal being welded.

2. Tack the plates together in the tee and lap configurations.

3. Place metal flat on the weld table. Start at one end using either a push or dragging technique to make a single pass weld along the entire 12 in. (305 mm) of the plate.

4. Use a weave pattern that follows the contour of the joint.

5. Inspect the appearance of the bead and make any necessary changes in the weld parameters, gun angle, or weave pattern to correct for discrepancies.

6. Repeat the weld until the bead can be made straight, uniform, free of any visual defects, and with good penetration.

7. Turn off the welding machine and shielding gas, and clean up your work area when finished with this practice.

INSTRUCTOR'S COMMENTS _____

_____

_____

■ PRACTICE 11-24

Name _____ Date _____

Class _____ Instructor _____ Grade _____

OBJECTIVE: After completing this practice you should be able to make a tee joint and a lap joint in the horizontal (2F) position using pulsed-arc globular metal transfer.

EQUIPMENT AND MATERIALS NEEDED FOR THIS PRACTICE

1. A properly set-up and adjusted GMA welding machine.

2. Proper safety protection (welding hood, safety glasses, wire brush, chipping hammer, leather gloves, pliers, long-sleeved shirt, long pants, and leather boots or shoes). Refer to Chapter 2 in the text for more specific safety information.

3. 0.035-in. and/or 0.045-in. (0.9-mm and/or 1.2-mm) diameter welding wire.

4. Two or more pieces of mild steel plate, 12-in. (305-mm) long by 1 1/2-in. (38-mm) wide by 1/4-in. (6-mm) thick. Edge preparation is not required.

INSTRUCTIONS

1. Have your instructor show you how to correctly set the controls on the welding machine so that you will be welding with a pulsed-arc globular metal transfer. The current and wire-feed speed as well as the frequency, amplitude, and pulse width must be correct for the metal being welded.

2. Tack the plates together in the tee and lap configurations.

3. Secure the pieces in the horizontal position on the weld table. Start at one end using either a push or dragging technique to make a single pass weld along the entire 12 in. (305 mm) of the plate.

4. The weave pattern must follow the plate surfaces and establish a shelf to support the weld.

5. To prevent undercutting, the beads must be small and quickly made.

6. Inspect the appearance of the bead and make any necessary changes in the weld parameters, gun angle, or weave pattern to correct for discrepancies.

7. Repeat the weld until the bead can be made straight, uniform, and free of any visual defects.

8. Turn off the welding machine and shielding gas, and clean up your work area when finished with this practice.

INSTRUCTOR'S COMMENTS _____

_____

_____

■ PRACTICE 11-25

Name _____ Date _____

Class _____ Instructor _____ Grade _____

OBJECTIVE: After completing this practice you should be able to make a stringer bead in the flat (1G) position using axial spray metal transfer.

EQUIPMENT AND MATERIALS NEEDED FOR THIS PRACTICE

1. A properly set-up and adjusted GMA welding machine.

2. Proper safety protection (welding hood, safety glasses, wire brush, chipping hammer, leather gloves, pliers, long-sleeved shirt, long pants, and leather boots or shoes). Refer to Chapter 2 in the text for more specific safety information.

3. 0.035-in. and/or 0.045-in. (0.9-mm and/or 1.2-mm) diameter welding wire.

4. One or more pieces of mild steel plate, 12-in. (305-mm) long by 1 1/2-in. (38-mm) wide by 1/4-in. (6-mm) thick.

INSTRUCTIONS

1. Have your instructor show you how to correctly set the controls on the welding machine so that you will be welding with axial spray transfer. The current and wire-feed speed must be correct for the metal being welded.

2. Be sure that the shielding gas you are using is either pure argon or has a high percentage of argon. Spray transfer only works correctly if the shielding gas being used is pure or nearly pure argon.

3. Place the metal flat on the weld table and start at one end using either a push or dragging technique to make a bead along the entire 12 in. (305 mm) of the plate.

4. Inspect the appearance of the bead and make any necessary changes in the weld parameters or gun angle to correct for discrepancies.

5. Repeat the weld until the bead can be made straight, uniform, free of any visual defects, and with good penetration.

6. Turn off the welding machine and shielding gas, and clean up your work area when finished with this practice.

INSTRUCTOR'S COMMENTS _____

_____

_____

■ PRACTICE 11-26

Name _____ Date _____

Class _____ Instructor _____ Grade _____

OBJECTIVE: After completing this practice you should be able to make a butt, lap, and tee joint in the flat position using axial spray metal transfer.

EQUIPMENT AND MATERIALS NEEDED FOR THIS PRACTICE

1. A properly set-up and adjusted GMA welding machine.

2. Proper safety protection (welding hood, safety glasses, wire brush, chipping hammer, leather gloves, pliers, long-sleeved shirt, long pants, and leather boots or shoes). Refer to Chapter 2 in the text for more specific safety information.

3. 0.035-in. and/or 0.045-in. (0.9-mm and/or 1.2-mm) diameter welding wire.

4. Two or more pieces of mild steel plate, 12-in. (305-mm) long by 1 1/2-in. (38-mm) wide by 1/4-in. (6-mm) thick.

INSTRUCTIONS

1. Have your instructor show you how to correctly set the controls on the welding machine so that you will be welding with axial spray transfer. The current and wire-feed speed must be correct for the metal being welded.

2. Be sure that the shielding gas you are using is either pure argon or has a high percentage of argon. Spray transfer only works correctly if the shielding gas being used is pure or nearly pure argon.

3. Tack the pieces together in the butt, lap, and tee configurations.

4. Place the metal flat on the weld table and start at one end using either a push or dragging technique to make a bead along the entire 12 in. (305 mm) of the plate.

5. Inspect the appearance of the bead and make any necessary changes in the weld parameters or gun angle to correct for discrepancies.

6. Repeat the weld until the bead can be made straight, uniform, free of any visual defects, and with good penetration.

7. Turn off the welding machine and shielding gas, and clean up your work area when finished with this practice.

INSTRUCTOR'S COMMENTS _____

_____

_____

■ PRACTICE 11-27

Name _____ Date _____

Class _____ Instructor _____ Grade _____

OBJECTIVE: After completing this practice you should be able to make a butt joint and tee joint in the flat position using axial spray metal transfer. All welds must pass the bend test.

EQUIPMENT AND MATERIALS NEEDED FOR THIS PRACTICE

1. A properly set-up and adjusted GMA welding machine.

2. Proper safety protection (welding hood, safety glasses, wire brush, chipping hammer, leather gloves, pliers, long-sleeved shirt, long pants, and leather boots or shoes). Refer to Chapter 2 in the text for more specific safety information.

3. 0.035-in. and/or 0.045-in. (0.9-mm and/or 1.2-mm) diameter welding wire.

4. Two or more pieces of mild steel plate, 12-in. (305-mm) long by 1 1/2-in. (38-mm) wide by 1/4-in. (6-mm) thick.

INSTRUCTIONS

1. Have your instructor show you how to correctly set the controls on the welding machine so that you will be welding with axial spray transfer. The current and wire-feed speed must be correct for the metal being welded.

2. Be sure that the shielding gas you are using is either pure argon or has a high percentage of argon. Spray transfer only works correctly if the shielding gas being used is pure or nearly pure argon.

3. Tack the pieces together in the butt and tee configurations.

4. Place the metal flat on the weld table and start at one end using either a push or dragging technique to make a bead along the entire 12 in. (305 mm) of the plate.

5. Cut 1-in. (25-mm) sections from your welded piece and subject them to the bend test.

6. After bending, inspect the specimens for defects.

7. Repeat the weld until the bead can consistently pass the bend test.

8. Turn off the welding machine and shielding gas, and clean up your work area when finished with this practice.

INSTRUCTOR'S COMMENTS _____

_____

_____

CHAPTER 11: GAS METAL ARC WELDING QUIZ

Name _____ Date _____

Class _____ Instructor _____ Grade _____

Instructions: Carefully read Chapter 11 in the text and answer the following questions.

A. IDENTIFICATION

In the space provided, identify the items shown in the illustrations.

1. List the parts of the GMA welding gun.

   A. _____

   B. _____

   C. _____

   D. _____

B. MATCHING

2. Match the following terms with the illustration at right.

   ____ gas diffuser

   ____ gas diffuser

   ____ contact tube (tip)

   ____ the wire picks up the welding current in this area.

   ____ conduit liner

   ____ liner setscrew

C. SHORT ANSWER

Write a brief answer in the space provided.

3. What problem(s) results from a failure to keep the wire in the correct spot on the pool?

_____

_____

4. What will cause a wire to feed erratically?

_____

_____

4. Why are brief stops and starts used?

_____

_____

5. What changes in the weld bead occur when the gun angle is changed?

_____

_____

D. DEFINITIONS

6. Define the following terms:

flow rate _____

wire-feed speed _____

feed rollers _____

cast _____

conduit liner _____

INSTRUCTOR'S COMMENTS _____

_____

_____

Chapter 12

Flux Cored Arc Welding Equipment, Setup, and Operation

CHAPTER 12: FLUX CORED ARC WELDING EQUIPMENT, SETUP, AND OPERATION QUIZ

Name _____ Date _____

Class _____ Instructor _____ Grade _____

Instructions: Carefully read Chapter 12 in the text and answer the following questions.

A. IDENTIFICATION

In the space provided, identify the items shown in the illustration.

1. Identify the various components of self-shielded flux cored arc welding.

_____ powdered metal flux and slag-forming materials

_____ slag

_____ molten weld pool

_____ insulator

_____ base metal

_____ molten slag

_____ contact tube

_____ weld metal

_____ arc shielding composed of vaporized compounds

_____ arc and metal transfer

_____ flux-filled tubular wire electrode

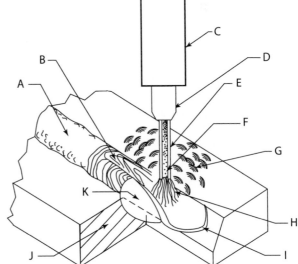

2. Identify the components of dual-shielded flux cored arc welding.

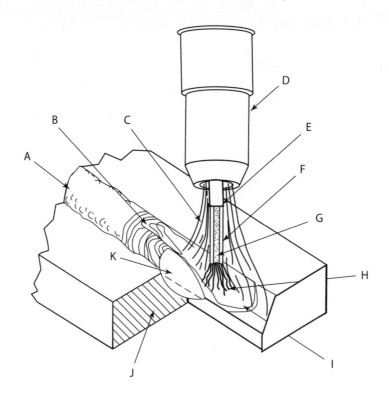

_____ molten weld pool

_____ slag

_____ wire guide and contact tube

_____ arc and metal transfer

_____ powdered metal flux and slag-forming materials

_____ weld metal

_____ flux-filled tubular wire electrode

_____ molten slag

_____ shielding gas

_____ base metal

_____ gas nozzle

B. FILL IN THE BLANK

Fill in the blank with the correct word. Answers may be more than one word.

3. FCAW _____ flux makes it the best choice for making welds outdoors or in windy conditions.

4. _____ electrodes can have more allowing elements and fluxing agents added to the flux core since not as much gas forming elements are needed.

5. Because the arc voltage remains constant, changes in the wire feed speed will automatically cause a change in the _____.

6. The CP welding machine's _____ naturally increases to whatever level is needed to maintain the same arc length.

7. The major atmospheric contaminations come from _____ and _____, the major elements in air.

8. The flux core additives that serve as _____, _____, and _____ either protect the molten weld pool or help to remove impurities from the base metal.

9. Some elements in the flux that rapidly expand and, called _____, push the surrounding air away from the molten weld pool.

10. FCA welding guns are available as _____-cooled or _____-cooled.

11. Water pressures higher than _____ may cause the water hoses to burst.

12. These systems use a vacuum to pull the smoke back into a specially designed _____ on the welding gun.

13. The addition of deoxidizers and other fluxing agents permits high-quality welds to be made on plates with _____ and _____.

14. Lay FCA welding wire on the floor and observe that it forms a circle, the diameter of the circle is known as the _____ of the wire.

15. T-1 fluxes are rutile-based fluxes and are _____.

16. T-5 fluxes are lime-based fluxes and are _____.

17. The addition of ferrite-forming elements can control the _____ and _____ of a weld.

18. The AWS designation was changed from the letter T for _____ to the letter C for _____.

19. Hydrogen entrapment can cause weld beads to _____ or become _____.

20. Shielding gas cylinders are supplied with _____ psi of pressure.

21. Changes in the gun angle will affect the weld bead _____ and _____.

22. The characteristic of _____ transfer is a smooth arc, through which hundreds of small droplets per second are transferred through the arc from the electrode to the weld pool.

C. SHORT ANSWER

Write a brief answer in the space provided that will answer the question or complete the statement.

23. Is there any problem with using a shielding gas on an FCAW electrode?

_____

_____

24. What elements can be placed in the flux to improve the hardness of weld?

_____

_____

25. Why are two sets of feed rollers recommended when using the FCAW process?

_____

_____

26. What is the primary purpose of the flux used in the FCAW process?

_____

_____

D. ESSAY

Provide complete answers for all of the following questions.

27. What effect on the weld does electrode extension have?

_____

_____

28. What effect on the weld does gun angle have?

_____

_____

29. List three things that the electrode flux provides to the weld.

a. _____

b. _____

c. _____

30. List three useful things that a slag covering helps.

a. _____

b. _____

c. _____

31. What is meant when it is said that a plate can be welded in the "as-cut" condition?

_____

_____

32. How is the flux packed inside the FCA welding wire?

_____

_____

33. Why are some FCA welding filler electrodes not recommended for multiple pass welds?

_____

_____

34. List three limitations to the FCAW process.

a. _____

b. _____

c. _____

35. Why is it important to remove the slag from a weld bead before the parts are painted?

_____

_____

INSTRUCTOR'S COMMENTS _____

_____

_____

Chapter 13

Flux Cored Arc Welding

ELECTRODE		WELDING POWER			SHIELDING GAS		BASE METAL	
Type	Size	Amps	Wire–Feed Speed IPM (cm/min)	Volts	Type	Flow	Type	Thickness
E70T-1	0.035 in. (0.9 mm)	130 to 150	288 to 380 (732 to 975)	22 to 25	none	n/a	Low-Carbon Steel	1/4 in. to 1/2 in. (6 mm to 13 mm)
E70T-1	0.045 in. (1.2 mm)	150 to 210	200 to 300 (508 to 762)	28 to 29	none	n/a	Low-Carbon Steel	1/4 in. to 1/2 in. (6 mm to 13 mm)
E70T-5	0.035 in. (0.9 mm)	130 to 200	288 to 576 (732 to 1463)	20 to 28	75% Argon 25% CO_2	30 cfh	Low-Carbon Steel	1/4 in. to 1/2 in. (6 mm to 13 mm)
E70T-5	0.045 in. (1.2 mm)	150 to 250	200 to 400 (508 to 1016)	23 to 29	75% Argon 25% CO_2	35 cfh	Low-Carbon Steel	1/4 in. to 1/2 in. (6 mm to 13 mm)

Table 13-1   FCA welding parameters that can be used if specific settings are not available from the electrode manufacturer.

Electrode		Welding Power			Shielding Gas		Base Metal	
Type	Size	Amps	Wire-Feed Speed IPM (cm/min)	Volts	Type	Flow	Type	Thickness
E70T-1	0.035 in. (0.9 mm)	130 to 150	288 to 380 (732 to 975)	22 to 25	None	n/a	Low carbon steel	1/4 in. to 3/4 in. (6 mm to 19 mm)
E70T-1	0.045 in. (1.2 mm)	150 to 210	200 to 300 (508 to 762)	28 to 29	None	n/a	Low carbon steel	1/4 in. to 3/4 in. (6 mm to 19 mm)
E70T-1	0.052 in. (1.4 mm)	150 to 300	150 to 350 (381 to 889)	25 to 33	None	n/a	Low carbon steel	1/4 in. to 3/4 in. (6 mm to 19 mm)
E70T-1	1/16 in. (1.6 mm)	200 to 400	150 to 300 (381 to 762)	27 to 33	None	n/a	Low carbon steel	1/4 in. to 3/4 in. (6 mm to 19 mm)
E70T-5	0.035 in. (0.9 mm)	130 to 200	288 to 576 (732 to 1463)	20 to 28	75% argon 25% CO_2	30 cfh	Low carbon steel	1/4 in. to 3/4 in. (6 mm to 19 mm)
E70T-5	0.045 in. (1.2 mm)	150 to 250	200 to 400 (508 to 1016)	23 to 29	75% argon 25% CO_2	35 cfh	Low carbon steel	1/4 in. to 3/4 in. (6 mm to 19 mm)
E70T-5	0.052 in. (1.4 mm)	150 to 300	150 to 350 (381 to 889)	21 to 32	75% argon 25% CO_2	35 cfh	Low carbon steel	1/4 in. to 3/4 in. (6 mm to 19 mm)
E70T-5	1/16 in. (1.6 mm)	180 to 400	145 to 350 (368 to 889)	21 to 34	75% argon 25% CO_2	40 cfh	Low carbon steel	1/4 in. to 3/4 in. (6 mm to 19 mm)

Table 13-2 FCA welding parameters that can be used if specific settings are not available from the electrode manufacturer.

Electrode		Welding Power			Shielding Gas		Base Metal	
Type	Size	Amps	Wire-Feed Speed IPM (cm/min)	Volts	Type	Flow	Type	Thickness
E70T-1	0.030 in. (0.8 mm)	40 to 145	90 to 340 (228 to 864)	20 to 27	None	n/a	Low carbon steel	16 gauge to 18 gauge
E70T-1	0.035 in. (0.9 mm)	130 to 200	288 to 576 (732 to 1463)	20 to 28	None	n/a	Low carbon steel	16 gauge to 18 gauge
E70T-5	0.035 in. (0.9 mm)	90 to 200	190 to 576 (483 to 1463)	16 to 29	75% argon 25% CO_2	35 cfh	Low carbon steel	16 gauge to 18 gauge

Table 13-3 FCA welding parameters that can be used if specific settings are not available from the electrode manufacturer.

■ PRACTICE 13-1

Name _____ Date _____

Class _____ Instructor _____ Grade _____

OBJECTIVE: After completing this practice, you should be able to safely set up FCAW equipment.

EQUIPMENT AND MATERIALS NEEDED FOR THIS PRACTICE

1. Semiautomatic welding power source approved for FCA welding.

2. Welding gun.

3. Electrode feed unit, electrode supply.

4. Shielding gas supply, shielding gas flowmeter regulator.

5. Electrode conduit, power and work leads.

6. Shielding gas hoses (if required).

7. Assorted hand tools, spare parts, and any other required materials.

INSTRUCTIONS

> **NOTE: Some manufacturers include detailed setup instructions with their equipment. If such instructions are available for your equipment, follow them. However, if no instructions are available, then follow the instructions listed below.**

1. If the shielding gas is to be used and it comes from a cylinder, the cylinder must be chained securely in place before the valve protection cap is removed. Stand to one side of the valve opening and quickly crack the valve to blow out any dirt in the valve before the flowmeter regulator is attached. Attach the correct hose from the regulator to the "gas-in" connection on the electrode feed unit or machine.

2. Install the reel of electrode (welding wire) on the holder and secure it. Check the feed roller size to ensure that it matches the wire size. The conduit liner size should be checked to be sure that it is compatible with the wire size. Connect the conduit to the feed unit. The conduit or an extension should be aligned with the groove in the roller and set as close to the roller as possible without touching. Misalignment at this point can contribute to a bird's nest.

3. Be sure that the power is off before attaching the welding cables. The electrode and work leads should be attached to the proper terminals. The electrode lead should be attached to electrode or positive (+) terminal. If necessary, it is also attached to the power cable part of the gun lead. The work lead should be attached to work or negative (−) terminal.

4. The shielding "gas-out" side of the solenoid is then also attached to the gun lead. If a separate splice is required from the gun switch circuit to the feed unit, it should be connected at this time. Check to see that the welding contactor circuit is connected from the wire-feed unit to the power source.

5. The welding cable liner or wire conduit must be securely attached to the gas diffuser and contact tube (tip). The contact tube must be the correct size to match the electrode wire size being used. If a shielding gas is to be used, a gas nozzle would be attached to complete the assembly. If a gas nozzle is not needed for a shielding gas, it may still be installed. Because it is easy for a student to touch the work with the contact tube during welding, an electrical short may occur. This short out of the contact tube will immediately destroy the tube. Although the gas nozzle may interfere with some visibility, it may be worth the trouble for a welder in training.

INSTRUCTOR'S COMMENTS _____

_____

_____

■ PRACTICE 13-2

Name _____ Date _____

Class _____ Instructor _____ Grade _____

OBJECTIVE: After completing this practice, you should be able to thread FCAW wire.

EQUIPMENT AND MATERIALS NEEDED FOR THIS PRACTICE

1. Semiautomatic welding power source approved for FCA welding.

2. Welding gun.

3. Electrode feed unit, electrode supply.

4. Shielding gas supply, shielding gas flowmeter regulator.

5. Electrode conduit, power and work leads.

6. Shielding gas hoses (if required).

7. Assorted hand tools, spare parts, and any other required materials.

INSTRUCTIONS

> **NOTE: Some manufacturers include detailed setup instructions with their equipment. If such instructions are available for your equipment, follow them. However, if no instructions are available, then follow the instructions listed below.**

1. Check to see that the unit is assembled correctly according to the manufacturer's specifications. Switch on the power and check the gun switch circuit by depressing the switch. The power source relays, feed relays, gas solenoid, and feed motor should all activate.

2. Cut off the end of the electrode wire if it is bent. When you are working with the wire, be sure to hold it tightly. The wire will become tangled if it is released. The wire has a natural curl called cast. Straighten out about 12 in. (300 mm) of the cast to make threading easier.

3. Separate the wire-feed rollers and push the wire first through the guides, then between the rollers, and finally into the conduit liner. Reset the rollers so that there is a slight amount of compression on the wire. Set the wire-feed speed control to a slow speed. Hold the welding gun so that the electrode conduit and cable are as straight as possible.

4. Remove the contact tube on the gun.

5. Press the gun switch. The wire should start feeding into the liner. Watch to make certain that the wire feeds smoothly. Release the gun switch as soon as the end comes through the gun. If the wire stops feeding before it reaches the end of the conduit, stop and check the system. If no obvious problem can be found, mark the wire with tape and remove it from the gun. It then can be held next to the electrode conduit and cable system to determine the location of the problem.

6. Replace the contact tube on the gun.

7. With the wire feed running, adjust the feed roller compression so that the wire reel can be stopped easily by a slight pressure. Too light a roller pressure will cause the wire to feed erratically. Too high a pressure can crush some wires, causing some flux to be dropped inside the wire liner. If this happens, you will have a continual problem with the wire not feeding smoothly or jamming.

8. With the wire feed running, adjust the spool drag so that the reel stops when the feed stops. The reel should not coast to a stop because the wire can be snagged easily. Also, when the feed restarts, a jolt occurs when the slack in the wire is taken up. This jolt can be enough to momentarily stop the wire, possibly causing a discontinuity in the weld or may cause the wire to break.

9. When the test runs are completed, the wire can either be rewound or cut off. Some wire-feed units have a retract button. This allows the feed driver to reverse and retract the wire automatically. To rewind the wire on units without this retract feature, release the rollers and turn them backward by hand. If the machine will not allow the feed rollers to be released without upsetting the tension, you must cut the wire. Some wire reels have covers to prevent the collection of dust, dirt, and metal filings on the wire.

INSTRUCTOR'S COMMENTS _____

_____

_____

■ PRACTICE 13-3

Name _____ Date _____

Class _____ Instructor _____ Grade _____

OBJECTIVE: After completing this practice, you should be able to make a stringer bead weld in the flat position.

EQUIPMENT AND MATERIALS NEEDED FOR THIS PRACTICE

1. A properly set-up and adjusted FCA welding machine.

2. Table 13-1 or specific settings from the electrode manufacturer.

3. Proper safety protection.

4. E70T-1 and/or E70T-5 electrodes with 0.035-in. and/or 0.045-in. (0.9-mm and/or 1.2-mm) diameter.

5. One or more pieces of mild steel plate, 12-in. (305-mm) long by 3-in. (76-mm) wide and 1/4-in. (6-mm) thick.

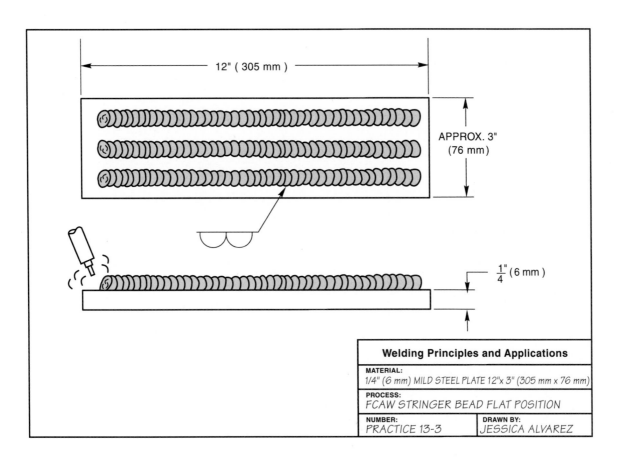

Welding Principles and Applications	
MATERIAL: 1/4" (6 mm) MILD STEEL PLATE 12"x 3" (305 mm x 76 mm)	
PROCESS: FCAW STRINGER BEAD FLAT POSITION	
NUMBER: PRACTICE 13-3	**DRAWN BY:** JESSICA ALVAREZ

INSTRUCTIONS

1. Starting at one end of the plate, and using a dragging technique, make a weld bead along the entire 12 in. (305 mm) length of the metal. Use 1 in. (25 mm) of stickout.

2. After the weld is complete, check its appearance. Make any needed changes to correct the weld.

3. Repeat the weld and make additional adjustments.

4. After the machine is set properly, start to work on improving the straightness and uniformity of the weld.

5. Use weave patterns of different widths and straight stringers without weaving.

6. Repeat with both classifications of electrodes as needed until beads can be made straight, uniform, and free from any visual defects.

7. Turn off the welding machine and shielding gas, and clean up your work area when you are finished welding.

INSTRUCTOR'S COMMENTS _____

_____

_____

■ PRACTICE 13-4

Name _____ Date _____

Class _____ Instructor _____ Grade _____

OBJECTIVE: After completing this practice, you should be able to make a groove weld in the flat position.

EQUIPMENT AND MATERIALS NEEDED FOR THIS PRACTICE

1. A properly set-up and adjusted FCA welding machine.

2. Table 13-1 or specific settings from the electrode manufacturer.

3. Proper safety protection.

4. E70T-1 and/or E70T-5 electrodes with 0.035-in. and/or 0.045-in. (0.9-mm and/or 1.2-mm) diameter.

5. Two or more pieces of mild steel plate, 12-in. (305-mm) long by 3-in. (76-mm) wide and 1/4-in. (6-mm) thick.

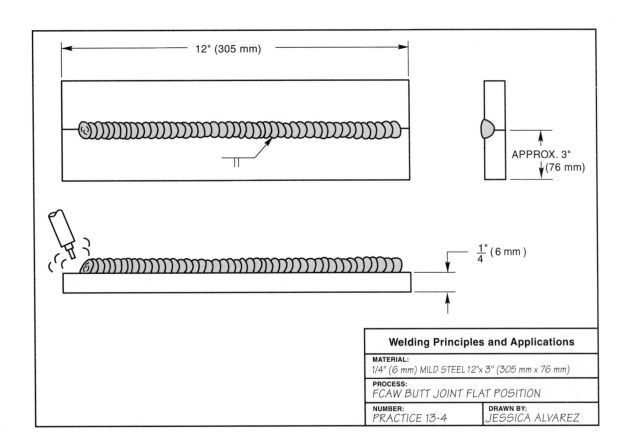

12" (305 mm)

APPROX. 3" (76 mm)

$\frac{1}{4}$" (6 mm)

Welding Principles and Applications

MATERIAL:
1/4" (6 mm) MILD STEEL 12"x 3" (305 mm x 76 mm)

PROCESS:
FCAW BUTT JOINT FLAT POSITION

NUMBER:
PRACTICE 13-4

DRAWN BY:
JESSICA ALVAREZ

INSTRUCTIONS

1. Tack weld the plates together and place them in the flat position to be welded.

2. Starting at one end, run a bead along the joint. Watch the molten weld pool and bead for signs that a change in technique may be required.

3. In order to produce a uniform weld, make any needed changes as the weld progresses.

4. Repeat with both classifications of electrodes as needed until consistently defect-free welds can be made in the 1/4-in. (6-mm) thick plate.

5. Turn off the welding machine and shielding gas, and clean up your work area when you are finished welding.

INSTRUCTOR'S COMMENTS _____

_____

_____

■ PRACTICE 13-5

Name _____ Date _____

Class _____ Instructor _____ Grade _____

OBJECTIVE: After completing this practice, you should be able to make a groove butt weld in the flat position.

EQUIPMENT AND MATERIALS NEEDED FOR THIS PRACTICE

1. A properly set-up and adjusted FCA welding machine.

2. Table 13-2 or specific settings from the electrode manufacturer.

3. Proper safety protection.

4. E70T-1 and/or E70T-5 electrodes with 0.035-in. and/or through 1/16-in. (0.9-mm and/or through 1.6-mm) diameter.

5. 12-in. (305-mm) long by 3-in. (76-mm) wide and 3/8-in. (9.5-mm) thick or thicker beveled plate.

6. One piece of mild steel plate, 14-in. (355-mm) long, 1-in. (25-mm) wide, and 1/4-in. (6-mm) thick for a backing strip.

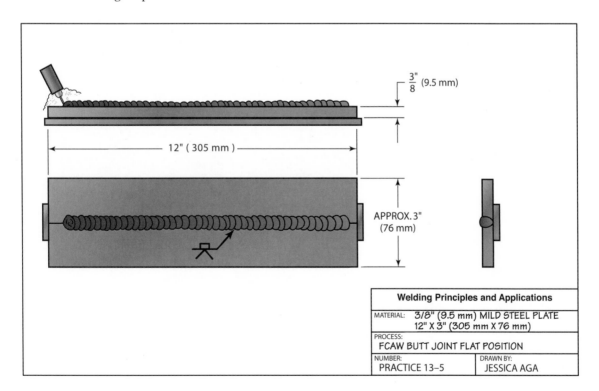

Welding Principles and Applications

MATERIAL: 3/8" (9.5 mm) MILD STEEL PLATE
12" X 3" (305 mm X 76 mm)

PROCESS:
FCAW BUTT JOINT FLAT POSITION

NUMBER: PRACTICE 13–5

DRAWN BY: JESSICA AGA

INSTRUCTIONS

1. Tack weld the backing strip to the plates. There should be approximately 1/8 in. (3 mm) root gap between the plates. The beveled surface can be made with or without a root face.

2. Place the test plates in position at a comfortable height and location. Be sure that you will have complete and free movement along the full length of the weld joint. Make a practice pass along the joint with the welding gun without power to make sure nothing will interfere with your making the weld. Be sure the welding cable is free and will not get caught on anything during the welding.

3. Start the weld outside of the groove on the backing strip tab. This is done so that the arc is smooth and the molten weld pool size is established at the beginning of the groove. Continue the weld out on to the tab at the other end of the groove. This is to insure that the end of the groove is completely filled with weld.

4. Repeat with both classifications of electrodes as needed until consistently defect-free welds can be made.

5. Turn off the welding machine and shielding gas, and clean up your work area when you are finished welding.

INSTRUCTOR'S COMMENTS _____

_____

_____

■ PRACTICE 13-6

Name _____ Date _____

Class _____ Instructor _____ Grade _____

OBJECTIVE: After completing this practice, you should be able to make fillet welds in both lap and tee joints in the flat position.

EQUIPMENT AND MATERIALS NEEDED FOR THIS PRACTICE

1. A properly set-up and adjusted FCA welding machine.

2. Table 13-1 or specific settings from the electrode manufacturer.

3. Proper safety protection.

4. E70T-1 and/or E70T-5 electrodes with 0.035-in. and/or 0.045-in. (0.9-mm and/or 1.2-mm) diameter.

5. Two or more pieces of mild steel plate, 12-in. (305-mm) long by 3-in. (76-mm) wide and 3/8-in. (9.5-mm) thick.

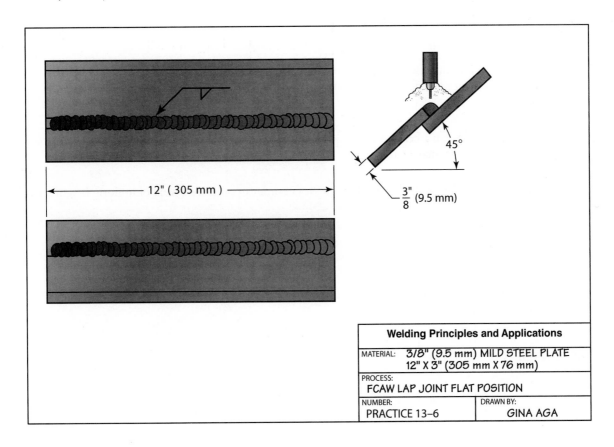

Welding Principles and Applications	
MATERIAL: 3/8" (9.5 mm) MILD STEEL PLATE 12" X 3" (305 mm X 76 mm)	
PROCESS: FCAW LAP JOINT FLAT POSITION	
NUMBER: PRACTICE 13–6	DRAWN BY: GINA AGA

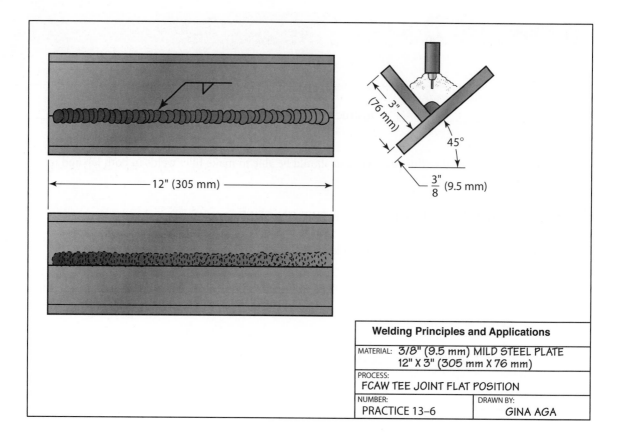

12" (305 mm)

3" (76 mm)

45°

3/8" (9.5 mm)

Welding Principles and Applications

MATERIAL: 3/8" (9.5 mm) MILD STEEL PLATE
12" X 3" (305 mm X 76 mm)

PROCESS:
FCAW TEE JOINT FLAT POSITION

NUMBER: PRACTICE 13–6

DRAWN BY: GINA AGA

INSTRUCTIONS

1. Tack weld the plates together and place them in position to be welded. When making the lap or tee joints in the flat position, the plates must be at a 45° angle so that the surface of the weld will be flat.

2. Starting at one end, run a bead along the joint. Watch the molten weld pool and bead for signs that a change in technique may be required.

3. Make any needed changes as the weld progresses in order to produce a weld of uniform height and width.

4. Repeat with both classifications of electrodes as needed until consistently defect-free welds can be made.

5. Turn off the welding machine and shielding gas, and clean up your work area when you are finished welding.

INSTRUCTOR'S COMMENTS _____

■ PRACTICE 13-7

Name _____ Date _____

Class _____ Instructor _____ Grade _____

OBJECTIVE: After completing this practice you should be able to make butt joint welds at a full vertical up angle.

EQUIPMENT AND MATERIALS NEEDED FOR THIS PRACTICE

1. A properly set-up and adjusted FCA welding machine.

2. Table 13-1 or specific settings from the electrode manufacturer.

3. Proper safety protection.

4. E70T-1 and/or E70T-5 electrodes with 0.035-in. and/or 0.045-in. (0.9-mm and/or 1.2-mm) diameter.

5. Two or more pieces of mild steel plate, 12-in. (305-mm) long by 3-in. (76-mm) wide by 1/4-in. (6-mm) thick (or thinner) plate.

INSTRUCTIONS

1. Tack weld the plates together in the butt configuration.

2. Start with the plate at a 45° angle. When making these joints, you will increase the plate angle gradually as you develop skill until you are making satisfactory welds in the full vertical up position.

3. Starting at the bottom, run a bead up the joint. Watch the molten weld pool and bead for signs that a change in technique may be required.

4. Repeat with both classifications of electrodes as needed until defect-free welds can be made.

5. Turn off the welding machine and shielding gas, and clean up your work area when finished with this practice.

INSTRUCTOR'S COMMENTS _____

_____

_____

■ PRACTICE 13-8

Name _____ Date _____

Class _____ Instructor _____ Grade _____

OBJECTIVE: After completing this practice you should be able to make groove joint welds in the vertical up position.

EQUIPMENT AND MATERIALS NEEDED FOR THIS PRACTICE

1. A properly set-up and adjusted FCA welding machine.

2. Table 13-1 or specific settings from the electrode manufacturer.

3. Proper safety protection.

4. E70T-1 and/or E70T-5 electrodes with 0.035-in. and/or 0.045-in. (0.9-mm and/or 1.2-mm) diameter.

5. Two or more pieces of mild steel plate, 12-in. (305-mm) long by 3-in. (76-mm) wide by 1/4-in. (6-mm) thick (or thinner) plate.

INSTRUCTIONS

1. Tack weld the plates together in the groove joint configuration.

2. Start with the plate at a 45° angle. When making these joints, you will increase the plate angle gradually as you develop skill until you are making satisfactory welds in the full vertical up position.

3. Starting at the bottom, run a bead up the joint. Watch the molten weld pool and bead for signs that a change in technique may be required.

4. Repeat with both classifications of electrodes as needed until defect-free welds can be made.

5. Turn off the welding machine and shielding gas, and clean up your work area when finished with this practice.

INSTRUCTOR'S COMMENTS _____

_____

_____

■ PRACTICE 13-9

Name _____ Date _____

Class _____ Instructor _____ Grade _____

OBJECTIVE: After completing this practice you should be able to make groove joint welds in the vertical up position with a backing strip.

EQUIPMENT AND MATERIALS NEEDED FOR THIS PRACTICE

1. A properly set-up and adjusted FCA welding machine.

2. Table 13-1 or specific settings from the electrode manufacturer.

3. Proper safety protection.

4. E70T-1 and/or E70T-5 electrodes with 0.035-in. and/or 0.045-in. (0.9-mm and/or 1.2 mm) diameter.

5. Two or more pieces of mild steel plate, 12-in. (305-mm) long by 3-in. (76-mm) wide by 3/8-in. (9.5-mm) thick beveled plate and a 14-in. (355-mm) long by 1-in. (25-mm) wide by 1/4-in. (6-mm) thick backing strip.

INSTRUCTIONS

1. Tack weld the plates in a groove joint configuration with the backing strip in place.

2. Start with the plate at a 45° angle. When making these joints, you will increase the angle until satisfactory welds can be made in a full vertical up angle.

3. Starting at the bottom, run a bead up the joint. Watch the molten weld pool and bead for signs that a change in technique may be required.

4. Repeat with both classifications of electrodes as needed until defect-free welds can be made.

5. Turn off the welding machine and shielding gas, and clean up your work area when finished with this practice.

INSTRUCTOR'S COMMENTS _____

_____

_____

■ PRACTICE 13-10

Name _____ Date _____ Electrode Used _____

Class _____ Instructor _____ Grade _____

OBJECTIVE: After completing this practice you should be able to make butt joint welds at a 45° vertical up position.

EQUIPMENT AND MATERIALS NEEDED FOR THIS PRACTICE

1. A properly set-up and adjusted FCA welding machine.

2. Table 13-2 or specific settings from the electrode manufacturer.

3. Proper safety protection.

4. E70T-1 and/or E70T-5 electrodes with 0.035-in. and/or through 1/16-in. (0.9-mm and/or through 1.6-mm) diameter.

5. Two or more pieces of mild steel plate, 7-in. (178-mm) long by 3-in. (76-mm) wide by 3/4-in. (19-mm) thick (or thicker) beveled plate and a 9-in. (230-mm) long by 1-in. (25-mm) wide by 1/4-in. (6-mm) thick backing strip.

INSTRUCTIONS

1. Tack weld the plates together in the butt joint configuration with the backing strip in place.

2. Start with the plate at a 45° angle. When making these joints, you will increase the plate angle gradually as you develop skill until you are making satisfactory welds in the full vertical up position.

3. Starting at the bottom, run a bead up the joint. Watch the molten weld pool and bead for signs that a change in technique may be required.

4. Repeat with both classifications of electrodes as needed until defect-free welds can be made.

5. Turn off the welding machine and shielding gas, and clean up your work area when finished with this practice.

INSTRUCTOR'S COMMENTS _____

_____

_____

■ PRACTICE 13-11

Name _____ Date _____

Class _____ Instructor _____ Grade _____

OBJECTIVE: After completing this practice you should be able to make a lap weld joint and a tee joint in the vertical up position.

EQUIPMENT AND MATERIALS NEEDED FOR THIS PRACTICE

1. A properly set-up and adjusted FCA welding machine.

2. Table 13-1 or specific settings from the electrode manufacturer.

3. Proper safety protection.

4. E70T-1 and/or E70T-5 electrodes with 0.035-in. and/or through 0.045-in. (0.9-mm and/or through 1.2-mm) diameter.

5. Two or more pieces of mild steel plate, 12-in. (305-mm) long by 3-in. (76-mm) wide by 3/8-in. (9.5-mm) thick.

INSTRUCTIONS

1. Tack weld the plates together in lap joint and tee joint configurations.

2. Start with the plate at a 45° angle to the table. You will gradually increase this angle as your skill develops until you are making satisfactory welds in the full vertical up position.

3. Starting at the bottom, run a bead up the joint. Watch the molten weld pool and bead for signs that a change in technique may be required.

4. Repeat with both joint types and with both classifications of electrodes as needed until defect-free welds can be made.

5. Turn off the welding machine and shielding gas, and clean up your work area when finished with this practice.

INSTRUCTOR'S COMMENTS _____

_____

_____

■ PRACTICE 13-12

Name _____ Date _____

Class _____ Instructor _____ Grade _____

OBJECTIVE: After completing this practice you should be able to make a lap joint and a tee joint in the horizontal position.

EQUIPMENT AND MATERIALS NEEDED FOR THIS PRACTICE

1. A properly set-up and adjusted FCA welding machine.

2. Table 13-1 or specific settings from the electrode manufacturer.

3. Proper safety protection.

4. E70T-1 and/or E70T-5 electrodes with 0.035-in. and/or 0.045-in. (0.9-mm and/or 1.2-mm) diameter.

5. Two or more pieces of mild steel plate, 12-in. (305-mm) long by 3-in. (76-mm) wide by 3/8-in. (9.5-mm) thick beveled plate.

INSTRUCTIONS

1. Tack weld the plates together in the lap and tee joint configurations.

2. Place the plate in the horizontal 2F position.

3. Make the root weld small so that you do not trap slag under the overlap on the lower edge of the weld.

4. Clean and wire brush each pass before the next pass is added.

5. Make all the succeeding welds small to help control the contour.

6. Repeat with both classifications of electrodes and with both joint types as needed until defect-free welds can be made.

7. Turn off the welding machine and shielding gas, and clean up your work area when finished with this practice.

INSTRUCTOR'S COMMENTS _____

_____

_____

■ PRACTICE 13-13

Name _____ Date _____

Class _____ Instructor _____ Grade _____

OBJECTIVE: After completing this practice you should be able to make a stringer bead at a 45° angle.

EQUIPMENT AND MATERIALS NEEDED FOR THIS PRACTICE

1. A properly set-up and adjusted FCA welding machine.

2. Table 13-1 or specific settings from the electrode manufacturer.

3. Proper safety protection.

4. E70T-1 and/or E70T-5 electrodes with 0.035-in. and/or 0.045-in. (0.9-mm and/or 1.2-mm) diameter.

5. One or more pieces of mild steel plate, 12-in. (305-mm) long by 3-in. (76-mm) wide by 1/4-in. (6-mm) thick or thicker.

INSTRUCTIONS

1. Start with the plate at a 45° angle to the table. You will gradually increase this angle as your skill develops until you are making satisfactory welds on the vertical face of the plate in the horizontal position.

2. Repeat with both classifications of electrodes as needed until defect-free welds can be made.

3. Turn off the welding machine and shielding gas, and clean up your work area when finished with this practice.

INSTRUCTOR'S COMMENTS _____

_____

_____

■ PRACTICE 13-14

Name _____ Date _____

Class _____ Instructor _____ Grade _____

OBJECTIVE: After completing this practice you should be able to make a butt (groove) joint in the horizontal position.

EQUIPMENT AND MATERIALS NEEDED FOR THIS PRACTICE

1. A properly set-up and adjusted FCA welding machine.

2. Table 13-1 or specific settings from the electrode manufacturer.

3. Proper safety protection.

4. E70T-1 and/or E70T-5 electrodes with 0.035-in. and/or 0.045-in. (0.9-mm and/or 1.2-mm) diameter.

5. Two or more pieces of mild steel plate, 12-in. (305-mm) long by 3-in. (76-mm) wide by 1/4-in. (6-mm) thick or thicker.

INSTRUCTIONS

1. Tack weld the plates together leaving a slight space between them of approximately one-half the thickness of the metal.

2. Position the plate horizontally.

3. Using a slightly upward gun angle begin to weld at one end of the groove. Travel at a relatively fast speed to prevent the bead from sagging down. If sagging occurs, travel faster.

4. Repeat with both classifications of electrodes as needed until defect-free welds can be made.

5. Turn off the welding machine and shielding gas, and clean up your work area when finished with this practice.

INSTRUCTOR'S COMMENTS _____

_____

_____

■ PRACTICE 13-15

Name _____ Date _____

Class _____ Instructor _____ Grade _____

OBJECTIVE: After completing this practice you should be able to make a butt joint in the horizontal position using a backing strip.

EQUIPMENT AND MATERIALS NEEDED FOR THIS PRACTICE

1. A properly set-up and adjusted FCA welding machine.

2. Table 13-2 or specific settings from the electrode manufacturer.

3. Proper safety protection.

4. E70T-1 and/or E70T-5 electrodes with 0.035-in. and/or through 1/16-in. (0.9-mm and/or through 1.6-mm) diameter.

5. Two or more pieces of mild steel plate, 7-in. (178-mm) long by 3-in. (76-mm) wide by 3/4-in. (19-mm) thick or thicker beveled plates and a 9-in. (230-mm) long by 1-in. (25-mm) wide by 1/4-in. (6-mm) thick backing strip.

INSTRUCTIONS

1. Tack weld the plates and the backing strip together. Leave a slight space between the plates and allow the backing strip to extend 1 in. (25-mm) beyond the plates at each end.

2. Position the plate horizontally.

3. Using a slightly upward gun angle begin to weld at one end of the groove. Travel at a relatively fast speed to prevent the bead from sagging down. Multiple passes will be required. Clean and wire brush the beads between passes.

4. Repeat with both classifications of electrodes as needed until defect-free welds can be made.

5. Turn off the welding machine and shielding gas, and clean up your work area when finished with this practice.

INSTRUCTOR'S COMMENTS _____

_____

_____

■ PRACTICE 13-16

Name _____ Date _____

Class _____ Instructor _____ Grade _____

OBJECTIVE: After completing this practice you should be able to make a butt (groove) joint in the overhead 4G position.

EQUIPMENT AND MATERIALS NEEDED FOR THIS PRACTICE

1. A properly set-up and adjusted FCA welding machine.

2. Table 13-1 or specific settings from the electrode manufacturer.

3. Proper safety protection. (Extra protection is needed for this practice. Wear a leather jacket and cap.)

4. E70T-1 and/or E70T-5 electrodes with 0.035-in. and/or 0.045-in. (0.9-mm and/or 1.2-mm) diameter.

5. Two or more pieces of mild steel plate, 12-in. (305-mm) long by 3-in. (76-mm) wide by 1/4-in. (19-mm) thick.

INSTRUCTIONS

1. Tack weld the plates together leaving a slight space between the plates of approximately one half the thickness of the metal.

2. Position the plate in the overhead position. Get yourself into a comfortable position.

3. Use a lower current setting. Maintain a small weld pool and travel quickly to prevent dripping of the molten metal.

4. Repeat with both classifications of electrodes as needed until defect-free welds can be made.

5. Turn off the welding machine and shielding gas, and clean up your work area when finished with this practice.

INSTRUCTOR'S COMMENTS _____

_____

_____

■ PRACTICE 13-17

Name _____ Date _____

Class _____ Instructor _____ Grade _____

OBJECTIVE: After completing this practice you should be able to make a butt joint in the overhead 4G position with a backing strip.

EQUIPMENT AND MATERIALS NEEDED FOR THIS PRACTICE

1. A properly set-up and adjusted FCA welding machine.

2. Table 13-1 or specific settings from the electrode manufacturer.

3. Proper safety protection. (Extra protection is needed for this practice. Wear a Leather jacket and cap.)

4. E70T-1 and/or E70T-5 electrodes with 0.035-in. and/or 0.045-in. (0.9-mm and/or 1.2-mm) diameter.

5. Two or more pieces of mild steel plate, 12-in. (305-mm) long by 3-in. (76-mm) wide by 3/8-in. (9.5 mm) thick beveled plate and a 14-in. (355-mm) long by 1-in. (25-mm) wide by 1/4-in (6-mm) thick backing strip.

INSTRUCTIONS

1. Tack weld the plates and the backing strip together leaving a slight space between the plates and `allow the backing strip to extend 1 in. (25 mm) beyond the end of the plates.

2. Position the plate in the overhead position. Get yourself into a comfortable position.

3. Use a lower current setting. Maintain a small weld pool and travel quickly to prevent dripping of the molten metal.

4. Multiple passes will be required. Clean and wire brush the beads between passes.

5. Repeat with both classifications of electrodes as needed until defect-free welds can be made.

6. Turn off the welding machine and shielding gas, and clean up your work area when finished with this practice.

INSTRUCTOR'S COMMENTS _____

_____

_____

■ PRACTICE 13-18

Name _____ Date _____

Class _____ Instructor _____ Grade _____

OBJECTIVE: After completing this practice you should be able to make lap and tee joints in the overhead 4F position.

EQUIPMENT AND MATERIALS NEEDED FOR THIS PRACTICE

1. A properly set-up and adjusted FCA welding machine.

2. Table 13-1 or specific settings from the electrode manufacturer.

3. Proper safety protection. (Extra protection is needed for this practice. Wear a Leather jacket and cap.)

4. E70T-1 and/or E70T-5 electrodes with 0.035-in. and/or 0.045-in. (0.9-mm and/or 1.2-mm) diameter.

5. Two or more pieces of mild steel plate, 12-in. (305-mm) long by 3-in. (76-mm) wide by 3/8-in. (9.5-mm) thick beveled plate.

INSTRUCTIONS

1. Tack weld the plates together in the lap and tee joint configuration.

2. Position the plate in the overhead position. Get yourself into a comfortable position.

3. Use a lower current setting. Maintain a small weld pool and travel quickly to prevent dripping.

4. Multiple passes will be required. Clean and wire brush the beads between passes.

5. Repeat with both classifications of electrodes as needed until defect-free welds can be made.

6. Turn off the welding machine and shielding gas, and clean up your work area when finished with this practice.

INSTRUCTOR'S COMMENTS _____

_____

_____

■ PRACTICE 13-19

Name _____ Date _____

Class _____ Instructor _____ Grade _____

OBJECTIVE: After completing this practice you should be able to make a butt joint in the flat 1G position.

EQUIPMENT AND MATERIALS NEEDED FOR THIS PRACTICE

1. A properly set-up and adjusted FCA welding machine.

2. Table 13-3 or specific settings from the electrode manufacturer.

3. Proper safety protection.

4. E70T-1 and/or E70T-5 electrodes with 0.030-in. and/or 0.035-in. (0.8-mm and/or 0.9-mm) diameter.

5. Two or more pieces of mild steel sheet, 12-in. (305-mm) long by 3-in. (76-mm) wide by 16-gauge to 18-gauge thick.

INSTRUCTIONS

1. Tack weld the plates together in the butt joint configuration without a root gap.

2. Position the plate in the flat position.

3. Use a low current setting and beware of burnthrough. If burnthrough occurs, the welder can be pulsed on and off to fill the hole.

4. Repeat with both classifications of electrodes as needed until defect-free welds can be made.

5. Turn off the welding machine and shielding gas, and clean up your work area when finished with this practice.

INSTRUCTOR'S COMMENTS _____

_____

_____

■ PRACTICE 13-20

Name _____ Date _____

Class _____ Instructor _____ Grade _____

OBJECTIVE: After completing this practice you should be able to make lap and tee joints in the flat 1F position.

EQUIPMENT AND MATERIALS NEEDED FOR THIS PRACTICE

1. A properly set-up and adjusted FCA welding machine.

2. Table 13-3 or specific settings from the electrode manufacturer.

3. Proper safety protection.

4. E70T-1 and/or E70T-5 electrodes with 0.030-in. and/or 0.035-in. (0.8-mm and/or 0.9-mm) diameter.

5. Two or more pieces of mild steel sheet, 12-in. (305-mm) long by 3-in. (76-mm) wide by 16-gauge to 18-gauge thick.

INSTRUCTIONS

1. Tack weld the plates together in the lap and tee joint configurations.

2. Position the plate in the flat position.

3. Use a low current setting and beware of burnthrough. If burnthrough occurs, the welder can be pulsed on and off to fill the hole.

4. Repeat with both classifications of electrodes as needed until defect-free welds can be made.

5. Turn off the welding machine and shielding gas, and clean up your work area when finished with this practice.

INSTRUCTOR'S COMMENTS _____

_____

_____

■ PRACTICE 13-21

Name _____ Date _____

Class _____ Instructor _____ Grade _____

OBJECTIVE: After completing this practice you should be able to make a butt joint in the vertical down 3G position.

EQUIPMENT AND MATERIALS NEEDED FOR THIS PRACTICE

1. A properly set-up and adjusted FCA welding machine.

2. Table 13-3 or specific settings from the electrode manufacturer.

3. Proper safety protection.

4. E70T-1 and/or E70T-5 electrodes with 0.030-in. and/or 0.035-in. (0.8-mm and/or 0.9-mm) diameter.

5. Two or more pieces of mild steel sheet, 12-in. (305-mm) long by 3-in. (76-mm) wide by 16-gauge to 18-gauge thick.

INSTRUCTIONS

1. Tack weld the plates together in the butt joint configuration without a root gap.

2. Position the plate in the vertical position.

3. Begin the bead at the top and weld downward. This works best on thin-gauge metal.

4. Use a low current setting and beware of burnthrough. If burnthrough occurs, the welder can be pulsed on and off to fill the hole.

5. Repeat with both classifications of electrodes as needed until defect-free welds can be made.

6. Turn off the welding machine and shielding gas, and clean up your work area when finished with this practice.

INSTRUCTOR'S COMMENTS _____

_____

_____

■ PRACTICE 13-22

Name _____ Date _____

Class _____ Instructor _____ Grade _____

OBJECTIVE: After completing this practice you should be able to make lap and tee joints in the vertical down 3F position.

EQUIPMENT AND MATERIALS NEEDED FOR THIS PRACTICE

1. A properly set-up and adjusted FCA welding machine.

2. Table 13-3 or specific settings from the electrode manufacturer.

3. Proper safety protection.

4. E70T-1 and/or E70T-5 electrodes with 0.030-in. and/or 0.035-in. (0.8-mm and/or 0.9-mm) diameter.

5. Two or more pieces of mild steel sheet, 12-in. (305-mm) long by 3-in. (76-mm) wide by 16-gauge to 18-gauge thick.

INSTRUCTIONS

1. Tack weld the plates together in the lap and tee joint configurations.

2. Position the plate in the vertical position.

3. Begin the bead at the top and weld downward. This works best on thin-gauge metal.

4. Use a low current setting and beware of burnthrough. If burnthrough occurs, the welder can be pulsed on and off to fill the hole.

5. Repeat with both classifications of electrodes as needed until defect-free welds can be made.

6. Turn off the welding machine and shielding gas, and clean up your work area when finished with this practice.

INSTRUCTOR'S COMMENTS _____

_____

_____

■ PRACTICE 13-23

Name _____ Date _____

Class _____ Instructor _____ Grade _____

OBJECTIVE: After completing this practice you should be able to make lap and tee joints in the horizontal 2F position.

EQUIPMENT AND MATERIALS NEEDED FOR THIS PRACTICE

1. A properly set-up and adjusted FCA welding machine.

2. Table 13-3 or specific settings from the electrode manufacturer.

3. Proper safety protection.

4. E70T-1 and/or E70T-5 electrodes with 0.030-in. and/or 0.035-in. (0.8-mm and/or 0.9-mm) diameter.

5. Two or more pieces of mild steel sheet, 12-in. (305-mm) long by 3-in. (76-mm) wide by 16-gauge to 18-gauge thick.

INSTRUCTIONS

1. Tack weld the plates together in the lap and tee joint configurations.

2. Position the plate in the horizontal position.

3. Angle the gun slightly upward and travel relatively fast to avoid sagging of the weld pool.

4. Use a low current setting and beware of burnthrough. If burnthrough occurs, the welder can be pulsed on and off to fill the hole.

5. Repeat with both classifications of electrodes as needed until defect-free welds can be made.

6. Turn off the welding machine and shielding gas, and clean up your work area when finished with this practice.

INSTRUCTOR'S COMMENTS _____

_____

_____

■ PRACTICE 13-24

Name _____ Date _____

Class _____ Instructor _____ Grade _____

OBJECTIVE: After completing this practice you should be able to make a butt joint in the horizontal 2G position.

EQUIPMENT AND MATERIALS NEEDED FOR THIS PRACTICE

1. A properly set-up and adjusted FCA welding machine.

2. Table 13-3 or specific settings from the electrode manufacturer.

3. Proper safety protection.

4. E70T-1 and/or E70T-5 electrodes with 0.030-in. and/or 0.035-in. (0.8-mm and/or 0.9-mm) diameter.

5. Two or more pieces of mild steel sheet, 12-in. (305-mm) long by 3-in. (76-mm) wide by 16-gauge to 18-gauge thick.

INSTRUCTIONS

1. Tack weld the plates together in the butt joint configuration with no root gap.

2. Position the plate in the horizontal position.

3. Angle the gun slightly upward and travel relatively fast to avoid sagging of the weld pool.

4. Use a low current setting and beware of burnthrough. If burnthrough occurs, the welder can be pulsed on and off to fill the hole.

5. Repeat with both classifications of electrodes as needed until defect-free welds can be made.

6. Turn off the welding machine and shielding gas, and clean up your work area when finished with this practice.

INSTRUCTOR'S COMMENTS _____

_____

_____

■ PRACTICE 13-25

Name _____ Date _____

Class _____ Instructor _____ Grade _____

OBJECTIVE: After completing this practice you should be able to make a butt joint in the overhead 4G position.

EQUIPMENT AND MATERIALS NEEDED FOR THIS PRACTICE

1. A properly set-up and adjusted FCA welding machine.

2. Table 13-3 or specific settings from the electrode manufacturer.

3. Proper safety protection. (Extra protection is required for this practice. Wear a Leather jacket and cap.)

4. E70T-1 and/or E70T-5 electrodes with 0.030-in. and/or 0.035-in. (0.8-mm and/or 0.9-mm) diameter.

5. Two or more pieces of mild steel sheet, 12-in. (305-mm) long by 3-in. (76-mm) wide by 16-gauge to 18-gauge thick.

INSTRUCTIONS

1. Tack weld the plates together in the butt joint configuration with no root gap.

2. Position the plate in the overhead position. Get yourself into a comfortable position.

3. Travel relatively fast to avoid dripping of the weld pool.

4. Use a low current setting and beware of burnthrough. If burnthrough occurs, the welder can be pulsed on and off to fill the hole.

5. Repeat with both classifications of electrodes as needed until defect-free welds can be made.

6. Turn off the welding machine and shielding gas, and clean up your work area when finished with this practice.

INSTRUCTOR'S COMMENTS _____

_____

_____

■ PRACTICE 13-26

Name _____ Date _____

Class _____ Instructor _____ Grade _____

OBJECTIVE: After completing this practice you should be able to make a lap and tee joints in the overhead 4F position.

EQUIPMENT AND MATERIALS NEEDED FOR THIS PRACTICE

1. A properly set-up and adjusted FCA welding machine.

2. Table 13-3 or specific settings from the electrode manufacturer.

3. Proper safety protection. (Extra protection is required for this practice. Wear a Leather jacket and cap.)

4. E70T-1 and/or E70T-5 electrodes with 0.030-in. and/or 0.035-in. (0.8-mm and/or 0.9-mm) diameter.

5. Two or more pieces of mild steel sheet, 12-in. (305-mm) long by 3-in. (76-mm) wide by 16-gauge to 18-gauge thick.

INSTRUCTIONS

1. Tack weld the plates together in the lap and tee joint configurations.

2. Position the plate in the overhead position. Get yourself into a comfortable position.

3. Travel relatively fast to avoid dripping of the weld pool.

4. Use a low current setting and beware of burnthrough. If burnthrough occurs, the welder can be pulsed on and off to fill the hole.

5. Repeat with both classifications of electrodes as needed until defect-free welds can be made.

6. Turn off the welding machine and shielding gas, and clean up your work area when finished with this practice.

INSTRUCTOR'S COMMENTS _____

_____

_____

CHAPTER 13: FLUX CORED ARC WELDING QUIZ

Name _____ Date _____

Class _____ Instructor _____ Grade _____

Instructions: Carefully read Chapter 13 in the text and answer the following questions.

SHORT ANSWER

Write a brief answer in the space provided that will answer the question or complete the statement.

1. How should a gas cylinder be secured before the cap is removed?

   _____

   _____

2. Why are larger pieces of metal used for most FCA welding?

   _____

   _____

3. What can cause bird nesting?

   _____

   _____

4. How tight should the feed roller pressure be set?

   _____

   _____

5. What is the purpose of the root pass?

   _____

   _____

6. What is the purpose of the filler pass?

   _____

   _____

7. What is the purpose of the cover pass?

   _____

   _____

8. When is it best to use stringer beads as opposed to weave beads?

   _____

   _____

9. How large should a fillet weld be?

   _____

   _____

10. How can some tee joints be prepared for better weld penetration?

_____

_____

11. What would a notch at the leading edge of a fillet weld pool indicate?

_____

_____

12. Why should the electrode never be allowed to strike the metal ahead of the molten weld pool?

_____

_____

13. How can burnthrough be repaired in thin-gauged metals?

_____

_____

INSTRUCTOR'S COMMENTS _____

_____

_____

Chapter 14

Gas Metal Arc and Flux Cored Arc Welding of Pipe

■ PRACTICE 14-1

Name _____ Date _____

Class _____ Instructor _____ Grade _____

OBJECTIVE: After completing this practice, you will make a pipe-to-plate fillet welded joint in the 1F flat rolled position using each of the three wire welding processes.

EQUIPMENT AND MATERIALS NEEDED FOR THIS PRACTICE

1. Semiautomatic welding power system approved for GMA and FCA welding.

2. Assorted hand tools, spare parts, and any other required materials.

3. Proper PPE.

4. E70T-1 and/or E70T-5 and ER70S-2 and/or ER70S-3 electrodes with 0.035-in. and/or 0.045-in. (0.9-mm and/or 1.2-mm) diameter.

5. One piece of schedule 40 mild steel pipe 3 in. to 6 in. (76 mm to 150 mm) in diameter, and

6. One 4 in. or 7 in. (100 mm or 170 mm) ¼ in. (6 mm) mild steel plate.

INSTRUCTIONS

1. Mark a straight line around the pipe using a pipe wrap-around or other pipe layout tool. Using all of the appropriate PPE for a flame cutting torch or plasma cutting torch and following all of the equipment manufacturer's safety and operating guidelines, make a cut along the line. Clean any slag off, and grind the end if necessary to have the pipe stand vertically on the plate.

2. Tack weld the plate to the pipe with 4 equally spaced approximately 1 in. (25 mm) long welds. Grind the ends of the weld to a featheredge using a thin disk-grinding wheel.

3. The welds on this practice will have three passes. The first pass will be made with the short-circuiting metal transfer process, and the last two passes will be made using the spray metal transfer process.

4. Mount the pipe and plate at a 45° angle so that it can be turned between welds and so that the surface of the fillet weld will remain in the flat position. Position the pipe so that the tack welds are at approximately the 1:30 and 10:30 o'clock positions. Start the first weld on the 1:30 o'clock position, and weld toward the 10:30 o'clock position. The gun should be a slight forehand angle; transition to a perpendicular angle around the 12:00 o'clock position, then on to a slight backhand position as it moves toward the 10:30 o'clock position. The weld bead should be approximately 3/16 in. (4.8 mm).

5. Visually inspect the root surface to see if you have complete fusion and no overlap. Discuss with your instructor what you need to do to improve your root weld.

6. Rotate the pipe so you can make the next root weld. Make the necessary changes in the machine settings before starting the next weld. Repeat this process until you have made the root pass all the way around the pipe.

7. Chip and wire brush the root, and grind any areas that need additional cleanup.

8. The next two weld passes will be made using the spray metal transfer method. Using the same gun angle as you used for the root pass, make a filler pass over the root weld. Use a slight weaving pattern to make the weld bead wide enough to cover the root weld. To reduce the need for grinding speed up your travel rate slightly just before stopping the weld. This will result in a slight tapering down of the weld bead and can eliminate the need to grind the end of the weld. However, you should grind the starting point of this first weld so that when you get all the way around, you can tie the ending of the weld smoothly into the starting point of the weld pass.

9. Visually inspect the root surface to see if you have complete fusion and no overlap. Discuss with your instructor what you need to do to improve your root weld.

10. Rotate the pipe so you can make the next root weld. Make the necessary changes in the machine settings before starting the next weld. Repeat this process until you have made the filler pass all the way around the pipe.

11. Chip and wire brush the root, and grind any areas that need additional cleanup.

12. The cover pass will be made with the same technique as the filler pass; however, a slightly wider weave pattern will be needed.

13. Visually inspect the weld, and repeat this weld as needed with all three processes until you can consistently make welds free of defects. Turn off the welding machine, and clean up your work area when you are finished welding.

INSTRUCTOR'S COMMENTS _____

_____

_____

■ PRACTICE 14-2

Name _____ Date _____

Class _____ Instructor _____ Grade _____

OBJECTIVE: After completing this practice, you are going to make a fillet weld in the 2F position using each of the three wire welding processes.

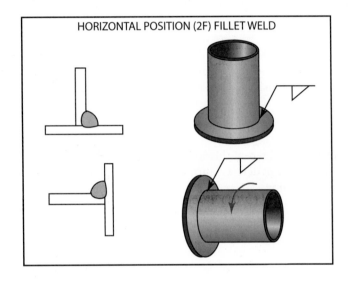

HORIZONTAL POSITION (2F) FILLET WELD

EQUIPMENT AND MATERIALS NEEDED FOR THIS PRACTICE

1. Semiautomatic welding power source approved for GMA and FCA welding.

2. Assorted hand tools, spare parts, and any other required materials.

3. Proper PPE.

4. E70T-1 and/or E70T-5 and ER70S-2 and/or ER70S-3 electrodes with 0.035-in. and/or 0.045-in. (0.9-mm and/or 1.2-mm) diameter.

5. One piece of schedule 40 mild steel pipe 3 in. to 6 in. (76 mm to 150 mm) in diameter, and

6. One 4 in. or 7 in. (100 mm or 170 mm) ¼ in. (6 mm) mild steel plate.

INSTRUCTIONS

1. Using a 15° to 20° backhand gun angle and a 35° to 40° work angle, start the weld on one of the tack welds, and make a stringer weld all the way to the center of the next tack weld. As you get to the tack weld, speed up your travel rate to taper the size of the weld down to make it easier to restart the next weld bead. Chip and wire brush the weld, and visually inspect it for uniformity and any undercut or overlap. The weld bead should have equal legs of approximately 3/16 in. (4.8 mm). The weld should be equal on both the plate and side of the pipe.

2. If there is undercut along the top toe of the weld or if the weld leg is not as large on the top side, you will need to decrease the work angle to direct more metal onto the pipe side. You may also want to make a slight J weave pattern to help correct this problem. Discuss with your instructor what you need to do to improve your root penetration before making the weld on the opposite side.

3. Make any corrections to your technique and complete the weld all the way around the pipe.

4. The second weld pass will be made using the spray metal transfer method. This weld bead will be placed around the lower side of the first weld so that approximately 2/3 to ¾ of the root weld face is covered by this weld pass. Chip and wire brush the weld each time you stop to reposition yourself. Visually inspect the weld, and discuss with your instructor what you need to do to improve your weld. Complete the weld all the way around the pipe.

5. The third weld pass will complete the cover pass, and it will also be made using the spray metal transfer method. This weld should be made approximately ½ to 2/3 of the way up on the second weld's face. Chip and wire brush the weld each time you stop to reposition yourself. Visually inspect the weld, and discuss with your instructor what you need to do to improve your weld. Complete the weld all the way around the pipe.

6. Visually inspect the weld, and repeat this weld as needed with all three processes until you can consistently make welds free of defects. Turn off the welding machine, and clean up your work area when you are finished welding.

INSTRUCTOR'S COMMENTS _____

_____

_____

■ PRACTICE 14-3

Name _____   Date _____

Class _____ Instructor _____ Grade _____

OBJECTIVE: After completing this practice, you are going to make a fillet weld in the 5F fixed position using each of the three wire welding processes.

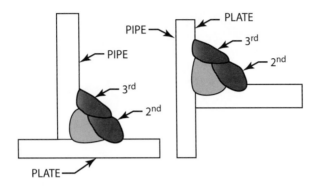

EQUIPMENT AND MATERIALS NEEDED FOR THIS PRACTICE

1. Semiautomatic welding power source approved for GMA and FCA welding.

2. Assorted hand tools, spare parts, and any other required materials.

3. Proper PPE.

4. E70T-1 and/or E70T-5 and ER70S-2 and/or ER70S-3 electrodes with 0.035-in. and/or 0.045-in. (0.9-mm and/or 1.2-mm) diameter.

5. One piece of schedule 40 mild steel pipe 3 in. to 6 in. (76 mm to 150 mm) in diameter, and

6. One 4 in. or 7 in. (100 mm or 170 mm) ¼ in. (6 mm) mild steel plate.

INSTRUCTIONS

1. The weld will be made in an uphill progression, starting on the tack weld at the 6:00 o'clock position. Most welders stop on top of the tack weld at the 9:00 o'clock position. They do this so they can get in a better position to complete the weld.

2. The gun angle relative to the pipe surface will be constantly changing as the weld progresses upward. Figure 14-1 shows how the gun angle starts at a slight forehand angle at 6:00 o'clock and becomes perpendicular to the pipe surface around 7:00 o'clock. It transitions to a steep forehand angle of around 30° between 7:00 and 9:00 o'clock. Between 9:00 and 11:00 o'clock, the angle transitions back to perpendicular and remains perpendicular through the 12:00 o'clock position where the weld ends.

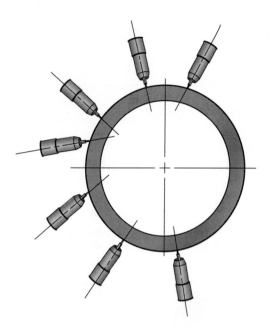

3. The weld bead should be made as an equal legged 3/16 in. (4.8 mm) fillet weld. Chip and wire brush the weld; and after you visually inspect it, discuss with your instructor what you need to do to improve your root penetration before making the weld on the opposite side.

4. Visually inspect the weld, and repeat this weld as needed with all three processes until you can consistently make welds free of defects. Turn off the welding machine, and clean up your work area when you are finished welding.

INSTRUCTOR'S COMMENTS _____

_____

_____

■ PRACTICE 14-4

Name _____ Date _____

Class_____ Instructor _____ Grade _____

OBJECTIVE: After completing this practice, you will make a pipe butt joint in the 1G horizontal rolled position using each of the three wire welding processes.

EQUIPMENT AND MATERIALS NEEDED FOR THIS PRACTICE

1. Semiautomatic welding power source approved for GMA and FCA welding.

2. Assorted hand tools, spare parts, and any other required materials.

3. Proper PPE.

4. E70T-1 and/or E70T-5 and ER70S-2 and/or ER70S-3 electrodes with 0.035-in. and/or 0.045-in. (0.9-mm and/or 1.2-mm) diameter, and

5. Two or more piece of schedule 40 mild steel pipe 6 in. to 8 in. (150 mm to 200 mm) in diameter.

INSTRUCTIONS

1. Tack weld two pieces of pipe together with four 1-in. (25 mm) long welds at the 12:00, 3:00, 6:00, and 9:00 o'clock positions. Grind the ends of the tack welds before placing the pipe horizontally on the welding table or pipe stand with the top tack welds between the 10:30 and 1:30 o'clock positions.

2. Start the root pass weld at the 1:30 o'clock position and weld to the 10:30 o'clock position with a 14° to 20° forehand angle in the direction of travel. The root pass is made as a stringer bead with the GMA welding process unless a backing ring is used; then it can be made with the FCA welding process. As you pass over the top of the pipe and start downhill, change to a slight backhand angle that will increase to approximately a 45° angle. Watch the leading edge of the molten weld pool to see if you have a keyhole, which indicates you are getting 100% penetration.

3. Visually inspect the root surface to see if you had 100% penetration. Mark any areas that did not have complete penetration so you can make changes in the next section to be welded. Discuss with your instructor what you need to do to improve your root penetration.

4. Rotate the pipe so you can make the next root weld. Make the necessary changes in the machine settings before starting the next weld. Repeat this process until you have made the root pass all the way around the pipe.

5. Clean the root, and grind any areas that need additional cleanup.

6. **NOTE:** The small weld coupon may become too hot to allow you to make a good weld. Be sure to give the coupon time to cool between weld passes. You may want to have two or three coupons being worked on at the same time so you can alternate between the three processes as a way of giving the coupon time to cool.

7. The hot pass is made with the same procedure of welding from 1:30 to 10:30 o'clock positions as the root pass. Watch the sides of the groove and the leading edge of the molten weld pool to make sure it is being fused.

8. Clean the weld face, and visually inspect it for uniformity and sidewall fusion. Grind any trapped slag, and taper the end of the weld to make it easier to restart the weld bead. Discuss with your instructor what you need to do to improve your root penetration.

9. Rotate the pipe so you can make the next hot pass weld. Make the necessary changes in the machine settings before starting the next weld. Repeat this process until you have made the hot pass all the way around the pipe.

10. Grind any areas of the hot pass that have buildup greater than 1/8 in. (3 mm) higher than the surrounding weld face before starting the filler passes.

11. The first filler pass is made along one side so that its toe is about two-thirds of the way across the hot pass. If the toe of the filler pass is made so it meets the opposite sidewall, the V it forms will be likely to trap slag, and it is hard to get the next filler pass to penetrate the narrow gap it forms.

12. The end of the filler passes can be tapered down by slightly increasing the welding travel speed and reducing the weave pattern. This will make it easier to restart the next weld bead. When you have reached the 10:30 o'clock position, stop welding and clean off the flux, and visually inspect the weld. Discuss with your instructor what you need to do to improve your root penetration.

13. Rotate the pipe so you can make the next filler pass. Make the necessary changes in the machine settings before starting the next weld. Repeat this process until you have filled the weld groove to within approximately 1/16 to 1/8-in. (1.6 mm to 3 mm) below the pipe's outer surface. Clean the weld, and grind any starts or stops that are more than 1/16 in. (1.6 mm) higher than the surrounding weld surface. Also grind any areas of the weld that are overfilled that would affect the uniformity of the cover weld pass.

14. Watch the edge of the groove as the first cover pass weld is made so that it does not overlap the edge more than 1/16 in. (1.6 mm). Bracing yourself on the table or bracing your hand on the pipe will help you make the cover pass more consistent. The transition for the pipe surface to the weld must be smooth and uniform, so do not spend too much time along the side of the groove because that could cause excessive reinforcement and an abrupt toe of the weld.

15. Clean each weld before the next weld is started. When the cover weld has been completed, clean the weld, do not grind it, and inspect it for discontinuities or defects.

16. Repeat this weld as needed with all three processes until you can consistently make welds free of defects. Turn off the welding machine, and clean up your work area when you are finished welding.

INSTRUCTOR'S COMMENTS _____

_____

_____

■ PRACTICE 14-5

Name _____ Date _____

Class _____ Instructor _____ Grade _____

OBJECTIVE: After completing this practice, you are going to make a V-grooved weld in the 2G vertical fixed position using each of the three wire welding processes.

EQUIPMENT AND MATERIALS NEEDED FOR THIS PRACTICE

1. Semiautomatic welding power source approved for GMA and FCA welding.

2. Assorted hand tools, spare parts, and any other required materials.

3. Proper PPE.

4. E70T-1 and/or E70T-5 and ER70S-2 and/or ER70S-3 electrodes with 0.035-in. and/or 0.045-in. (0.9-mm and/or 1.2-mm) diameter.

5. Two or more piece of schedule 40 mild steel pipe 6 in. to 8 in. (150 mm to 200 mm) in diameter.

INSTRUCTIONS

1. Prepare the pipe and tack it together as before, then place it vertically on the table or on a pipe stand.

2. You will use a 14° to 20° forehand angle in the direction of travel for the root pass and the same forehand angle with a 5° to 10° upward work angle for all of the additional weld passes, Figure 14-2. This upward work angle will help to prevent the weld bead's bottom edge from sagging.

3. Keeping the weld beads small makes it easier to control the molten weld pool. Place the weld beads in the groove as illustrated in Figure 14-3.

4. Repeat this weld as needed with all three processes until you can consistently make welds free of defects. Turn off the welding machine, and clean up your work area when you are finished welding.

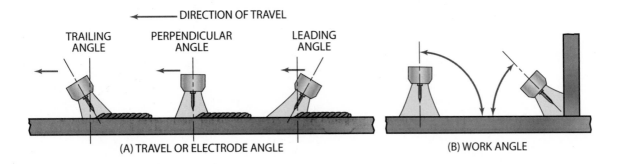

DIRECTION OF TRAVEL

TRAILING ANGLE   PERPENDICULAR ANGLE   LEADING ANGLE

(A) TRAVEL OR ELECTRODE ANGLE   (B) WORK ANGLE

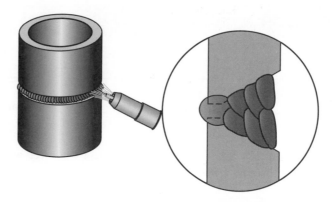

INSTRUCTOR'S COMMENTS _____

_____

_____

■ PRACTICE 14-6

Name _____ Date _____

Class _____ Instructor _____ Grade _____

OBJECTIVE: After completing this practice, you are going to make a V-grooved weld in the 5G fixed position using each of the three wire welding processes.

EQUIPMENT AND MATERIALS NEEDED FOR THIS PRACTICE

1. Semiautomatic welding power source approved for GMA and FCA welding.

2. Assorted hand tools, spare parts, and any other required materials.

3. Proper PPE.

4. E70T-1 and/or E70T-5 and ER70S-2 and/or ER70S-3 electrodes with 0.035-in. and/or 0.045-in. (0.9-mm and/or 1.2-mm) diameter.

5. Two or more piece of schedule 40 mild steel pipe 6 in. to 8 in. (150 mm to 200 mm) in diameter.

INSTRUCTIONS

1. Prepare the pipe and tack it together as before, and place it horizontally on a pipe stand. Any time the weld has to be stopped, the weld must be cleaned and the end of the weld bead feathered even if the weld stops on a tack weld.

2. These welds will be made in an uphill progression, starting near the 6:00 o'clock position. Most welders stop on top of the tack weld at the 9:00 o'clock position. They do this so they can get in a better position to complete the weld.

3. This weld transitions from the overhead position to the vertical position and then to the flat position. So the gun angle relative to the pipe surface must be constantly changing as the weld progresses upward. Refer back to Figure 14-1 in Practice 14-3 to see how the gun angle starts at a slight forehand angle at 6:00 o'clock and becomes perpendicular to the pipe surface around 7:00 o'clock. It transitions to a steep forehand angle of around 30° between 7:00 and 9:00 o'clock. Between 9:00 and 11:00 o'clock, the angle transitions back to perpendicular and remains perpendicular through the 12:00 o'clock position where the weld ends.

4. Clean the weld, and feather the starting and stopping points. Chip and wire brush the root pass before visually inspecting it for any problem areas with lack of fusion, excessive buildup, burnthrough, etc. Discuss with your instructor what you need to do to improve your root penetration before making the weld on the opposite side.

5. Grind the root pass to remove any trapped slag and deep undercut before starting the hot pass. The beginning and ending points of all of the weld should be staggered so they are not all in the same area and overlap. Slightly increasing the voltage or decreasing the wire feed speed can help to make the weld pool a little more fluid and help increase the hot pass penetration. Use the same technique of transitioning the electrode angle as you used on the root pass to make the hot pass. Use a

stringer technique with little or no weave for most of the weld. A slight weave may be needed to fill any area that was ground out to remove root pass undercut.

6. Chip and wire brush the root pass before visually inspecting the hot pass for any problem areas with lack of fusion, excessive buildup, etc. Discuss with your instructor what you need to do to improve your root penetration before making the weld on the opposite side.

7. When the hot pass is completed, clean and wire brush it before showing it to your instructor for their advice.

8. Grind any areas of the hot pass that have a weld buildup greater than 1/8 in. (3 mm) before starting the filler passes.

9. Slightly reduce the voltage or increase the wire feed speed so that the weld has good fusion but not deep penetration since the filler passes do not need to have deep penetration.

10. The first filler pass can be made by rocking the welding gun so that the arc is directed from side to side or by using a slight side-to-side weave, Figure 14-4. This can make a concave weld that can make it easier to prevent slag entrapment and/or undercut along the toe of the weld. If this technique is used, it is important not to make the weave too large because that can cause excessive buildup and overlap.

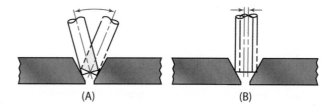

(A)   (B)

11. Whether you used a slight weave or stringer bead for the first filler pass, all of the following passes should be made as stringer beads. Remember to grind the ends of the welds before restarting the next weld. Also clean and grind any parts of a weld bead that has undercut, overlap, or slag trapped before making the next weld pass.

12. The groove must be left with 1/16 in. to 1/8 in. (1.6 mm to 3 mm) below the surface so that there will be room for the cover pass. If necessary, grind away any areas of the filler passes that are too high or uneven that would affect the appearance of the cover passes.

13. Again, follow the edge of the V-groove for the first cover pass. Keep the weld bead small and uniform. Overlap each weld pass about 1/3 of the way. Judge the width of the weld beads so that the last weld pass will not overlap the edge of the V-groove more than 1/8 in. (3 mm). You may have a problem if the last weld pass needs to be much wider than the other cover pass welds, because trying to make it larger can cause overlap or excessive buildup.

14. Visually inspect the weld, and repeat this weld as needed with all three processes until you can consistently make welds free of defects. Turn off the welding machine, and clean up your work area when you are finished welding.

INSTRUCTOR'S COMMENTS _____

_____

_____

■ PRACTICE 14-7

Name _____ Date _____

Class_____ Instructor _____ Grade _____

OBJECTIVE: After completing this practice, you are going to make a V-grooved weld in the 6G fixed position using each of the three wire welding processes.

EQUIPMENT AND MATERIALS NEEDED FOR THIS PRACTICE

1. Semiautomatic welding power source approved for GMA and FCA welding.

2. Assorted hand tools, spare parts, and any other required materials.

3. Proper PPE.

4. E70T-1 and/or E70T-5 and ER70S-2 and/or ER70S-3 electrodes with 0.035-in. and/or 0.045-in. (0.9-mm and/or 1.2-mm) diameter.

5. Two or more piece of schedule 40 mild steel pipe 6 in. to 8 in. (150 mm to 200 mm) in diameter.

INSTRUCTIONS

1. Prepare the pipe and tack it together as before, and place it at a 45° angle on a pipe stand. Any time the weld has to be stopped, the weld must be cleaned and the end of the weld bead feathered even if the weld stops on a tack weld.

2. The root pass is made without changing the work angle, because getting good root penetration is more important than trying to prevent the weld face from sagging to the downhill side of the groove. Any sagging can be removed by post weld grinding.

3. Once the root pass is made, the remaining welds in the 6G position become more challenging because both the electrode angle and the work angle must be constantly changed to maintain a consistent weld bead. That is because parts of the weld are much like the overhead portion of the 5G weld. Other parts are more like the 2G weld.

4. Slightly increase the voltage or decrease the wire feed speed to help increase the weld penetration for the hot pass. Start the hot pass weld on the opposite side of the bottom that the root pass was started. Use a slight forehand angle and around a 5° upward work angle when starting to weld. Transition to a perpendicular work angel at the 6:00 o'clock position with the same 5° upward work angle. As the weld progresses toward the 9:00 o'clock position, increase the forward gun angle to around 30° and keep the same 5° work angle. Between the 9:00 and 11:00 o'clock positions, transition to a perpendicular gun angle while keeping enough work angle to force the weld face to be uniform and not sagging toward the downhill side of the joint.

5. Clean and visually inspect the hot pass. Discuss with your instructor what you need to do to improve your hot pass before making the weld on the opposite side.

6. The filler passes will be made with a similar technique as the hot pass but with a little less penetration, so set the voltage and/or amperage for a less penetrating weld. The first filler welds will be made on the downhill side of the groove and cover about two-thirds of the hot pass weld surface. Each of the next filler welds will be made on the upper side of the previous filler welds, Figure 14-5. Remember to stagger the starts and stops as you make the filler welds.

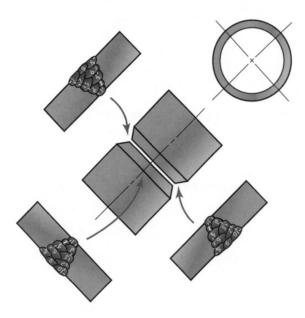

7. Grind any areas of the filler pass that are not uniform in buildup or are excessively built up before starting the cover passes. The cover passes are made with the same technique as the filler passes but with possibly a little greater work angle of between 5° and 10°. Keep the cover passes small so they can be controlled and kept uniform.

8. Visually inspect the weld, and repeat this weld as needed with all three processes until you can consistently make welds free of defects. Turn off the welding machine, and clean up your work area when you are finished welding.

INSTRUCTOR'S COMMENTS _____

_____

_____

CHAPTER 14: GAS METAL ARC AND FLUX CORED ARC WELDING OF PIPE QUIZ

Name _____ Date _____

Class_____ Instructor _____ Grade _____

Instructions: Carefully read Chapter 14 in the text and answer the following questions.

SKETCH

Sketch and label the following.

1. The joint geometry without a backing.

2. Cover pass as a single weld pass and multiple weld passes.

3. Travel or electrode angle and work angle.

SHORT ANSWER

Write a brief answer in the space provided that will answer the question or complete the statement.

4. What is a high-low gauge used for?

   _____

   _____

5. Why are the ends of tack welds ground to a featheredge?

   _____

   _____

6. Why should a thin grinding disk be used to grind the tack welds?

   _____

   _____

7. What is the easiest way of controlling the root pass penetration?

_____

_____

8. At what point should the filler pass be stopped?

_____

_____

9. What are the acceptable visual inspection tolerances for discontinuities welds?

_____

_____

10. What are some examples of uses for pipe to plate welds?

_____

_____

11. How can a straight line be drawn around a pipe?

_____

_____

12. What problems can be caused if the contact tip gets too close to the weld while using the spray metal transfer process?

_____

_____

13. How large should a fillet weld be?

_____

_____

14. How much of the hot pass should a filler pass cover on a 1G weld?

_____

_____

15. How much of the side of the surface next to a V-groove should the cover weld cover?

_____

_____

INSTRUCTOR'S COMMENTS _____

_____

_____

Chapter 15

Gas Metal Arc and Flux Cored Arc Welding AWS SENSE Certification

Electrode			Welding Power		Shielding Gas		Base Metal	
Type	Size	Amps	Wire-Feed Speed IPM (cm/min)	Volts	Type	Flow	Type	Thickness
E70S-X	0.035 in. (0.9 mm)	90 to 120	180 to 300 (457 to 762)	15 to 19	CO_2 or 75% Ar/25% CO_2	30 to 50	Low carbon steel	1/4 in. to 1/2 in. (6 mm to 13 mm)
E70S-X	0.045 in. (1.2 mm)	130 to 200	125 to 200 (318 to 508)	17 to 20	CO_2 or 75% Ar/25% CO_2	30 to 50	Low carbon steel	1/4 in. to 1/2 in. (6 mm to 13 mm)

Electrode			Welding Power		Shielding Gas		Base Metal	
Type	Size	Amps	Wire-Feed Speed IPM (cm/min)	Volts	Type	Flow	Type	Thickness
E70S-X	0.035 in. (0.9 mm)	180 to 230	400 to 550 (1016 to 1397)	25 to 27	Ar plus 2% O_2 or 90% Ar/10% CO_2	30 to 50	Low carbon steel	1/4 in. to 1/2 in. (6 mm to 13 mm)
E70S-X	0.045 in. (1.2 mm)	260 to 340	300 to 500 (762 to 1270)	25 to 30	Ar plus 2% O_2 or 90% Ar/10% CO_2	30 to 50	Low carbon steel	1/4 in. to 1/2 in. (6 mm to 13 mm)

Electrode			Welding Power		Shielding Gas		Base Metal	
Type	Size	Amps	Wire-Feed Speed IPM (cm/min)	Volts	Type	Flow	Type	Thickness
E71T-1	0.035 in. (0.9 mm)	130 to 150	288 to 380 (732 to 975)	22 to 25	CO_2 or 75% Ar/25% CO_2	30 to 50	Low carbon steel	1/4 in. to 1/2 in. (6 mm to 13 mm)
E71T-1	0.045 in. (1.2 mm)	150 to 210	200 to 300 (508 to 762)	28 to 29	CO_2 or 75% Ar/25% CO_2	30 to 50	Low carbon steel	1/4 in. to 1/2 in. (6 mm to 13 mm)

ELECTRODE TYPE	DIAMETER	VOLTS	AMPS	WIRE FEED SPEED IPM (cm/min)	ELECTRODE STICKOUT (INCH)
SELF-SHIELDED E70T-11 or E71T-11	0.030	15	40	69 (175)	3/8
		16	100	175 (445)	3/8
		16	160	440 (1118)	3/8
	0.035	15	80	81 (206)	3/8
		17	120	155 (394)	3/8
		17	200	392 (996)	3/8
	0.045	15	95	54 (137)	1/2
		17	150	118 (300)	1/2
		18	225	140 (356)	1/2

■ PRACTICE 15-1

Name _____ Date _____

Class _____ Instructor _____ Grade _____

OBJECTIVE: After completing this practice, you will make welded butt joints, tee joints, and lap joints in all four positions on 10 gauge to 14 gauge thickness carbon sheet steel.

EQUIPMENT AND MATERIALS NEEDED FOR THIS PRACTICE

1. Semiautomatic welding power system approved for GMA.

2. Assorted hand tools, spare parts, and any other required materials.

3. Proper PPE.

4. ER70S-2 and/or ER70S-3 electrodes with 0.035-in. and/or 0.045-in. (0.9-mm and/or 1.2-mm) diameter.

5. 75% Ar + 25% CO_2 or Ar + 2% to 5% O_2 shielding gas.

6. Two or more pieces of 6 in. long 1−1/2 in. wide (150 mm by 38 mm) 10 to 14 gauge carbon sheet steel.

INSTRUCTIONS

1. One at a time tack weld the sheets together to form the joints and place them in the desired position on a welding stand.

2. Starting at one end, run a 1/8 in. (3 mm) bead along the joint. Watch the molten weld pool and bead for signs it is increasing or decreasing in size. Make any necessary changes in your technique to keep the weld consistent.

3. When the welds have cooled use a hack saw or thermal cutting torch to cut out 1 in. (25 mm) wide strips as shown in Figure 15-1.

4. Using a vice and a hammer bend the strip of welded joint as shown in Figure 15-2.

5. Check for complete root fusion on the lap and tee joints and 100% penetration on the butt joint.

6. Repeat each type of joint in each position as needed until consistently good beads are obtained. Turn off the welding machine and shielding gas and clean up your work area when you are finished welding.

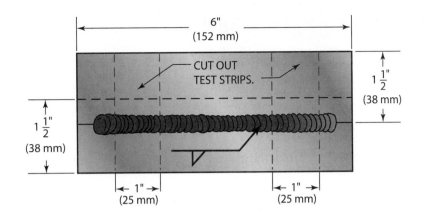

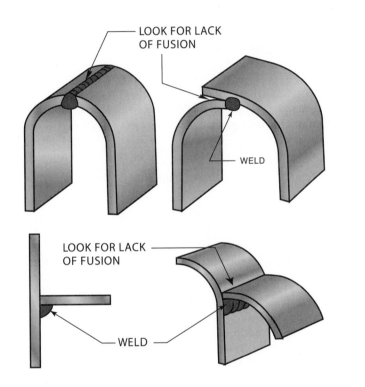

INSTRUCTOR'S COMMENTS _____

_____

_____

■ PRACTICE 15-2 AWS SENSE

Name _____ Date _____

Class _____ Instructor _____ Grade _____

GAS METAL ARC WELDING—SHORT-CIRCUIT METAL TRANSFER (GMAW-S)

WORKMANSHIP SAMPLE

Welding Procedure Specification (WPS) No.: Practice 15-2.

TITLE:

Welding GMAW-S of carbon steel plate to carbon steel plate.

SCOPE:

This procedure is applicable for square 2G, 3G, and 4G groove and 3F, 3F and 4F fillet welds within the range of 10 gauge (3.4 mm) through 14 gauge (1.9 mm) for the AWS SENSE Level I short-circuit metal transfer GMA welding.

WELDING MAY BE PERFORMED IN THE FOLLOWING POSITIONS:

All.

BASE METAL:

The base metal shall conform to carbon steel M-1, P-1, and S-1 Group 1 or 2.

BACKING MATERIAL SPECIFICATION:

None.

FILLER METAL:

The filler metal shall conform to AWS specification no. E70S-X from AWS specification A5.18. This filler metal falls into F-number F-6 and A-number A-1.

SHIELDING GAS:

The shielding gas, or gases, shall conform to the following compositions and purity: 25% CO_2 + 75% Ar at 30 to 50 cfh or Ar + 2% to 5% O_2 at 30 to 50 cfh.

JOINT DESIGN AND TOLERANCES:

Refer to the drawing in Figure 15-3 for the joint layout specifications.

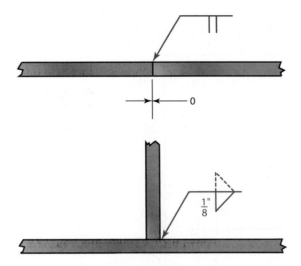

PREPARATION OF BASE METAL:

All parts may be mechanically cut or machine PAC unless specified as manual PAC.

All hydrocarbons and other contaminants, such as cutting fluids, grease, oil, and primers, must be cleaned off all parts and filler metals before welding. This cleaning can be done with any suitable solvents or detergents. The groove face and inside and outside plate surface within 1 in. (25 mm) of the joint must be mechanically cleaned of slag, rust, and mill scale. Cleaning must be done with a wire brush or grinder down to bright metal.

ELECTRICAL CHARACTERISTICS:

Set the voltage, amperage, wire-feed speed, and shielding gas flow according to Table 15-1.

PREHEAT:

The parts must be heated to a temperature higher than 50°F (10°C) before any welding is started.

BACKING GAS:

N/A.

SAFETY:

Proper protective clothing and equipment must be used. The area must be free of all hazards that may affect the welder or others in the area. The welding machine, welding leads, work clamp, electrode holder, and other equipment must be in safe working order.

WELDING TECHNIQUE:

Using a 1/2-in. (13-mm) or larger gas nozzle for all welding, first tack weld the plates together according to the drawing, Figure 15-4. Use the E70S-X filler metal to fuse the plates together. Clean any silicon slag, being sure to remove any trapped silicon slag along the sides of the weld.

All welds are to be made with one pass. All welds must be placed in the orientation shown in the drawing.

INTERPASS TEMPERATURE:

The plate should not be heated to a temperature higher than 350°F (175°C) during the welding process. After each weld pass is completed, allow it to cool but never to a temperature below 50°F (10°C). The weldment must not be quenched in water.

CLEANING:

The weld beads may be cleaned by a hand wire brush, a chipping hammer, or a punch and hammer. All weld cleaning must be performed with the test plate in the welding position.

VISUAL INSPECTION:

Visually inspect the weld for uniformity and discontinuities.

- There shall be no cracks, no incomplete fusion.
- There shall be no incomplete joint penetration in groove welds except as permitted for partial joint penetration welds.
- The Test Supervisor shall examine the weld for acceptable appearance and shall be satisfied that the welder is skilled in using the process and procedure specified for the test.
- Undercut shall not exceed the lesser of 10% of the base metal thickness or 1/32 in. (0.8 mm).
- Where visual examination is the only criterion for acceptance, all weld passes are subject to visual examination at the discretion of the Test Supervisor.
- The frequency of porosity shall not exceed one in each 4 in. (100 mm) of weld length, and the maximum diameter shall not exceed 3/32 in. (2.4 mm).
- Welds shall be free from overlap.

SKETCHES:

GMAW Short-Circuit Metal Transfer Workmanship Sample drawing.

INSTRUCTOR'S COMMENTS _____

_____

_____

■ PRACTICE 15-3

Name _____ Date _____

Class _____ Instructor _____ Grade _____

OBJECTIVE: After completing this practice, you will make a V-groove welds in the flat position using the spray metal transfer technique.

EQUIPMENT AND MATERIALS NEEDED FOR THIS PRACTICE

1. Semiautomatic welding power system approved for GMA.

2. Assorted hand tools, spare parts, and any other required materials.

3. Proper PPE.

4. ER70S-2 and/or ER70S-3 electrodes with 0.035-in. and/or 0.045-in. (0.9-mm and/or 1.2-mm) diameter.

5. 75% Ar + 25% CO_2 or Ar + 2% to 5% O_2 shielding gas.

6. Two or more pieces of 6 in. long 1−1/2 in. wide and 3/8 thick (150 mm × 38 mm × 9.5 mm) steel plate.

INSTRUCTIONS

1. The 30° bevel angles are to be flame or plasma cut on the edges of the plate before the parts are assembled.

2. Grind a 1/16-in. (1.6 mm) to 1/8-in. (3 mm) root face on the beveled edge.

3. The groove face and inside and outside plate surface within 1 in. (25 mm) of the joint must be mechanically cleaned of slag, rust, and mill scale. Cleaning must be done with a wire brush or grinder down to bright metal.

4. Set the voltage, amperage, wire-feed speed, and shielding gas flow according to Table 15-2.

5. Tack weld the plates with a root gap between the plates that measures approximately 1/16 in. to 1/8 in. (1.6 mm to 3 mm).

6. Place the plates in position at a comfortable height and location. Be sure you have complete and free movement along the full length of the weld joint. It is often a good idea to make a practice pass along the joint with the welding gun without power to make sure nothing will interfere with your making the weld. Be sure the welding cable is free and will not get caught on anything during the weld.

7. Grind the ends of the tack welds with a thin grinding disk.

8. Using 20° to 30° backhand technique start the weld at the very end of the groove. Allow the weld pool to expand to approximately ¼ -in. to 3/8 in. (6 mm to 9.5 mm) wide before starting to travel along the groove. Watch the leading edge of the weld pool for a keyhole that indicates that you are getting 100% penetration.

9. When the weld reaches the opposed end pause long enough for the weld crater to fill up, so the end of the weld is all the way at the edge of the plates.

10. Clean off all of the slag with a wire brush, chipping hammer, or punch.

11. It should take 3 or more weld passes to fill the groove. Make sure not to make the weld any wider than 1/8 in. (3 mm) than the groove and no more than 1/8 in. (3 mm) of reinforcement.

INSTRUCTOR'S COMMENTS _____

_____

_____

■ PRACTICE 15-4

Name _____ Date _____

Class _____ Instructor _____ Grade _____

OBJECTIVE: After completing this practice, you will make a tee joint with fillet welds in the horizontal position using the spray metal transfer technique.

EQUIPMENT AND MATERIALS NEEDED FOR THIS PRACTICE

1. Semiautomatic welding power system approved for GMA.

2. Assorted hand tools, spare parts, and any other required materials.

3. Proper PPE.

4. ER70S-2 and/or ER70S-3 electrodes with 0.035-in. and/or 0.045-in. (0.9-mm and/or 1.2-mm) diameter.

5. 75% Ar + 25% CO_2 or Ar + 2% to 5% O_2 shielding gas.

6. Two or more pieces of 6 in. long 1−1/2 in. wide and 3/8 thick (150 mm × 38 mm × 9.5 mm) steel plate.

INSTRUCTIONS

1. The plate surfaces within 1 in. (25 mm) of the joint must be mechanically cleaned of slag, rust, and mill scale. Cleaning must be done with a wire brush or grinder down to bright metal.

2. Set the voltage, amperage, wire-feed speed, and shielding gas flow according to Table 15-2.

3. Tack weld the vertical piece in the center of the base plate. Make sure it is square and in the center of the base plate. Make tack welds on both sides of the tee joint.

4. Place the plates in position at a comfortable height and location. Be sure you have complete and free movement along the full length of the weld joint.

5. Using 20° to 30° backhand technique start the weld at the very end of the joint. Allow the weld pool to expand to ¼-in. (6 mm) in size before starting to travel along the joint. Watch the leading edge of the weld pool to make sure you are getting complete root fusion.

6. When the weld reaches the opposite end pause long enough for the weld crater to fill up, so the end of the weld is all the way at the edge of the plates.

7. Clean off all of the slag with a wire brush, chipping hammer, or punch.

8. Make any adjustments in your machine setting and techniques your instructor might suggest before making the second weld on the opposite side.

INSTRUCTOR'S COMMENTS _____

_____

_____

■ PRACTICE 15-5 AWS SENSE

Name _____ Date _____

Class _____ Instructor _____ Grade _____

GAS METAL ARC WELDING (GMAW) SPRAY TRANSFER WORKMANSHIP SAMPLE

Welding Procedure Specification (WPS) No.: Practice 15-5.

TITLE:

Welding GMAW of carbon steel plate to carbon steel plate.

SCOPE:

This procedure is applicable for V-groove and fillet welds on 3/8 in. (9.5 mm) thick plate for the AWS SENSE Level I flat groove and horizontal fillet welds using spray transfer GMA welding.

WELDING MAY BE PERFORMED IN THE FOLLOWING POSITIONS:

1G and 2F.

BASE METAL:

The base metal shall conform to carbon steel M-1, P-1, and S-1, Group 1 or 2.

BACKING MATERIAL SPECIFICATION:

None.

FILLER METAL:

The filler metal shall conform to AWS specification no. 0.035-in. (0.90-mm) to 0.045-in. (1.2-mm) diameter E70S-X from AWS specification A5.18. This filler metal falls into F-number F-6 and A-number A-1.

SHIELDING GAS:

The shielding gas, or gases, shall conform to the following compositions and purity: 25% CO_2 + 75% Ar at 30 to 50 cfh or Ar + 2% to 5% O_2 at 30 to 50 cfh.

JOINT DESIGN AND TOLERANCES:

Refer to the drawing in **Figure 15-5** for the joint layout specifications.

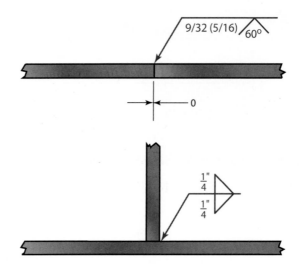

PREPARATION OF BASE METAL:

The 30° bevel angles are to be flame or plasma cut on the edges of the plate before the parts are assembled. The beveled surface must be smooth and free of notches. Any roughness or notches deeper than 1/64 in. (0.4 mm) must be ground smooth.

All hydrocarbons and other contaminants, such as cutting fluids, grease, oil, and primers, must be cleaned off all parts and filler metals before welding. This cleaning can be done with any suitable solvents or detergents. The groove face and inside and outside plate surface within 1 in. (25 mm) of the joint must be mechanically cleaned of slag, rust, and mill scale. Cleaning must be done with a wire brush or grinder down to bright metal.

ELECTRICAL CHARACTERISTICS:

Set the voltage, amperage, wire-feed speed, and shielding gas flow according to Table 15-2.

PREHEAT:

The parts must be heated to a temperature higher than 50°F (10°C) before any welding is started.

BACKING GAS:

N/A.

SAFETY:

Proper protective clothing and equipment must be used. The area must be free of all hazards that may affect the welder or others in the area. The welding machine, welding leads, work clamp, electrode holder, and other equipment must be in safe working order.

WELDING TECHNIQUE:

Using a 3/4-in. (19-mm) or larger gas nozzle for all welding, first tack weld the plates together, Figure 15-6. Use the E70S-X arc welding electrodes to make the welds.

Using the E70S-X arc welding electrodes, make a series of stringer filler welds, no thicker than 1/4 in. (6.4 mm) in the groove until the joint is filled.

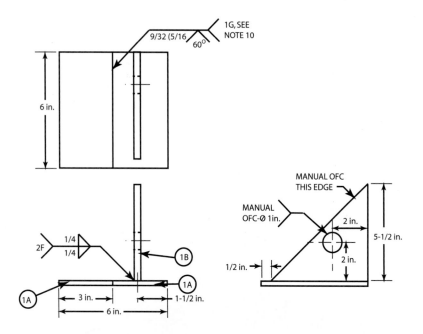

INTERPASS TEMPERATURE:

The plate should not be heated to a temperature higher than 350°F (175°C) during the welding process. After each weld pass is completed, allow it to cool but never to a temperature below 50°F (10°C). The weldment must not be quenched in water.

CLEANING:

Any slag must be cleaned off between passes. The weld beads may be cleaned by a hand wire brush, a chipping hammer, a punch and hammer, or a needle scaler. All weld cleaning must be performed with the test plate in the welding position.

VISUAL INSPECTION:

Visually inspect the weld for uniformity and discontinuities.

1. There shall be no cracks, no incomplete fusion.

2. There shall be no incomplete joint penetration in groove welds except as permitted for partial joint penetration welds.

3. The Test Supervisor shall examine the weld for acceptable appearance and shall be satisfied that the welder is skilled in using the process and procedure specified for the test.

4. Undercut shall not exceed the lesser of 10% of the base metal thickness or 1/32 in. (0.8 mm).

5. Where visual examination is the only criterion for acceptance, all weld passes are subject to visual examination at the discretion of the Test Supervisor.

6. The frequency of porosity shall not exceed one in each 4 in. (100 mm) of weld length and the maximum diameter shall not exceed 3/32 in. (2.4 mm).

7. Welds shall be free from overlap.

SKETCHES:

Gas Metal Arc Welding Spray Transfer (GMAW)

Workmanship Sample drawing, Figure 15-7.

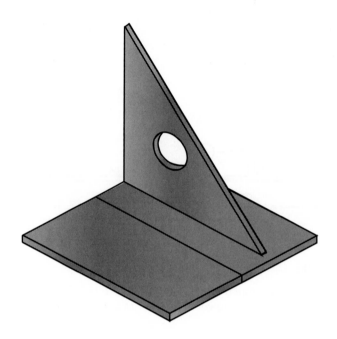

INSTRUCTOR'S COMMENTS _____

_____

_____

■ PRACTICE 15-6 AWS SENSE

Name _____ Date _____

Class _____ Instructor _____ Grade _____

GAS METAL ARC WELDING (GMAW) SPRAY TRANSFER WORKMANSHIP SAMPLE

Welding Procedure Specification (WPS) No.: Practice 15-6.

TITLE:

Welding GMAW of carbon steel plate to carbon steel plate.

SCOPE:

This procedure is applicable for V-groove weld on 1 in. (25 mm) thick plate with backing for the AWS SENSE Level II flat groove weld using spray transfer GMA welding.

WELDING MAY BE PERFORMED IN THE FOLLOWING POSITIONS:

1G and 2F.

BASE METAL:

The base metal shall conform to carbon steel M-1, P-1, and S-1, Group 1 or 2.

BACKING MATERIAL SPECIFICATION:

The backing metal strip shall conform to carbon steel M-1, P-1, and S-1, Group 1 or 2.

FILLER METAL:

The filler metal shall conform to AWS specification no. 0.035-in. (0.90-mm) to 0.045-in. (1.2-mm) diameter E70S-X from AWS specification A5.18. This filler metal falls into F-number F-6 and A-number A-1.

SHIELDING GAS:

The shielding gas, or gases, shall conform to the following compositions and purity: 25% CO_2 + 75% Ar at 30 to 50 cfh or Ar + 2% to 5% O_2 at 30 to 50 cfh.

JOINT DESIGN AND TOLERANCES:

Refer to the drawing in Figure 15-8 for the joint layout specifications.

Joint Details:

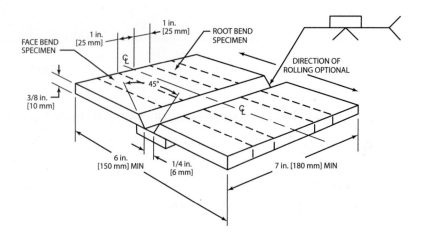

PREPARATION OF BASE METAL:

The 22 1/2° to 30° bevel angles are to be flame or plasma cut on the edges of the plate before the parts are assembled. The beveled surface must be smooth and free of notches. Any roughness or notches deeper than 1/64 in. (0.4 mm) must be ground smooth.

All hydrocarbons and other contaminants, such as cutting fluids, grease, oil, and primers, must be cleaned off all parts and filler metals before welding. This cleaning can be done with any suitable solvents or detergents. The groove face and inside and outside plate surface within 1 in. (25 mm) of the joint must be mechanically cleaned of slag, rust, and mill scale. Cleaning must be done with a wire brush or grinder down to bright metal.

ELECTRICAL CHARACTERISTICS:

Set the voltage, amperage, wire-feed speed, and shielding gas flow according to Table 15-2.

PREHEAT:

The parts must be heated to a temperature higher than 50°F (10°C) before any welding is started.

BACKING GAS:

N/A.

SAFETY:

Proper protective clothing and equipment must be used. The area must be free of all hazards that may affect the welder or others in the area. The welding machine, welding leads, work clamp, electrode holder, and other equipment must be in safe working order.

WELDING TECHNIQUE:

Using a 3/4-in. (19-mm) or larger gas nozzle for all welding, first tack weld the plates together. Use the E70S-X arc welding electrodes to make the welds.

Using the E70S-X arc welding electrodes, make a series of stringer filler welds, no thicker than 1/4 in. (6.4 mm) in the groove until the joint is filled.

INTERPASS TEMPERATURE:

The plate should not be heated to a temperature higher than 350°F (175°C) during the welding process. After each weld pass is completed, allow it to cool but never to a temperature below 50°F (10°C). The weldment must not be quenched in water.

CLEANING:

Any slag must be cleaned off between passes. The weld beads may be cleaned by a hand wire brush, a chipping hammer, a punch and hammer, or a needle scaler. All weld cleaning must be performed with the test plate in the welding position.

VISUAL INSPECTION:

There shall be no cracks, no incomplete fusion. There shall be no incomplete joint penetration in groove welds except as permitted for partial joint penetration welds.

- The Test Supervisor shall examine the weld for acceptable appearance and shall be satisfied that the welder is skilled in using the process and procedure specified for the test.
- Undercut shall not exceed the lesser of 10% of the base metal thickness or 1/32 in. (0.8 mm).
- Where visual examination is the only criterion for acceptance, all weld passes are subject to visual examination at the discretion of the Test Supervisor.
- The frequency of porosity shall not exceed one in each 4 in. (100 mm) of weld length, and the maximum diameter shall not exceed 3/32 in. (2.4 mm).
- Welds shall be free from overlap.

BEND TEST:

The weld is to be mechanically tested only after it has passed the visual inspection. Be sure that the test specimens are properly marked to identify the welder, the position, and the process.

SPECIMEN PREPARATION:

For 1/2-in. (13-mm) test plates, two side bend specimens are to be located in accordance with the requirements in Figure 15-9.

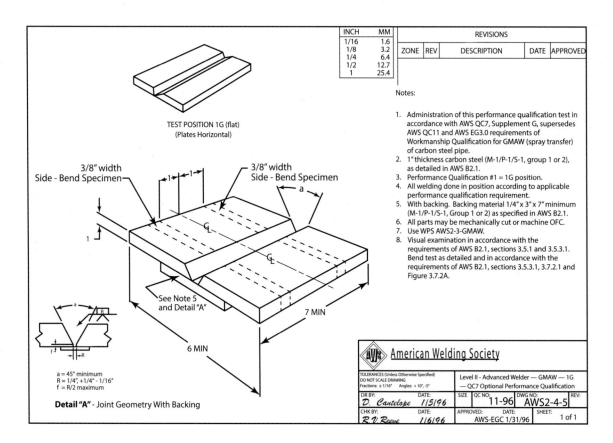

INCH	MM
1/16	1.6
1/8	3.2
1/4	6.4
1/2	12.7
1	25.4

REVISIONS

ZONE	REV	DESCRIPTION	DATE	APPROVED

TEST POSITION 1G (flat)
(Plates Horizontal)

3/8" width
Side - Bend Specimen

3/8" width
Side - Bend Specimen

See Note 5
and Detail "A"

7 MIN

6 MIN

a = 45° minimum
R = 1/4", +1/4" - 1/16"
f = R/2 maximum

Detail "A" - Joint Geometry With Backing

Notes:

1. Administration of this performance qualification test in accordance with AWS QC7, Supplement G, supersedes AWS QC11 and AWS EG3.0 requirements of Workmanship Qualification for GMAW (spray transfer) of carbon steel pipe.
2. 1" thickness carbon steel (M-1/P-1/S-1, group 1 or 2), as detailed in AWS B2.1.
3. Performance Qualification #1 = 1G position.
4. All welding done in position according to applicable performance qualification requirement.
5. With backing. Backing material 1/4" x 3" x 7" minimum (M-1/P-1/S-1, Group 1 or 2) as specified in AWS B2.1.
6. All parts may be mechanically cut or machine OFC.
7. Use WPS AWS2-3-GMAW.
8. Visual examination in accordance with the requirements of AWS B2.1, sections 3.5.1 and 3.5.3.1. Bend test as detailed and in accordance with the requirements of AWS B2.1, sections 3.5.3.1, 3.7.2.1 and Figure 3.7.2A.

American Welding Society

TOLERANCES (Unless Otherwise Specified)
DO NOT SCALE DRAWING
Fractions: ± 1/16"   Angles: ± 10', -5°

Level II - Advanced Welder — GMAW — 1G
— QC7 Optional Performance Qualification

| DR BY: D. Cantelope | DATE: 1/5/96 | SIZE | QC NO: 11-96 | DWG NO: AWS2-4-5 | REV: |
| CHK BY: R V. Reeve | DATE: 1/6/96 | APPROVED: AWS-EGC | DATE: 1/31/96 | SHEET: 1 of 1 |

- *Transverse side bend.* The weld is perpendicular to the longitudinal axis of the specimen and is bent so that the side of the weld face becomes the tension surface of the specimen.

ACCEPTANCE CRITERIA FOR BEND TEST:

For acceptance, the convex surface of the side of the weld bend specimen shall meet both of the following requirements:

- No single indication shall exceed 1/8 in. (3.2 mm) measured in any direction on the surface.
- The sum of the greatest dimensions of all indications on the surface that exceed 1/32 in. (0.8 mm) but are less than or equal to 1/8 in. (3.2 mm) shall not exceed 3/8 in. (9.5 mm).

SKETCHES:

Gas Metal Arc Welding Spray Transfer (GMAW) Workmanship Sample drawing.

INSTRUCTOR'S COMMENTS _____

_____

_____

■ PRACTICE 15-7

Name _____ Date _____

Class _____ Instructor _____ Grade _____

OBJECTIVE: After completing this practice, you will make a V-groove welds in the flat and vertical position using the FCAW-S and FCAW-G techniques.

EQUIPMENT AND MATERIALS NEEDED FOR THIS PRACTICE

1. Semiautomatic welding power system approved for GMA.

2. Assorted hand tools, spare parts, and any other required materials.

3. Proper PPE.

4. E70T-1 and/or E70T-5 electrodes with 0.035-in. and/or 0.045-in. (0.9-mm and/or 1.2-mm) diameter.

5. 75% Ar + 25% CO_2 or Ar + 2% to 5% O_2 shielding gas.

6. Two or more pieces of 6 in. long 1−1/2 in. wide and 3/8 thick (150 mm × 38 mm × 9.5 mm) steel plate.

INSTRUCTIONS

1. The 30° bevel angles are to be flame or plasma cut on the edges of the plate before the parts are assembled.

2. Grind a 1/16-in. (1.6-mm) to 1/8-in. (3-mm) root face on the beveled edge.

3. The groove face and inside and outside plate surface within 1 in. (25 mm) of the joint must be mechanically cleaned of slag, rust, and mill scale. Cleaning must be done with a wire brush or grinder down to bright metal.

4. Set the voltage, amperage, wire-feed speed, and shielding gas flow according to Table 15-3.

5. Tack weld the plates with a root gap between the plates that measures approximately 1/16 in. (1.6 mm).

6. Place the plates in position at a comfortable height and location. Be sure you have complete and free movement along the full length of the weld joint.

7. Using 20° to 30° backhand technique start the weld at the very end of the groove. Allow the weld pool to expand to fill the full width of the groove wide before starting to travel along the groove.

8. A slight weave pattern will help you control the molten weld pool in the 3G position.

9. When the weld reaches the opposed end pause long enough for the weld crater to fill up, so the end of the weld is all the way at the edge of the plates.

10. Clean off all of the slag with a wire brush, chipping hammer, or punch.

11. Make sure not to make the weld any wider than 1/8 in. (3 mm) than the groove and no more than 1/8 in. (3 mm) of reinforcement.

INSTRUCTOR'S COMMENTS _____

_____

_____

■ PRACTICE 15-8

Name _____ Date _____

Class _____ Instructor _____ Grade _____

OBJECTIVE: After completing this practice, you will make a beveled grooved butt joint in the horizontal position using both the FCAW-G and FCAW-S processes.

EQUIPMENT AND MATERIALS NEEDED FOR THIS PRACTICE

1. Semiautomatic welding power system approved for GMA.

2. Assorted hand tools, spare parts, and any other required materials.

3. Proper PPE.

4. E70T-1 and/or E70T-5 electrodes with 0.035-in. and/or 0.045-in. (0.9-mm and/or 1.2-mm) diameter.

5. 75% Ar + 25% CO_2 or Ar + 2% to 5% O_2 shielding gas.

6. Two or more pieces of 6 in. long 1−1/2 in. wide and 3/8 thick (150 mm × 38 mm × 9.5 mm) steel plate.

INSTRUCTIONS

1. The plate surfaces within 1 in. (25 mm) of the joint must be mechanically cleaned of slag, rust, and mill scale. Cleaning must be done with a wire brush or grinder down to bright metal.

2. Set the voltage, amperage, wire-feed speed, and shielding gas flow according to Table 15-3 for FCAW-G and Table 15-4 for FCAW-S.

3. Tack weld the plates with a root gap between the plates that measures approximately 1/16 in. (1.6 mm).

4. Place the plates in position at a comfortable height and location. Be sure you have complete and free movement along the full length of the weld joint.

5. Using 20° to 30° backhand stringer bead technique start the weld at the very end of the groove. Allow the root weld pool to expand to approximately ¼ in. (6 mm) in width before starting to travel along the groove.

6. When the weld reaches the opposite end pause long enough for the weld crater to fill up, so the end of the weld is all the way at the edge of the plates.

7. Clean off all of the slag with a wire brush, chipping hammer, or punch.

8. Use the same backhand technique but with a J weave pattern for the second weld. Place this weld bead on the bottom plate along its outside edge extending 2/3 of the way up the beveled surface.

9. Clean the weld.

10. The cover pass is made with the same backhand technique with a J weave pattern and extends from the top edge of the bevel to almost the bottom edge.

INSTRUCTOR'S COMMENTS _____

_____

_____

■ PRACTICE 15-9

Name _____ Date _____

Class _____ Instructor _____ Grade _____

OBJECTIVE: After completing this practice, you will make a beveled grooved butt joint in the overhead position using both the FCAW-G and FCAW-S processes.

EQUIPMENT AND MATERIALS NEEDED FOR THIS PRACTICE

1. Semiautomatic welding power system approved for GMA.

2. Assorted hand tools, spare parts, and any other required materials.

3. Proper PPE.

4. E70T-1 and/or E70T-5 electrodes with 0.035-in. and/or 0.045-in. (0.9-mm and/or 1.2-mm) diameter.

5. 75% Ar + 25% CO_2 or Ar + 2% to 5% O_2 shielding gas.

6. Two or more pieces of 6 in. long 1−1/2 in. wide and 3/8 thick (150 mm × 38 mm × 9.5 mm) steel plate.

INSTRUCTIONS

1. The plate surfaces within 1 in. (25 mm) of the joint must be mechanically cleaned of slag, rust, and mill scale. Cleaning must be done with a wire brush or grinder down to bright metal.

2. Set the voltage, amperage, wire-feed speed, and shielding gas flow according to Table 15-3 for FCAW-G and Table 15-4 for FCAW-S.

3. Tack weld the plates with a root gap between the plates that measures approximately 1/16 in. (1.6 mm).

4. Place the plates in position at a comfortable height and location. Be sure you have complete and free movement along the full length of the weld joint.

5. Using 10° to 15° backhand stringer bead technique start the weld at the very end of the groove. Allow the root weld pool to expand to approximately ¼ in. (6 mm) in width before starting to travel along the groove.

6. When the weld reaches the opposite end pause long enough for the weld crater to fill up, so the end of the weld is all the way at the edge of the plates.

7. Clean off all of the slag with a wire brush, chipping hammer, or punch.

8. Using 15° to 25° backhand technique with a zigzag or semi-circle for the second weld placing this weld bead on the top of the root pass.

9. Clean the weld.

10. The cover pass is made with a 15° to 25° backhand technique with a zigzag or semi-circle for the second weld placing this weld bead on the top of the filler pass. This weld can be made in two passes if you have trouble controlling the larger molten weld pool in the overhead position.

INSTRUCTOR'S COMMENTS _____

_____

_____

■ PRACTICE 15-10

Name _____ Date _____

Class _____ Instructor _____ Grade _____

OBJECTIVE: After completing this practice, you will make fillet welded lap joints and tee joints in all four positions using two or more pieces of mild steel plate welds with both the FCAW-G and FCAW-S processes.

EQUIPMENT AND MATERIALS NEEDED FOR THIS PRACTICE

1. Semiautomatic welding power system approved for GMA.

2. Assorted hand tools, spare parts, and any other required materials.

3. Proper PPE.

4. E70T-1 and/or E70T-5 electrodes with 0.035-in. and/or 0.045-in. (0.9-mm and/or 1.2-mm) diameter.

5. 75% Ar + 25% CO_2 or Ar + 2% to 5% O_2 shielding gas.

6. Two or more pieces of 6 in. long 1−1/2 in. wide and 3/8 thick (150 mm × 38 mm × 9.5 mm) steel plate.

INSTRUCTIONS

1. One at a time tack weld the sheets together to form the joints and place them in the desired position on a welding stand.

2. Starting at one end, run a 1/4 in. (6 mm) equal leg fillet weld along the joint. Watch the molten weld pool and bead for signs it is increasing or decreasing in size. Make any necessary changes in your technique to keep the weld consistent.

3. Check for complete root fusion on the lap and tee joints and 100% penetration on the butt joint.

INSTRUCTOR'S COMMENTS _____

_____

_____

■ PRACTICE 15-11

Name _____ Date _____

Class _____ Instructor _____ Grade _____

AWS SENSE ENTRY-LEVEL WELDER WORKMANSHIP SAMPLE FOR FLUX CORED ARC WELDING, GAS-SHIELDED (FCAW-G)

Welding Procedure Specification (WPS) No.: Practice 15-11.

TITLE:

Welding FCAW of plate to plate.

SCOPE:

This procedure is applicable for V-groove, bevel, and fillet welds within the range of 1/8 in. (3.2 mm) through 1 1/2 in. (38 mm) for the AWS SENSE Level I all position groove and fillet welds using the FCAW-G welding process.

WELDING MAY BE PERFORMED IN THE FOLLOWING POSITIONS:

All.

BASE METAL:

The base metal shall conform to carbon steel M-1, P-1, and S-1, Group 1 or 2.

BACKING MATERIAL SPECIFICATION:

None.

FILLER METAL:

The filler metal shall conform to AWS specification no. E71T-1 from AWS specification A5.20. This filler metal falls into F-number F-6 and A-number A-1.

SHIELDING GAS:

The shielding gas, or gases, shall conform to the following compositions and purity: 25% CO_2 + 75% Ar at 30 to 50 cfh.

JOINT DESIGN AND TOLERANCES:

Refer to the drawing and specifications in Figure 15-10 for the workmanship sample layout.

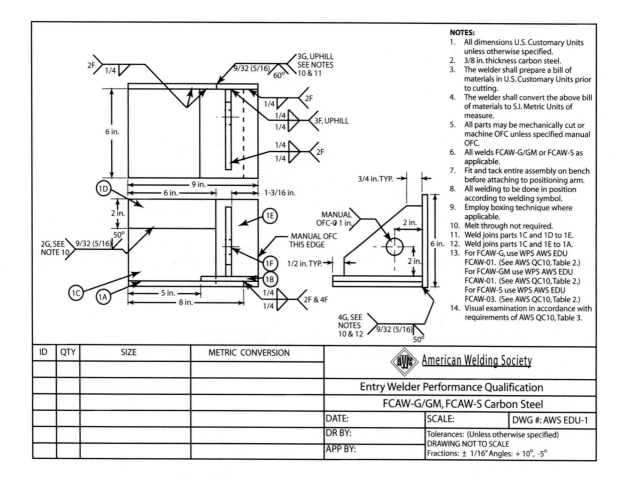

NOTES:
1. All dimensions U.S. Customary Units unless otherwise specified.
2. 3/8 in. thickness carbon steel.
3. The welder shall prepare a bill of materials in U.S. Customary Units prior to cutting.
4. The welder shall convert the above bill of materials to S.I. Metric Units of measure.
5. All parts may be mechanically cut or machine OFC unless specified manual OFC.
6. All welds FCAW-G/GM or FCAW-S as applicable.
7. Fit and tack entire assembly on bench before attaching to positioning arm.
8. All welding to be done in position according to welding symbol.
9. Employ boxing technique where applicable.
10. Melt through not required.
11. Weld joins parts 1C and 1D to 1E.
12. Weld joins parts 1C and 1E to 1A.
13. For FCAW-G, use WPS AWS EDU FCAW-01. (See AWS QC10, Table 2.)
    For FCAW-GM use WPS AWS EDU FCAW-01. (See AWS QC10, Table 2.)
    For FCAW-S use WPS AWS EDU FCAW-03. (See AWS QC10, Table 2.)
14. Visual examination in accordance with requirements of AWS QC10, Table 3.

ID	QTY	SIZE	METRIC CONVERSION			
				| **American Welding Society**		
				Entry Welder Performance Qualification		
				FCAW-G/GM, FCAW-S Carbon Steel		
				DATE:	SCALE:	DWG #: AWS EDU-1
				DR BY:	Tolerances: (Unless otherwise specified) DRAWING NOT TO SCALE	
				APP BY:	Fractions: ± 1/16" Angles: + 10°, -5°	

PREPARATION OF BASE METAL:

The bevels are to be flame or plasma cut on the edges of the plate before the parts are assembled. The beveled surface must be smooth and free of notches. Any roughness or notches deeper than 1/64 in. (0.4 mm) must be ground smooth.

All hydrocarbons and other contaminants, such as cutting fluids, grease, oil, and primers, must be cleaned off all parts and filler metals before welding. This cleaning can be done with any suitable solvents or detergents. The groove face and inside and outside plate surface within 1 in. (25 mm) of the joint must be mechanically cleaned of slag, rust, and mill scale. Cleaning must be done with a wire brush or grinder down to bright metal.

ELECTRICAL CHARACTERISTICS:

Set the voltage, amperage, wire-feed speed, and shielding gas flow according to Table 15-3.

PREHEAT:

The parts must be heated to a temperature higher than 50°F (10°C) before any welding is started.

BACKING GAS:

N/A.

SAFETY:

Proper protective clothing and equipment must be used. The area must be free of all hazards that may affect the welder or others in the area. The welding machine, welding leads, work clamp, electrode holder, and other equipment must be in safe working order.

WELDING TECHNIQUE:

Using a 1/2-in. (13-mm) or larger gas nozzle and a distance from contact tube to work of approximately 3/4 in. (19 mm) for all welding, first tack weld the plates together according to Figure 15-10. There should be a root gap of about 1/8 in. (3.2 mm) between the plates with V-grooved or beveled edges. Use an E71T-1 arc welding electrode to make a weld. If multiple pass welds are going to be made, a root pass weld should be made to fuse the plates together. Clean the slag from the root pass, being sure to remove any trapped slag along the sides of the weld.

Using an E71T-1 arc welding electrode, make a series of stringer or weave filler welds, no thicker than 1/4 in. (6.4 mm) in the groove until the joint is filled. The 1/4-in. (6.4-mm) fillet welds are to be made with one pass.

INTERPASS TEMPERATURE:

The plate should not be heated to a temperature higher than 350°F (175°C) during the welding process. After each weld pass is completed, allow it to cool but never to a temperature below 50°F (10°C). The weldment must not be quenched in water.

CLEANING:

The slag must be cleaned off between passes. The weld beads may be cleaned by a hand wire brush, a chipping hammer, a punch and hammer, or a needle scaler. All weld cleaning must be performed with the test plate in the welding position. A grinder may not be used to remove weld control problems such as undercut, overlap, or trapped slag.

INSPECTION:

Visually inspect the weld for uniformity and discontinuities. There shall be no cracks, no incomplete fusion, and no overlap. Undercut shall not exceed the lesser of 10% of the base metal thickness or 1/32 in. (0.8 mm). The frequency of porosity shall not exceed one in each 4 in. (100 mm) of weld length, and the maximum diameter shall not exceed 3/32 in. (2.4 mm).

SKETCHES:

Flux Cored Arc Welding (FCAW) Gas Shielded Workmanship Sample drawing.

INSTRUCTOR'S COMMENTS _____

_____

_____

■ PRACTICE 15-12

Name _____ Date _____

Class _____ Instructor _____ Grade _____

AWS SENSE ENTRY-LEVEL WELDER WORKMANSHIP SAMPLE FOR FLUX CORED ARC WELDING SELF-SHIELDED (FCAW-S)

Welding Procedure Specification (WPS) No.: Practice 15-12.

TITLE:

Welding FCAW of plate to plate.

SCOPE:

This procedure is applicable for V-groove, bevel, and fillet welds within the range of 1/8 in. (3.2 mm) through 1 1/2 in. (38 mm) for the AWS SENSE Level I all position groove and fillet welds using the FCAW-S welding process.

WELDING MAY BE PERFORMED IN THE FOLLOWING POSITIONS:

All.

BASE METAL:

The base metal shall conform to carbon steel M-1, P-1, and S-1, Group 1 or 2.

BACKING MATERIAL SPECIFICATION:

None.

FILLER METAL:

The filler metal shall conform to AWS specification no. 0.035 in. (0.90 mm) to 0.0415 in. (1.2 mm) diameter E71T-11 from AWS specification A5.20. This filler metal falls into F-number F-6 and A-number A-1.

SHIELDING GAS:

None.

JOINT DESIGN AND TOLERANCES:

Refer to the drawing and specifications in Figure 15-11 for the workmanship sample layout.

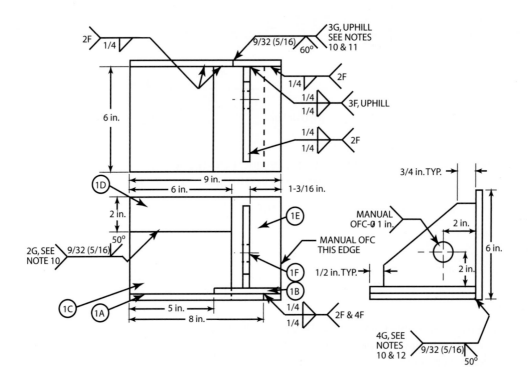

PREPARATION OF BASE METAL:

The bevels are to be flame or plasma cut on the edges of the plate before the parts are assembled. The beveled surface must be smooth and free of notches. Any roughness or notches deeper than 1/64 in. (0.4 mm) must be ground smooth.

All hydrocarbons and other contaminants, such as cutting fluids, grease, oil, and primers, must be cleaned off all parts and filler metals before welding. This cleaning can be done with any suitable solvents or detergents. The groove face and inside and outside plate surface within 1 in. (25 mm) of the joint must be mechanically cleaned of slag, rust, and mill scale. Cleaning must be done with a wire brush or grinder down to bright metal.

ELECTRICAL CHARACTERISTICS:

Set the voltage, amperage, and wire-feed speed flow according to Table 15-4.

PREHEAT:

The parts must be heated to a temperature higher than 50°F (10°C) before any welding is started.

BACKING GAS:

N/A.

SAFETY:

Proper protective clothing and equipment must be used. The area must be free of all hazards that may affect the welder or others in the area. The welding machine, welding leads, work clamp, electrode holder, and other equipment must be in safe working order.

WELDING TECHNIQUE:

Using a 1/2-in. (13-mm) or larger gas nozzle and a distance from contact tube to work of approximately 3/4 in. (19 mm) for all welding, first tack weld the plates together according to Figure 15-11. There should be about a root gap of about 1/8 in. (3.2 mm) between the plates with V-grooved or beveled edges. Use an E71T-11 arc welding electrode to make a weld. If multiple pass welds are going to be made, a root pass weld should be made to fuse the plates together. Clean the slag from the root pass, being sure to remove any trapped slag along the sides of the weld.

Using an E71T-11 arc welding electrode, make a series of stringer or weave filler welds, no thicker than 1/4 in. (6.4 mm) in the groove until the joint is filled. The 1/4-in. (6.4-mm) fillet welds are to be made with one pass.

INTERPASS TEMPERATURE:

The plate should not be heated to a temperature higher than 350°F (175°C) during the welding process. After each weld pass is completed, allow it to cool but never to a temperature below 50°F (10°C). The weldment must not be quenched in water.

CLEANING:

The slag must be cleaned off between passes. The weld beads may be cleaned by a hand wire brush, a chipping hammer, a punch and hammer, or a needle scaler. All weld cleaning must be performed with the test plate in the welding position. A grinder may not be used to remove weld control problems such as undercut, overlap, or trapped slag.

VISUAL INSPECTION CRITERIA FOR ENTRY WELDERS:

There shall be no cracks, no incomplete fusion. There shall be no incomplete joint penetration in groove welds except as permitted for partial joint penetration welds.

- The Test Supervisor shall examine the weld for acceptable appearance and shall be satisfied that the welder is skilled in using the process and procedure specified for the test.
- Undercut shall not exceed the lesser of 10% of the base metal thickness or 1/32 in. (0.8 mm).
- Where visual examination is the only criterion for acceptance, all weld passes are subject to visual examination at the discretion of the Test Supervisor.
- The frequency of porosity shall not exceed one in each 4 in. (100 mm) of weld length, and the maximum diameter shall not exceed 3/32 in. (2.4 mm).
- Welds shall be free from overlap.

SKETCHES:

Flux Cored Arc Welding (FCAW) Self-Shielded Workmanship Sample drawing.

INSTRUCTOR'S COMMENTS _____

_____

_____

■ PRACTICE 15-13

Name _____ Date _____

Class _____ Instructor _____ Grade _____

GAS METAL ARC WELDING—SHORT-CIRCUIT METAL TRANSFER (GMAW-S) WORKMANSHIP SAMPLE

Welding Procedure Specification (WPS) No.: Practice 15-13.

TITLE:

Welding GMAW-S of carbon steel plate to carbon steel plate.

SCOPE:

This procedure is applicable for square 2G and 5G groove on 4 in. to 6 in. (100 mm to 150 mm) schedule 40 carbon steel pipe with and without backing for the AWS SENSE Level II short-circuit metal transfer GMA pipe welding.

WELDING MAY BE PERFORMED IN THE FOLLOWING POSITIONS:

2G and 5G

BASE METAL:

The base metal shall conform to carbon steel M-1, P-1, and S-1 Group 1 or 2.

BACKING MATERIAL SPECIFICATION:

Backing metal shall conform to carbon steel M-1, P-1, and S-1 Group 1 or 2.

FILLER METAL:

The filler metal shall conform to AWS specification no. E70S-X from AWS specification A5.18. This filler metal falls into F-number F-6 and A-number A-1.

SHIELDING GAS:

The shielding gas, or gases, shall conform to the following compositions and purity: 25% CO_2 + 75% Ar at 30 to 50 cfh or Ar + 2% to 5% O_2 at 30 to 50 cfh.

JOINT DESIGN AND TOLERANCES:

Refer to the drawing in Figure 15-12 for the joint layout specifications.

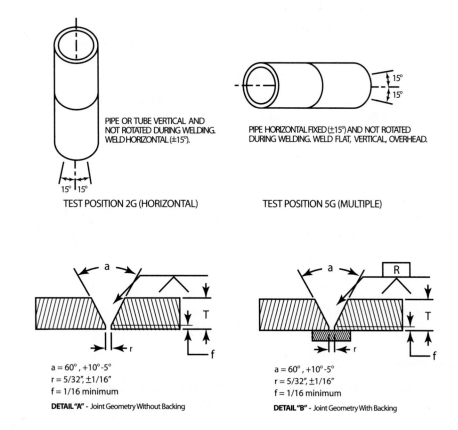

PIPE OR TUBE VERTICAL AND NOT ROTATED DURING WELDING. WELD HORIZONTAL (±15°).

15° 15°

TEST POSITION 2G (HORIZONTAL)

15°
15°

PIPE HORIZONTAL FIXED (±15°) AND NOT ROTATED DURING WELDING. WELD FLAT, VERTICAL, OVERHEAD.

TEST POSITION 5G (MULTIPLE)

a = 60°, +10°-5°
r = 5/32", ±1/16"
f = 1/16 minimum

DETAIL "A" - Joint Geometry Without Backing

a = 60°, +10°-5°
r = 5/32", ±1/16"
f = 1/16 minimum

DETAIL "B" - Joint Geometry With Backing

PREPARATION OF BASE METAL:

All parts may be mechanically cut or machine PAC unless specified as manual PAC.

All hydrocarbons and other contaminants, such as cutting fluids, grease, oil, and primers, must be cleaned off all parts and filler metals before welding. This cleaning can be done with any suitable solvents or detergents. The groove face and inside and outside plate surface within 1 in. (25 mm) of the joint must be mechanically cleaned of slag, rust, and mill scale. Cleaning must be done with a wire brush or grinder down to bright metal.

ELECTRICAL CHARACTERISTICS:

Set the voltage, amperage, wire-feed speed, and shielding gas flow according to Table 15-1.

PREHEAT:

The parts must be heated to a temperature higher than 50°F (10°C) before any welding is started.

BACKING GAS:

N/A.

SAFETY:

Proper protective clothing and equipment must be used. The area must be free of all hazards that may affect the welder or others in the area. The welding machine, welding leads, work clamp, electrode holder, and other equipment must be in safe working order.

WELDING TECHNIQUE:

Using a 1/2-in. (13-mm) or larger gas nozzle for all welding, first tack weld the plates together according to the drawing, Figure 15-12. Make four tack welds of approximately 1 in. (25 mm) long. Grind the ends of the tack welds to a featheredge with a narrow grinding disk.

Use the E70S-X filler metal to fuse the plates together. Clean any silicon slag, being sure to remove any trapped silicon slag along the sides of the weld.

All welds are to be made with one pass. All welds must be placed in the orientation shown in the drawing.

INTERPASS TEMPERATURE:

The plate should not be heated to a temperature higher than 350°F (175°C) during the welding process. After each weld pass is completed, allow it to cool but never to a temperature below 50°F (10°C). The weldment must not be quenched in water.

CLEANING:

The weld beads may be cleaned by a hand wire brush, a chipping hammer, a punch, hammer, and/or a narrow grinding disk. All weld cleaning must be performed with the test plate in the welding position.

VISUAL INSPECTION:

There shall be no cracks, no incomplete fusion. There shall be no incomplete joint penetration in groove welds except as permitted for partial joint penetration welds.

- The Test Supervisor shall examine the weld for acceptable appearance and shall be satisfied that the welder is skilled in using the process and procedure specified for the test.

- Undercut shall not exceed the lesser of 10% of the base metal thickness or 1/32 in. (0.8 mm).

- Where visual examination is the only criterion for acceptance, all weld passes are subject to visual examination at the discretion of the Test Supervisor.

- The frequency of porosity shall not exceed one in each 4 in. (100 mm) of weld length, and the maximum diameter shall not exceed 3/32 in. (2.4 mm).

- Welds shall be free from overlap.

BEND TEST:

The weld is to be mechanically tested only after it has passed the visual inspection. Be sure that the test specimens are properly marked to identify the welder, the position, and the process.

SPECIMEN PREPARATION:

Four specimens are to be located in accordance with the requirements in Figure 15-13. Two are to be prepared for a transverse face bend, and the other two are to be prepared for a transverse root bend, Figure 15-14.

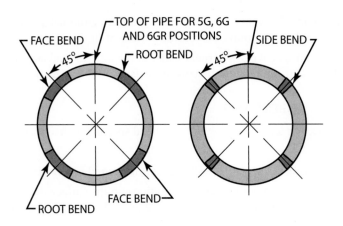

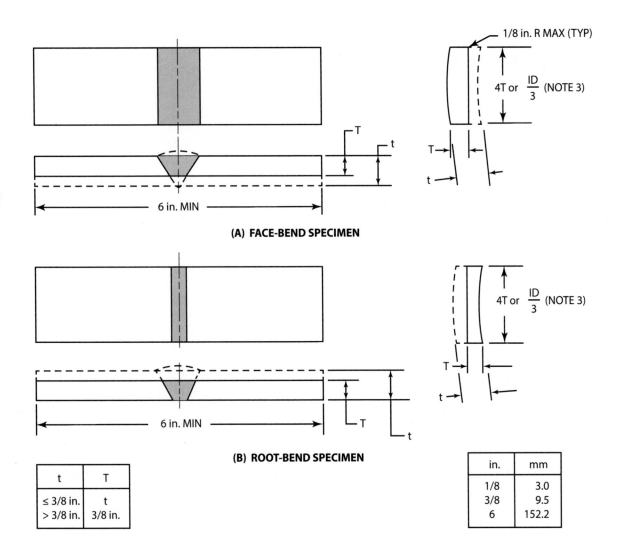

(A) FACE-BEND SPECIMEN

(B) ROOT-BEND SPECIMEN

t	T
≤ 3/8 in.	t
> 3/8 in.	3/8 in.

in.	mm
1/8	3.0
3/8	9.5
6	152.2

- Transverse face bend. The weld is perpendicular to the longitudinal axis of the specimen and is bent so that the weld face becomes the tension surface of the specimen.
- Transverse root bend. The weld is perpendicular to the longitudinal axis of the specimen and is bent so that the weld root becomes the tension surface of the specimen.

ACCEPTANCE CRITERIA FOR BEND TEST:

For acceptance, the convex surface of the face and root bend specimens shall meet both of the following requirements:

- No single indication shall exceed 1/8 in. (3.2 mm) measured in any direction on the surface.
- The sum of the greatest dimensions of all indications on the surface that exceed 1/32 in. (0.8 mm) but are less than or equal to 1/8 in. (3.2 mm) shall not exceed 3/8 in. (9.5 mm).

SKETCHES:

GMAW Short-Circuit Metal Transfer Workmanship Sample drawing.

INSTRUCTOR'S COMMENTS _____

_____

_____

■ PRACTICE 15-14

Name _____ Date _____

Class _____ Instructor _____ Grade _____

AWS SENSE ENTRY-LEVEL WELDER WORKMANSHIP SAMPLE FOR FLUX CORED ARC WELDING, GAS-SHIELDED (FCAW-G)

Welding Procedure Specification (WPS) No.: Practice 15-14.

TITLE:

Welding FCAW of plate to plate.

SCOPE:

This procedure is applicable for square 2G and 5G groove on 6 in. to 8 in. (150 mm to 200 mm) schedule 40 carbon steel pipe with and without backing for the AWS SENSE Level II short-circuit metal transfer FCAW-G pipe welding.

WELDING MAY BE PERFORMED IN THE FOLLOWING POSITIONS:

All.

BASE METAL:

The base metal shall conform to carbon steel M-1, P-1, and S-1, Group 1 or 2.

BACKING MATERIAL SPECIFICATION:

None.

FILLER METAL:

The filler metal shall conform to AWS specification no. E71T-1 from AWS specification A5.20. This filler metal falls into F-number F-6 and A-number A-1.

SHIELDING GAS:

The shielding gas, or gases, shall conform to the following compositions and purity: 25% CO_2 + 75% Ar at 30 to 50 cfh.

JOINT DESIGN AND TOLERANCES:

Refer to the drawing and specifications in Figure 15-15 for the workmanship sample layout.

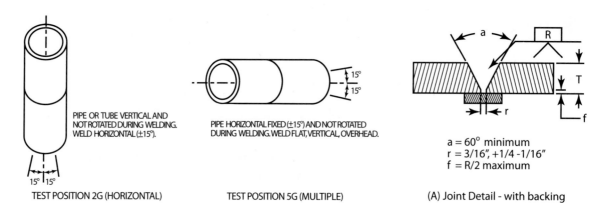

PIPE OR TUBE VERTICAL AND
NOT ROTATED DURING WELDING.
WELD HORIZONTAL (±15°).

PIPE HORIZONTAL FIXED (±15°) AND NOT ROTATED
DURING WELDING. WELD FLAT, VERTICAL, OVERHEAD.

a = 60° minimum
r = 3/16", +1/4 -1/16"
f = R/2 maximum

TEST POSITION 2G (HORIZONTAL)

TEST POSITION 5G (MULTIPLE)

(A) Joint Detail - with backing

PREPARATION OF BASE METAL:

The bevels are to be flame or plasma cut on the edges of the plate before the parts are assembled. The beveled surface must be smooth and free of notches. Any roughness or notches deeper than 1/64 in. (0.4 mm) must be ground smooth.

All hydrocarbons and other contaminants, such as cutting fluids, grease, oil, and primers, must be cleaned off all parts and filler metals before welding. This cleaning can be done with any suitable solvents or detergents. The groove face and inside and outside plate surface within 1 in. (25 mm) of the joint must be mechanically cleaned of slag, rust, and mill scale. Cleaning must be done with a wire brush or grinder down to bright metal.

ELECTRICAL CHARACTERISTICS:

Set the voltage, amperage, wire-feed speed, and shielding gas flow according to Table 15-3.

PREHEAT:

The parts must be heated to a temperature higher than 50°F (10°C) before any welding is started.

BACKING GAS:

N/A.

SAFETY:

Proper protective clothing and equipment must be used. The area must be free of all hazards that may affect the welder or others in the area. The welding machine, welding leads, work clamp, electrode holder, and other equipment must be in safe working order.

WELDING TECHNIQUE:

Using a 1/2-in. (13-mm) or larger gas nozzle and a distance from contact tube to work of approximately 3/4 in. (19 mm) for all welding, first tack weld the plates together according to Figure 15-15. There should be a root gap of about 1/8 in. (3.2 mm) between the plates with V-grooved or beveled edges. Use an E71T-1 arc welding electrode to make a weld. If multiple pass welds are going to be made, a root pass weld should be made to fuse the plates together. Clean the slag from the root pass, being sure to remove any trapped slag along the sides of the weld.

Using an E71T-1 arc welding electrode, make a series of stringer or weave filler welds, no thicker than 1/4 in. (6.4 mm) in the groove until the joint is filled. The 1/4-in. (6.4-mm) fillet welds are to be made with one pass.

INTERPASS TEMPERATURE:

The plate should not be heated to a temperature higher than 350°F (175°C) during the welding process. After each weld pass is completed, allow it to cool but never to a temperature below 50°F (10°C). The weldment must not be quenched in water.

CLEANING:

The slag must be cleaned off between passes. The weld beads may be cleaned by a hand wire brush, a chipping hammer, a punch and hammer, or a needle scaler. All weld cleaning must be performed with the test plate in the welding position. A grinder may not be used to remove weld control problems such as undercut, overlap, or trapped slag.

VISUAL INSPECTION:

There shall be no cracks, no incomplete fusion. There shall be no incomplete joint penetration in groove welds except as permitted for partial joint penetration welds.

- The Test Supervisor shall examine the weld for acceptable appearance and shall be satisfied that the welder is skilled in using the process and procedure specified for the test.
- Undercut shall not exceed the lesser of 10% of the base metal thickness or 1/32 in. (0.8 mm).
- Where visual examination is the only criterion for acceptance, all weld passes are subject to visual examination at the discretion of the Test Supervisor.
- The frequency of porosity shall not exceed one in each 4 in. (100 mm) of weld length, and the maximum diameter shall not exceed 3/32 in. (2.4 mm).
- Welds shall be free from overlap.

BEND TEST:

The weld is to be mechanically tested only after it has passed the visual inspection. Be sure that the test specimens are properly marked to identify the welder, the position, and the process.

SPECIMEN PREPARATION:

Four specimens are to be located in accordance with the requirements in Figure 15-13. Two are to be prepared for a transverse face bend, and the other two are to be prepared for a transverse root bend, Figure 15-14.

- Transverse face bend. The weld is perpendicular to the longitudinal axis of the specimen and is bent so that the weld face becomes the tension surface of the specimen.
- Transverse root bend. The weld is perpendicular to the longitudinal axis of the specimen and is bent so that the weld root becomes the tension surface of the specimen.

ACCEPTANCE CRITERIA FOR BEND TEST:

For acceptance, the convex surface of the face and root bend specimens shall meet both of the following requirements:

- No single indication shall exceed 1/8 in. (3.2 mm) measured in any direction on the surface.
- The sum of the greatest dimensions of all indications on the surface that exceed 1/32 in. (0.8 mm) but are less than or equal to 1/8 in. (3.2 mm) shall not exceed 3/8 in. (9.5 mm).

SKETCHES:

Flux Cored Arc Welding (FCAW) Gas Shielded Workmanship Sample drawing.

INSTRUCTOR'S COMMENTS _____

_____

_____

■ PRACTICE 15-15

Name _____ Date _____

Class _____ Instructor _____ Grade _____

AWS SENSE ENTRY-LEVEL WELDER WORKMANSHIP SAMPLE FOR FLUX CORED ARC WELDING SELF-SHIELDED (FCAW-S)

Welding Procedure Specification (WPS) No.: Practice 15-15.

TITLE:

Welding FCAW of plate to plate.

SCOPE:

This procedure is applicable for square 2G and 5G groove on 6 in. to 8 in. (150 mm to 200 mm) schedule 40 carbon steel pipe with and without backing for the AWS SENSE Level II short-circuit metal transfer FCAW-S pipe welding.

WELDING MAY BE PERFORMED IN THE FOLLOWING POSITIONS:

All.

BASE METAL:

The base metal shall conform to carbon steel M-1, P-1, and S-1, Group 1 or 2.

BACKING MATERIAL SPECIFICATION:

None.

FILLER METAL:

The filler metal shall conform to AWS specification no. 0.035 in. (0.90 mm) to 0.0415 in. (1.2 mm) diameter E71T-11 from AWS specification A5.20. This filler metal falls into F-number F-6 and A-number A-1.

SHIELDING GAS:

None.

JOINT DESIGN AND TOLERANCES:

Refer to the drawing and specifications in Figure 15-15 for the workmanship sample layout.

PREPARATION OF BASE METAL:

The bevels are to be flame or plasma cut on the edges of the plate before the parts are assembled. The beveled surface must be smooth and free of notches. Any roughness or notches deeper than 1/64 in. (0.4 mm) must be ground smooth.

All hydrocarbons and other contaminants, such as cutting fluids, grease, oil, and primers, must be cleaned off all parts and filler metals before welding. This cleaning can be done with any suitable solvents or detergents. The groove face and inside and outside plate surface within 1 in. (25 mm) of the joint must be mechanically cleaned of slag, rust, and mill scale. Cleaning must be done with a wire brush or grinder down to bright metal.

ELECTRICAL CHARACTERISTICS:

Set the voltage, amperage, and wire-feed speed flow according to Table 15-4.

PREHEAT:

The parts must be heated to a temperature higher than 50°F (10°C) before any welding is started.

BACKING GAS:

N/A.

SAFETY:

Proper protective clothing and equipment must be used. The area must be free of all hazards that may affect the welder or others in the area. The welding machine, welding leads, work clamp, electrode holder, and other equipment must be in safe working order.

WELDING TECHNIQUE:

Using a 1/2-in. (13-mm) or larger gas nozzle and a distance from contact tube to work of approximately 3/4 in. (19 mm) for all welding, first tack weld the plates together according to Figure 15-15. There should be about a root gap of about 1/8 in. (3.2 mm) between the plates with V-grooved or beveled edges. Use an E71T-11 arc welding electrode to make a weld. If multiple pass welds are going to be made, a root pass weld should be made to fuse the plates together. Clean the slag from the root pass, being sure to remove any trapped slag along the sides of the weld.

Using an E71T-11 arc welding electrode, make a series of stringer or weave filler welds, no thicker than 1/4 in. (6.4 mm) in the groove until the joint is filled. The 1/4-in. (6.4-mm) fillet welds are to be made with one pass.

INTERPASS TEMPERATURE:

The plate should not be heated to a temperature higher than 350°F (175°C) during the welding process. After each weld pass is completed, allow it to cool but never to a temperature below 50°F (10°C). The weldment must not be quenched in water.

CLEANING:

The slag must be cleaned off between passes. The weld beads may be cleaned by a hand wire brush, a chipping hammer, a punch and hammer, or a needle scaler. All weld cleaning must be performed with the test plate in the welding position. A grinder may not be used to remove weld control problems such as undercut, overlap, or trapped slag.

VISUAL INSPECTION CRITERIA FOR ENTRY WELDERS:

There shall be no cracks and no incomplete fusion. There shall be no incomplete joint penetration in groove welds except as permitted for partial joint penetration welds.

- The Test Supervisor shall examine the weld for acceptable appearance and shall be satisfied that the welder is skilled in using the process and procedure specified for the test.

- Undercut shall not exceed the lesser of 10% of the base metal thickness or 1/32 in. (0.8 mm).

- Where visual examination is the only criterion for acceptance, all weld passes are subject to visual examination at the discretion of the Test Supervisor.

- The frequency of porosity shall not exceed one in each 4 in. (100 mm) of weld length, and the maximum diameter shall not exceed 3/32 in. (2.4 mm).

- Welds shall be free from overlap.

BEND TEST:

The weld is to be mechanically tested only after it has passed the visual inspection. Be sure that the test specimens are properly marked to identify the welder, the position, and the process.

SPECIMEN PREPARATION:

For 3/8-in. (9.5-mm) test plates, two specimens are to be located in accordance with the requirements in Figure 15-13. One is to be prepared for a transverse face bend, and the other is to be prepared for a transverse root bend, Figure 15-14.

- Transverse face bend. The weld is perpendicular to the longitudinal axis of the specimen and is bent so that the weld face becomes the tension surface of the specimen.

- Transverse root bend. The weld is perpendicular to the longitudinal axis of the specimen and is bent so that the weld root becomes the tension surface of the specimen.

ACCEPTANCE CRITERIA FOR BEND TEST:

For acceptance, the convex surface of the face and root bend specimens shall meet both of the following requirements:

- No single indication shall exceed 1/8 in. (3.2 mm) measured in any direction on the surface.

- The sum of the greatest dimensions of all indications on the surface that exceed 1/32 in. (0.8 mm) but are less than or equal to 1/8 in. (3.2 mm) shall not exceed 3/8 in. (9.5 mm).

SKETCHES:

Flux Cored Arc Welding (FCAW) Self-Shielded Workmanship Sample drawing.

INSTRUCTOR'S COMMENTS _____

_____

_____

Problem	Remedy
Tungsten Inclusions	1. Use a larger tungsten electrode or reduce current. 2. Do not allow the tungsten to come into contact with the molten weld pool while welding. 3. Use high-frequency starting device. 4. Do not weld with tungsten electrodes that have become cracked or split on the ends.
Lack of Penetration	1. Increase current. 2. Decrease welding speed. 3. Decrease the size of the filler metal. 4. Adjust root opening.
Undercutting	1. Decrease current. 2. Make smaller weld beads. 3. Use a closer tungsten-to-work distance. 4. Change welding technique.

CHAPTER 16: GAS TUNGSTEN ARC WELDING EQUIPMENT, SETUP, OPERATION, AND FILLER METALS QUIZ

Name _____ Date _____

Class _____ Instructor _____ Grade _____

Instructions: Carefully read Chapter 16 in the text and answer the following questions.

A. SHORT ANSWER

Write a brief answer in the space provided that will answer the question or complete the statement.

1. List four methods of limiting tungsten erosion.

    a. _____

    b. _____

    c. _____

    d. _____

2. List nine classifications of tungsten electrodes that are commercially available.

    a. _____

    b. _____

    c. _____

    d. _____

    e. _____

    f. _____

    g. _____

    h. _____

    i. _____

3. List two types of GTAW torches.

    a. _____

    b. _____

4. Why are rubber hoses not recommended to be used for shielding gas hose from the GTAW process?

    _____

    _____

5. What are the advantages and disadvantages of using a small diameter nozzle versus a large diameter nozzle?

_____

_____

_____

_____

B. IDENTIFICATION

In the space provided, identify the items shown in the illustration.

6. Identify the following GTAW equipment by placing the correct number in the space provided from the drawing below.

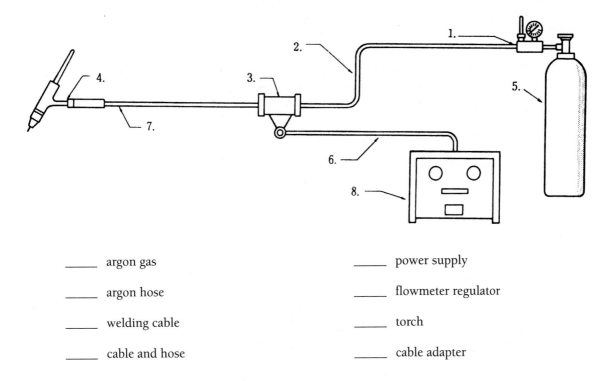

_____ argon gas   _____ power supply

_____ argon hose   _____ flowmeter regulator

_____ welding cable   _____ torch

_____ cable and hose   _____ cable adapter

7. From the illustration below, identify the following GTAW torch parts by placing the correct number in the space provided.

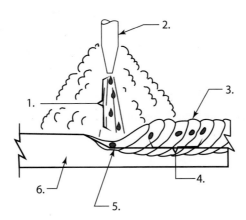

_____ tungsten inclusions in the weld metal

_____ electrode

_____ molten weld pool

_____ tungsten droplets transferring across the arc

_____ weld bead

_____ base metal

C. MATCHING

In the space provided to the left, write the letter from Column B that best answers the question or completes the statement in Column A.

Column A   Column B

_____ 8. The high melting temperature and good electrical conductivity   a. collet
        make _____ the best choice for a
        consumable electrode.   b. tungsten

_____ 9. A chucking device used to hold the tungsten in the torch.   c. flowmeter

_____ 10. A calibrated metering device used to regulate the flow of gases to   d. inert gas
        the torch and measured in cubic feet per hour (cfh).

_____ 11. A gas which will not combine with other elements used to surround
        the weld zone and prevent contamination of the weld by oxygen and
        nitrogen in the air.

D. FILL IN THE BLANK

Fill in the blank with the correct word. Answers may be more than one word.

12. The gas tungsten arc welding (GTAW) process is sometimes referred to as

    _____, or _____.

13. With _____ the filler metal is thin wire that is fed directly into the molten weld

    pool where it melts.

14. The _____ gas provides the needed arc characteristics and protects the molten

    weld pool.

15. The GTAW welds produced are sound, free of _____, and as

    _____ resistant as the parent metal.

16. Tungsten, atomic symbol _____.

17. The high melting temperature and good _____ make tungsten the best choice for a non-consumable electrode.

18. The heat of the arc is _____ away from the electrode's end so fast that it does not reach its melting temperature.

19. Tungsten inclusions are _____ spots that cause _____ to concentrate, possibly resulting in weld _____.

20. Most _____ require only ground surfaces of tungsten electrodes be used on all GTA welding.

21. _____ oxide (ZrO_2) also helps tungsten emit electrons freely.

22. _____ oxide electrodes have a current-carrying capacity similar to that of pure tungsten; however, they have a low arc-starting characteristic for improved arc starting and arc stability characteristic, similar to that of thoriated tungstens.at low current levels.

23. Because of the similarities in the performance of _____ electrodes to thorium electrodes _____ electrodes can often be substituted for thorium electrodes without having to requalify the welding procedure.

24. The EWG classification are a mixture of _____ such as cerium, lanthanum, yttrium and others are combined with approximately of 97% tungsten to form these electrodes.

25. _____ tungsten electrodes do not break evenly, they often split when you try to break off contamination.

26. When hand grinding with a _____ grinding stone will result in more tungsten breakage and a poorer finish.

27. When hand grinding, because of the hardness of the tungsten, it will become

    _____.

28. The tapered tungsten with a balled end, a shape sometimes used for _____.

29. The _____ -cooled torch, as compared to the air-cooled torch, operates at a lower temperature, resulting in a lower tungsten temperature and less erosion.

30. Without the cooling water flowing, the cable quickly _____ and melts through the hose.

31. The shielding gas hose must be made from a material such as _____ that will not contaminate the shielding gas.

32. Nozzles may be made from a ceramic such as alumina or silicon nitride (_____) or from fused quartz (_____).

33. The _____ reduces the turbulence in the shielding gas as it leaves the nozzle.

34. The rate of shielding gas flow is then read in units of CHF (_____), or L/min (_____)

35. The major differences among the currents are in their _____ and the _____ or degree of _____.

36. The high-frequency current can be created using a _____ or by _____.

37. The _____, cleaning portion of the current on these welders was much stronger than needed for welding on new or very clean aluminum plate or pipe.

38. The _____ squarewave combined the good surface wetting characteristics of the traditional AC sine wave with the improved molten weld pool control of the squarewave.

39. The _____ waveform reduces the heat input without sacrificing the arc control.

40. The _____ frequency setting produce a much less focused arc and the _____ frequency setting provide a much more concentrated stiffer more focused arc.

41. The shielding gases used for the GTA welding process are _____ (_____),_____ (_____),_____ (_____),_____ (_____), or a mixture of two or more of these gases.

42. Argon makes up approximately 1% of air and is a by-product of the _____.

43. Because argon is _____ than air, it effectively shields welds in deep grooves in the flat position.

44. Helium is a by-product of the natural _____.

45. The most common of these mixtures is _____% helium and

_____% argon.

46. The _____ allows a controlled surge of welding current as the arc is started to

establish a molten weld pool quickly.

47. The _____ time period serves to protect the molten weld pool, the filler rod, and

the tungsten electrode as they cool to a temperature at which they will not oxidize rapidly.

48. High shielding gas flow rates _____ shielding gases and may lead to

_____.

49. A variable resistor in the remote controls works in a manner similar to the _____

on a car to increase the power (current).

E. DRAW

In the space provided, make a pencil drawing to illustrate the question.

50. Sketch the three basic tungsten electrode end shapes and name them.

F. ESSAY

Provide complete answers for all of the following questions.

51. Describe the process for producing tungsten electrodes.

_____

_____

52. What type of equipment is recommended for grinding tungsten electrodes?

_____

_____

53. How should a tungsten electrode be held while shaping the end with a grinder?

_____

_____

54. Why is argon used as a shielding gas for GTA welding?

_____

_____

55. What effect does the addition of thoria have on the tungsten electrode?

_____

_____

56. What effect does the addition of zirconia have on the tungsten electrode?

_____

_____

57. What effect does the end shape of the tungsten have on its welding performance?

_____

_____

58. Why must tungsten be hot in order to have a stable arc?

_____

_____

59. What factor leads to increased tungsten erosion?

_____

_____

60. Why is high frequency used for some GTA welding?

_____

_____

61. Why is the hot start used by welders?

_____

_____

62. The postflow time should be how long?

_____

_____

INSTRUCTOR'S COMMENTS _____

_____

_____

Chapter 17

Gas Tungsten Arc Welding of Plate

OPERATING INSTRUCTIONS FOR THE GTAW PROCESS

1. With the power off, switch the machine to the GTA welding mode.

2. Select the desired type of current and amperage range:

Electrode Diameter		DCSP	DCRP	AC
in.	(mm)			
.04	(1)	15–60	Not recommended	10–50
1/16	(2)	70–100	10–20	50–90
3/32	(2.4)	90–200	15–30	80–130
1/8	(3)	150–350	25–40	100–200
5/32	(4)	300–450	40–55	160–300

3. Set the fine current adjustment to the proper range, depending on the size of the tungsten being used.

4. Place the high-frequency switch in the appropriate position, auto or start for DC or continuous for AC.

5. The remote control can be plugged in and the selector switch set if you are using the remote.

6. The collet and collet body should be installed and match the tungsten size being used.

7. Install the tungsten into the torch and tighten the back cap to hold the tungsten in place.

8. Select and install the desired nozzle size. Adjust the tungsten length so that it does not stick out more than the inside diameter of the nozzle.

9. Check the manufacturer's operating manual for the machine to ensure that all connections and settings are correct.

10. Turn on the power, depress the remote control, and check for water or gas leaks.

11. While postpurge is still engaged, set the gas flow by adjusting the valve on the flowmeter.

12. Position yourself so that you are comfortable and can see the torch, tungsten, and plate while the tungsten tip is held about 1/4 in. (6 mm) above the metal. Try to hold the torch at a vertical angle ranging from 0° to 15°. Too steep an angle will not give adequate gas coverage.

CAUTION **Avoid touching the metal table with any unprotected skin or jewelry. The high frequency can cause an uncomfortable shock.**

13. Lower your arc welding helmet and depress the remote control. A high-pitched, erratic arc should be immediately jumping across the gap between the tungsten and the plate. If the high-frequency arc is not established, lower the torch until it appears.

14. Slowly increase the current until the main welding arc appears.

15. Observe the color change of the tungsten as the arc appears.

16. Move the tungsten around in a small circle until a weld pool appears on the metal.

17. Slowly decrease the current and observe the change in the weld pool.

18. Reduce the current until the arc is extinguished.

19. Hold the torch in place over the weld until the postpurge stops.

20. Raise your hood and inspect the weld.

21. Turn off the welding machine and shielding gas, and clean up your work area when you are finished welding.

■ PRACTICE 17-1

Name _____ Date _____

Base Metal _____ Thickness _____ Filler Dia. _____ Joint _____

Class _____ Instructor _____ Grade _____

Welding Principles and Applications

MATERIAL:
1/8" X 6" MILD STEEL

PROCESS:
GTAW STRINGER BEAD FLAT POSITION

NUMBER:	DRAWN BY:
PRACTICE 17-1	WENDY JEFFUS

OBJECTIVE: After completing this practice, you should be able to make stringer beads in the flat position on 16-gauge and 1/8-in. (3-mm) thick carbon steel without filler metal using the GTAW process.

EQUIPMENT AND MATERIALS NEEDED FOR THIS PRACTICE

1. A properly set-up and adjusted GTA welding machine.

2. Proper safety protection (welding hood, safety glasses, pliers, gloves, long-sleeved shirt, long pants, and leather boots or shoes). Refer to Chapter 2 in the text for more specific safety information.

3. One or more pieces of mild steel, 3-in. (76-mm) wide by 6-in. (152-mm) long, 16-gauge, and 1/8-in. (3-mm) thick.

INSTRUCTIONS

1. Starting at one end of the piece of metal that is 1/8-in. (3-mm) thick, hold the torch as close as possible to a 90° angle.

2. Lower your hood, strike an arc, and establish a weld pool.

3. Move the torch in a circular oscillation pattern down the plate toward the other end.

4. If the size of the weld pool changes, speed up or slow down the travel rate to keep the weld pool the same width for the entire length of the plate.

5. Repeat the process using both thicknesses of metal until you can consistently make the weld pool visually defect free.

6. Turn off the welding machine and shielding gas, and clean up your work area when you are finished welding.

INSTRUCTOR'S COMMENTS _____

_____

_____

■ PRACTICE 17-2

Name _____ Date _____

Base Metal _____ Thickness _____ Filler Dia. _____ Joint _____

Class _____ Instructor _____ Grade _____

OBJECTIVE: After completing this practice, you should be able to make stringer beads in the flat position on 16-gauge and ¼-in. (6-mm) thick stainless steel without filler metal using the GTAW process.

EQUIPMENT AND MATERIALS NEEDED FOR THIS PRACTICE

1. A properly set-up and adjusted GTA welding machine.

2. Proper safety protection (welding hood, safety glasses, pliers, gloves, long-sleeved shirt, long pants, and leather boots or shoes). Refer to Chapter 2 in the text for more specific safety information.

3. One or more pieces of mild steel, 3-in. (76-mm) wide by 6-in. (152-mm) long, 16-gauge, and 1/8-in. (3-mm) thick.

INSTRUCTIONS

1. Starting at one end of the piece of metal that is 1/8-in. (3-mm) thick, hold the torch as close as possible to a 90° angle.

2. Lower your hood, strike an arc, and establish a weld pool.

3. Move the torch in a circular oscillation pattern down the plate toward the other end.

4. If the size of the weld pool changes, speed up or slow down the travel rate to keep the weld pool the same width for the entire length of the plate.

   To keep the formation of oxides on the bead to a minimum, a chill plate (a thick piece of metal used to absorb heat) may be required. Another method is to make the bead using as low a heat input as possible. When the weld is finished, the weld bead should be no darker than dark blue.

5. Repeat the process using both thicknesses of metal until you can consistently make the weld visually defect free.

6. Turn off the welding machine and shielding gas, and clean up your work area when you are finished welding.

INSTRUCTOR'S COMMENTS _____

_____

_____

■ PRACTICE 17-3

Name _____ Date _____

Base Metal _____ Thickness _____ Filler Dia. _____ Joint _____

Class _____ Instructor _____ Grade _____

OBJECTIVE: After completing this practice, you should be able to make stringer beads in the flat position on 16-gauge and 1/8-in. (3-mm) thick and ¼-in. (6-mm) thick aluminum without the filler metal using the GTAW process.

EQUIPMENT AND MATERIALS NEEDED FOR THIS PRACTICE

1. A properly set-up and adjusted GTA welding machine.

2. Proper safety protection (welding hood, safety glasses, pliers, gloves, long-sleeved shirt, long pants, and leather boots or shoes). Refer to Chapter 2 in the text for more specific safety information.

3. One or more pieces of aluminum, 3-in. (76-mm) wide by 6-in. (152-mm) long, 16-gauge, and 1/8-in. (3-mm) thick.

INSTRUCTIONS

1. Starting at one end of the piece of metal that is 1/8-in. (3-mm) thick, hold the torch as close as possible to a 90° angle.

2. Lower your hood, strike an arc, and establish a weld pool.

3. Move the torch in a circular oscillation pattern down the plate toward the other end.

4. If the size of the weld pool changes, speed up or slow down the travel rate to keep the weld pool the same width for the entire length of the plate.

   A high current setting will allow faster travel speeds. The faster speed helps to control excessive penetration. Hot cracking may occur on some types of aluminum after a surfacing weld. This is not normally a problem when filler metal is added. If hot cracking should occur during this practice, do not be concerned.

5. Repeat the process using both thicknesses of metal until you can consistently make the weld visually defect free.

6. Turn off the welding machine and shielding gas, and clean up your work area when you are finished welding.

INSTRUCTOR'S COMMENTS _____

_____

_____

■ PRACTICE 17-4

Name _____ Date _____

Base Metal _____ Thickness _____ Filler Dia. _____ Joint _____

Class _____ Instructor _____ Grade _____

Welding Principles and Applications

MATERIAL: 1/16" x 6" MILD STEEL & STAINLESS STEEL
1/8" x 6" MILD STEEL & STAINLESS STEEL
1/4" x 6" ALUMINUM

PROCESS:
GTAW STRINGER BEAD FLAT POSITION

| NUMBER: | DRAWN BY: |
| PRACTICE 17-4 | WENDY JEFFUS |

OBJECTIVE: After completing this practice, you should be able to weld stringer beads on carbon steel, stainless steel, and aluminum in the flat position using the correct filler metal using the GTAW process.

EQUIPMENT AND MATERIALS NEEDED FOR THIS PRACTICE

1. A properly set-up and adjusted GTA welding machine.

2. Proper safety protection.

3. Filler rods, 36-in. (0.9-m) long by 1/16-in. (1.6-mm), 3/32-in. (2.4-mm), and 1/8-in. (3-mm) diameter.

4. One or more pieces of mild steel, stainless steel, and aluminum, 3-in. (76-mm) wide by 6-in. (152-mm) long by 1/16-in. (2-mm) and 1/8-in. (3-mm) thick, and aluminum plate 1/4-in. (6-mm) thick.

INSTRUCTIONS

1. Starting with the metal that is 1/8-in. (3-mm) thick and the filler rod having a 3/32 in. (2.4 mm) diameter, strike an arc and establish a weld pool.

2. Move the torch in a circle as in the first three practices. When the torch is on one side, add filler rod to the other side of the molten weld pool. The end of the filler rod should be dipped into the front of the weld pool but should not be allowed to melt and drip into the weld pool.

3. Change to another size filler rod and determine its effect on the weld pool.

4. Maintain a smooth and uniform rhythm as filler metal is added. This will help to keep the bead uniform.

5. Vary the rhythms to determine which one is easiest for you. If the filler rod sticks, move the torch toward the rod until it melts free.

6. When the full 6-in. (152-mm) long weld bead is completed, cool and inspect it for uniformity and defects.

7. Repeat the process using all thicknesses and types of metal until you can consistently make the weld visually defect free.

8. Turn off the welding machine and shielding gas, and clean up your work area when you are finished welding.

INSTRUCTOR'S COMMENTS _____

_____

_____

■ PRACTICE 17-5

Name _____ Date _____

Base Metal _____ Thickness _____ Filler Dia. _____ Joint _____

Class _____ Instructor _____ Grade _____

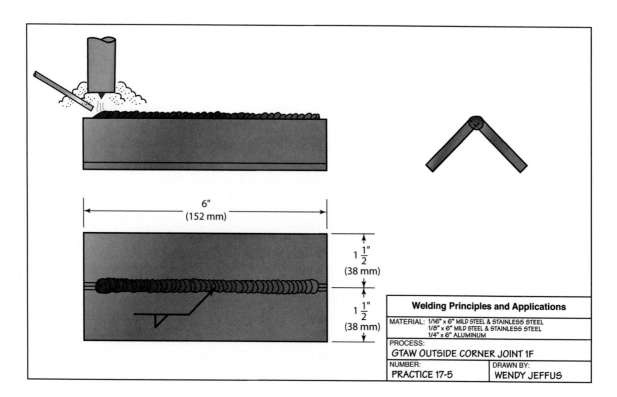

6"
(152 mm)

$1\frac{1}{2}$"
(38 mm)

$1\frac{1}{2}$"
(38 mm)

Welding Principles and Applications	
MATERIAL: 1/16" x 6" MILD STEEL & STAINLESS STEEL 1/8" x 6" MILD STEEL & STAINLESS STEEL 1/4" x 6" ALUMINUM	
PROCESS: GTAW OUTSIDE CORNER JOINT 1F	
NUMBER: PRACTICE 17-5	DRAWN BY: WENDY JEFFUS

OBJECTIVE: After completing this practice, you should be able to GTAW weld an outside corner joint on carbon steel, stainless steel, and aluminum in the flat position.

EQUIPMENT AND MATERIALS NEEDED FOR THIS PRACTICE

1. A properly set-up and adjusted GTA welding machine.

2. Proper safety protection.

3. Filler rods, 36-in. (0.9-m) long by 1/16-in. (1.6-mm), 3/32-in. (2.4-mm), and 1/8-in. (3-mm) diameter.

4. One or more pieces of mild steel, stainless steel, and aluminum, 1 1/2-in. (38-mm) wide by 6-in. (152-mm) long by 1/16-in. (2-mm) and 1/8-in. (3-mm) thick, and aluminum plate 1/4-in. (6-mm) thick.

INSTRUCTIONS

1. Place one of the pieces of metal flat on the table and hold or brace the other piece of metal vertically on it forming a corner joint.

2. Tack weld both ends of the plates together.

3. Set the plates up and add two or three more tack welds on the joint as required.

4. Clean the weld area before making the weld. Then, starting at one end, make a weld bead of uniform height and width, adding filler metal as needed to the outside corner.

5. Repeat each weld as needed with all types of base metal and filler rod sizes until all welds can be done with consistently high quality.

6. Turn off the welding machine and shielding gas, and clean up your work area when you are finished welding.

INSTRUCTOR'S COMMENTS _____

_____

_____

■ PRACTICE 17-6

Name _____ Date _____

Base Metal _____ Thickness _____ Filler Dia. _____ Joint _____

Class _____ Instructor _____ Grade _____

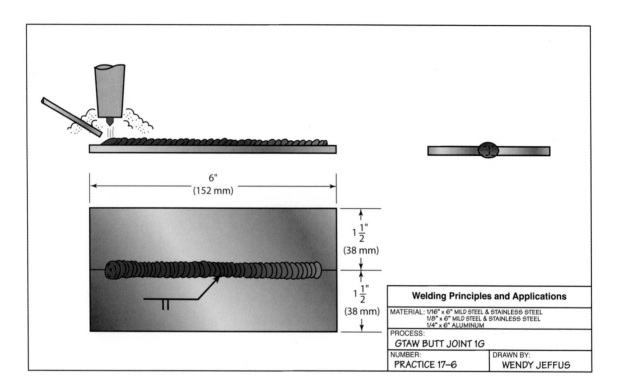

6"
(152 mm)

$1\frac{1}{2}$"
(38 mm)

$1\frac{1}{2}$"
(38 mm)

Welding Principles and Applications

MATERIAL: 1/16" x 6" MILD STEEL & STAINLESS STEEL
1/8" x 6" MILD STEEL & STAINLESS STEEL
1/4" x 6" ALUMINUM

PROCESS:
GTAW BUTT JOINT 1G

NUMBER:
PRACTICE 17–6

DRAWN BY:
WENDY JEFFUS

OBJECTIVE: After completing this practice, you should be able to weld a butt joint on carbon steel, stainless steel, and aluminum in the 1G position using the GTAW process.

EQUIPMENT AND MATERIALS NEEDED FOR THIS PRACTICE

1. A properly set-up and adjusted GTA welding machine.

2. Proper safety protection.

3. Filler rods, 36-in. (0.9-m) long by 1/16-in. (1.6-mm), 3/32-in. (2.4-mm), and 1/8-in. (3-mm) diameter.

4. Two or more pieces of mild steel, stainless steel, and aluminum, 1 1/2-in. (38-mm) wide by 6-in. (152-mm) long by 1/16-in. (2-mm) and 1/8-in. (3-mm) thick, and aluminum plate 1/4-in. (6-mm) thick.

INSTRUCTIONS

1. Place the metal flat on the table and tack weld both ends together.

2. Two or three additional tack welds can be made along the joint as needed.

3. Starting at one end, make a uniform width weld along the joint. Add filler metal as required to make a uniform height weld.

4. Repeat the process using all thicknesses and types of metal until you can consistently make the weld visually defect free and of high quality.

5. Turn off the welding machine and shielding gas, and clean up your work area when you are finished welding.

INSTRUCTOR'S COMMENTS _____

_____

_____

■ PRACTICE 17-7

Name _____ Date _____

Base Metal _____ Thickness _____ Filler Dia. _____ Joint _____

Class _____ Instructor _____ Grade _____

OBJECTIVE: After completing this practice you should be able to make a butt joint while controlling both distortion and penetration in the flat position using the GTAW process.

EQUIPMENT AND MATERIALS NEEDED FOR THIS PRACTICE

1. A properly set-up and adjusted GTA welding machine.

2. Proper safety protection.

3. Filler rods, 36-in. (0.9-m) long by 1/16-in. (1.6-mm), 3/32-in. (2.4-mm), and 1/8-in. (3-mm) diameter.

4. Two or more pieces of mild steel, stainless steel, and aluminum, 1 1/2-in. (38-mm) wide by 6-in. (152-mm) long by 1/16-in. (2-mm) and 1/8-in. (3-mm) thick, and aluminum plate 1/4-in. (6-mm) thick.

INSTRUCTIONS

1. Place the metal flat on the table and tack weld both ends together in the butt joint configuration.

2. Two or three additional tacks can be made along the joint as needed.

3. Using a back-stepping weld sequence, make a series of welds approximately 1-in. (25-mm) long along the joint.

4. Be sure to fill each weld crater adequately to reduce crater cracking.

5. Repeat the process using all thicknesses and types of metal until you can consistently make distortion-free and defect-free welds.

6. Turn off the welding machine and shielding gas, and clean up your work area when finished with this practice.

INSTRUCTOR'S COMMENTS _____

_____

_____

■ PRACTICE 17-8

Name _____ Date _____

Base Metal _____ Thickness _____ Filler Dia. _____ Joint _____

Class _____ Instructor _____ Grade _____

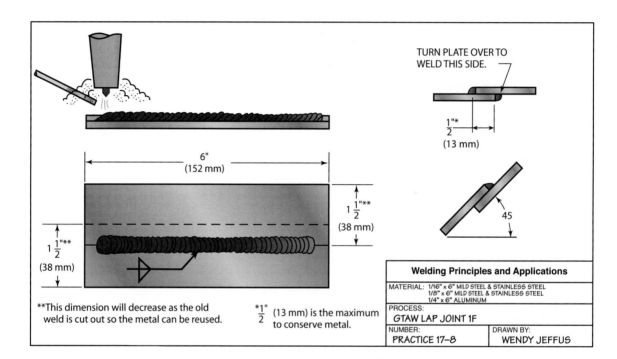

TURN PLATE OVER TO WELD THIS SIDE.

$\frac{1}{2}$"*
(13 mm)

6"
(152 mm)

$1\frac{1}{2}$"**
(38 mm)

$1\frac{1}{2}$"**
(38 mm)

45

**This dimension will decrease as the old weld is cut out so the metal can be reused.

*$\frac{1}{2}$" (13 mm) is the maximum to conserve metal.

Welding Principles and Applications

MATERIAL: 1/16" x 6" MILD STEEL & STAINLESS STEEL
1/8" x 6" MILD STEEL & STAINLESS STEEL
1/4" x 6" ALUMINUM

PROCESS:
GTAW LAP JOINT 1F

NUMBER: PRACTICE 17-8

DRAWN BY: WENDY JEFFUS

OBJECTIVE: After completing this practice, you should be able to weld lap joints on carbon steel, stainless steel, and aluminum in the 1F position using the GTAW process.

EQUIPMENT AND MATERIALS NEEDED FOR THIS PRACTICE

1. A properly set-up and adjusted GTA welding machine.

2. Proper safety protection.

3. Filler rods, 36-in. (0.9-m) long by 1/16-in. (1.6-mm), 3/32-in. (2.4-mm), and 1/8-in. (3-mm) diameter.

4. Two or more pieces of mild steel, stainless steel, and aluminum, 1 1/2-in. (38-mm) wide by 6-in. (152-mm) long by 1/16-in. (2-mm) and 1/8-in. (3-mm) thick, and aluminum plate 1/4-in. (6-mm) thick.

INSTRUCTIONS

1. Place the two pieces of metal flat on the table with an overlap of 1/4 in. (6 mm) to 3/8 in. (10 mm).

2. Hold the pieces of metal tightly together and tack weld them.

3. Starting at one end, make a uniform fillet weld along the joint.

4. Both sides of the joint should be welded.

5. Repeat the process using all thicknesses of metal until you can consistently make the weld visually defect free and of high quality.

6. Turn off the welding machine and shielding gas, and clean up your work area when you are finished welding.

INSTRUCTOR'S COMMENTS _____

_____

_____

■ PRACTICE 17-9

Name _____ Date _____

Base Metal _____ Thickness _____ Filler Dia. _____ Joint _____

Class _____ Instructor _____ Grade _____

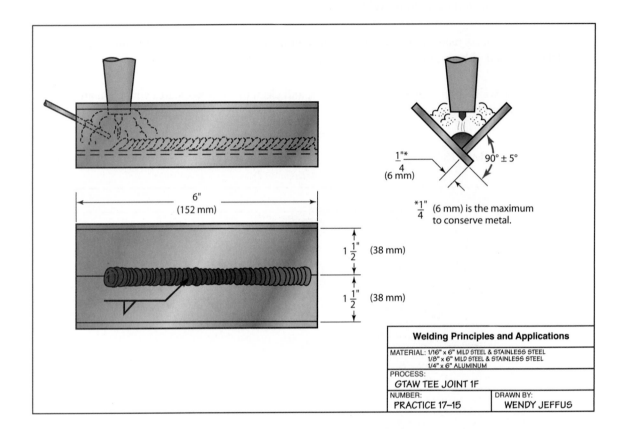

6"
(152 mm)

$1\frac{1}{2}$" (38 mm)

$1\frac{1}{2}$" (38 mm)

$\frac{1}{4}$"* (6 mm)

90° ± 5°

*$\frac{1}{4}$" (6 mm) is the maximum to conserve metal.

Welding Principles and Applications

MATERIAL: 1/16" x 6" MILD STEEL & STAINLESS STEEL
1/8" x 6" MILD STEEL & STAINLESS STEEL
1/4" x 6" ALUMINUM

PROCESS:
GTAW TEE JOINT 1F

NUMBER: PRACTICE 17–15

DRAWN BY: WENDY JEFFUS

OBJECTIVE: After completing this practice, you should be able to weld tee joints on carbon steel, stainless steel, and aluminum in the 1F position using the GTAW process.

EQUIPMENT AND MATERIALS NEEDED FOR THIS PRACTICE

1. A properly set-up and adjusted GTA welding machine.

2. Proper safety protection.

3. Filler rods, 36-in. (0.9-m) long by 1/16-in. (1.6-mm), 3/32-in. (2.4-mm), and 1/8-in. (3-mm) diameter.

4. Two or more pieces of mild steel, stainless steel, and aluminum, 1 1/2-in. (38-mm) wide by 6-in. (152-mm) long by 1/16 in. (2 mm) and 1/8-in. (3-mm) thick, and aluminum plate 1/4-in. (6-mm) thick.

INSTRUCTIONS

1. Place one of the pieces of metal flat on the table and hold or brace the other piece of metal vertically on it in the tee joint configuration.

2. Tack weld both ends of the plates together.

3. Set up the plates in the flat position and add two or three more tack welds to the joint as required.

4. On the metal that is 1/16 in. (1.5 mm) thick, it may not be possible to weld both sides, but on the thicker material a fillet weld can usually be made on both sides. The exception to this is if carbide precipitation occurs on the stainless steel during welding.

5. Starting at one end, make a uniform height and width weld, adding filler metal as needed.

6. Repeat the process using all thicknesses of metal until you can consistently make the weld visually defect free.

7. Turn off the welding machine and shielding gas, and clean up your work area when you are finished welding.

INSTRUCTOR'S COMMENTS _____

_____

_____

■ PRACTICE 17-10

Name _____ Date _____

Base Metal _____ Thickness _____ Filler Dia. _____ Joint _____

Class _____ Instructor _____ Grade _____

OBJECTIVE: After completing this practice you should be able to make a stringer bead at a 45° vertical up angle using mild steel, stainless steel, and aluminum with the GTAW process.

EQUIPMENT AND MATERIALS NEEDED FOR THIS PRACTICE

1. A properly set-up and adjusted GTA welding machine.

2. Proper safety protection.

3. Filler rods, 36-in. (0.9-m) long by 1/16-in. (1.6-mm), 3/32-in. (2.4-mm), and 1/8-in. (3-mm) diameter.

4. One or more pieces of mild steel, stainless steel, and aluminum, 1 1/2-in. (38-mm) wide by 6-in. (152-mm) long by 1/16-in. (2-mm) and 1/8-in. (3-mm) thick, and aluminum plate 1/4-in. (6-mm) thick.

INSTRUCTIONS

1. Position the workpiece at a 45° angle to the table.

2. Starting at the bottom and welding in an upward direction, add filler metal to the top edge of the weld pool, and move the torch in a circle or "C" pattern.

3. If the weld pool size starts to increase, the "C" pattern can be increased in length or the power can be decreased.

4. Watch the weld pool and establish a rhythm of torch movement and filler metal addition, which will keep the bead size and penetration uniform.

5. Repeat the process using all thicknesses and types of metal until you can consistently make the weld visually defect free.

6. Turn off the welding machine and shielding gas, and clean up your work area when finished with this practice.

INSTRUCTOR'S COMMENTS _____

_____

_____

■ PRACTICE 17-11

Name _____ Date _____

Base Metal _____ Thickness _____ Filler Dia. _____ Joint _____

Class _____ Instructor _____ Grade _____

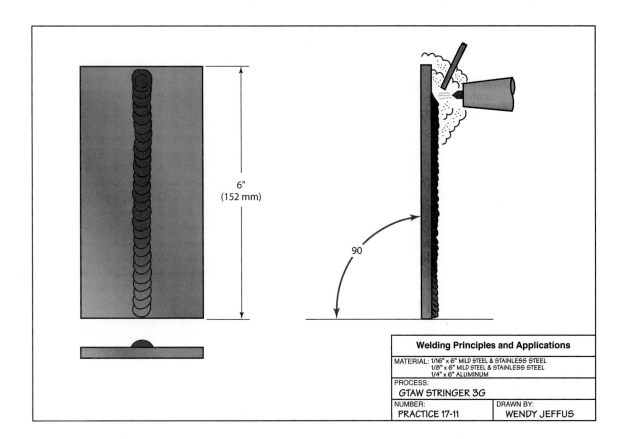

Welding Principles and Applications

MATERIAL: 1/16" x 6" MILD STEEL & STAINLESS STEEL
1/8" x 6" MILD STEEL & STAINLESS STEEL
1/4" x 6" ALUMINUM

PROCESS:
GTAW STRINGER 3G

NUMBER:	DRAWN BY:
PRACTICE 17-11	WENDY JEFFUS

OBJECTIVE: After completing this practice, you should be able to make stringer beads on carbon steel, stainless steel, and aluminum in the 3G position using the GTAW process.

EQUIPMENT AND MATERIALS NEEDED FOR THIS PRACTICE

1. A properly set-up and adjusted GTA welding machine.

2. Proper safety protection.

3. Filler rods, 36-in. (0.9-m) long by 1/16-in. (1.6 mm), 3/32-in. (2.4-mm), and 1/8-in. (3-mm) diameter.

4. One or more pieces of mild steel, stainless steel, and aluminum, 1 1/2-in. (38-mm) wide by 6-in. (152-mm) long by 1/16-in. (2-mm) and 1/8-in. (3-mm) thick, and aluminum plate 1/4-in. (6-mm) thick.

INSTRUCTIONS

1. Start with the plate at a 45° angle to the table.

2. Starting at the bottom and welding in an upward direction, add the filler metal at the top edge of the weld pool and move the torch in a circle or "C" pattern.

3. If the weld pool size starts to increase, the "C" pattern can be increased in length or the amperage can be decreased.

4. Watch the weld pool and establish a rhythm of torch movement and addition of filler rod to keep the weld uniform.

5. Gradually increase the angle as you develop skill until the weld is being made in the vertical up position.

6. Repeat the process using all thicknesses of metal until you can consistently make the weld visually defect free and the same height and width throughout the length of the bead.

7. Turn off the welding machine and shielding gas, and clean up your work area when you are finished welding.

INSTRUCTOR'S COMMENTS _____

_____

_____

■ PRACTICE 17-12

Name _____ Date _____

Base Metal _____ Thickness _____ Filler Dia. _____ Joint _____

Class _____ Instructor _____ Grade _____

OBJECTIVE: After completing this practice, you should be able to make butt joints on carbon steel, stainless steel, and aluminum in a 45° vertical angle position using the GTAW process.

EQUIPMENT AND MATERIALS NEEDED FOR THIS PRACTICE

1. A properly set-up and adjusted GTA welding machine.

2. Proper safety protection.

3. Filler rods, 36-in. (0.9-m) long by 1/16-in. (1.6-mm), 3/32-in. (2.4-mm), and 1/8-in. (3-mm) diameter.

4. Two or more pieces of mild steel, stainless steel, and aluminum, 1 1/2-in. (25-mm) wide by 6-in. (152-mm) long by 1/16-in. (2-mm) and 1/8-in. (3-mm) thick, and aluminum plate 1/4-in. (6-mm) thick.

INSTRUCTIONS

1. Tack weld the plates together and then place the plates on the welding table at a 45° vertical angle. Start the weld at the bottom and weld in an upward direction.

2. If the weld pool size starts to increase, the "C" pattern can be increased in length or the amperage can be decreased.

3. Watch the weld pool and establish a rhythm of torch movement and addition of filler rod to keep the weld uniform.

4. Repeat the process using all thicknesses and types of metal until you can consistently make the weld visually defect free and the same height and width throughout the length of the bead.

5. Turn off the welding machine and shielding gas, and clean up your work area when you are finished welding.

INSTRUCTOR'S COMMENTS _____

_____

_____

■ PRACTICE 17-13

Name _____ Date _____

Base Metal _____ Thickness _____ Filler Dia. _____ Joint _____

Class _____ Instructor _____ Grade _____

OBJECTIVE: After completing this practice, you should be able to make butt joints on carbon steel, stainless steel, and aluminum in the 3G position using the GTAW process.

EQUIPMENT AND MATERIALS NEEDED FOR THIS PRACTICE

1. A properly set-up and adjusted GTA welding machine.

2. Proper safety protection.

3. Filler rods, 36-in. (0.9-m) long by 1/16-in. (1.6-mm), 3/32-in. (2.4-mm), and 1/8-in. (3-mm) diameter.

4. Two or more pieces of mild steel, stainless steel, and aluminum, 1 1/2-in. (25-mm) wide by 6-in. (152-mm) long by 1/16-in. (2-mm) and 1/8-in. (3-mm) thick, and aluminum plate 1/4-in. (6-mm) thick.

INSTRUCTIONS

1. Tack weld the plates together and then place the plates on the welding table at a 45° vertical angle. Start the weld at the bottom and weld in an upward direction.

2. If the weld pool size starts to increase, the "C" pattern can be increased in length or the amperage can be decreased.

3. Watch the weld pool and establish a rhythm of torch movement and addition of filler rod to keep the weld uniform.

4. As you develop skill, continue increasing the angle until the weld is being made in the vertical up position.

5. Repeat the process using all thicknesses and types of metal until you can consistently make the weld visually defect free and the same height and width throughout the length of the bead.

6. Turn off the welding machine and shielding gas, and clean up your work area when you are finished welding.

INSTRUCTOR'S COMMENTS _____
_____
_____

■ PRACTICE 17-14

Name _____ Date _____

Base Metal _____ Thickness _____ Filler Dia. _____ Joint _____

Class _____ Instructor _____ Grade _____

OBJECTIVE: After completing this practice you should be able to make fillet weld lap joints on carbon steel, stainless steel, and aluminum in the 45° vertical up position using the GTAW process.

EQUIPMENT AND MATERIALS NEEDED FOR THIS PRACTICE

1. A properly set-up and adjusted GTA welding machine.

2. Proper safety protection.

3. Filler rods, 36-in. (0.9-m) long by 1/16-in. (1.6-mm), 3/32 in. (2.4-mm), and 1/8-in. (3-mm) diameter.

4. Two or more pieces of mild steel, stainless steel, and aluminum, 1 1/2-in. (38-mm) wide by 6-in. (152-mm) long by 1/16-in. (2-mm) and 1/8-in. (3-mm) thick, and aluminum plate 1/4-in. (6-mm) thick.

INSTRUCTIONS

1. Tack weld the plates together in the lap joint configuration with an overlap of 1/4 in. (6 mm) to 3/8 in. (10 mm). Then place them on the welding table at a 45° vertical angle. Start the weld at the bottom and weld in an upward direction.

2. Maintain a uniform rhythm so that a nice looking weld bead is formed.

3. It may be necessary to move the torch in and around the base of the weld pool to ensure adequate root fusion. The filler metal should be added along the top edge of the weld pool to ensure adequate fusion.

4. Repeat the process using all thicknesses and types of metal until you can consistently make the weld visually defect free.

5. As you develop skill, increase the angle until the weld is being made in the vertical up position.

6. Turn off the welding machine and shielding gas, and clean up your work area when you are finished.

INSTRUCTOR'S COMMENTS _____

_____

_____

■ PRACTICE 17-15

Name _____ Date _____

Base Metal _____ Thickness _____ Filler Dia. _____ Joint _____

Class _____ Instructor _____ Grade _____

OBJECTIVE: After completing this practice you should be able to make fillet weld lap joints on carbon steel, stainless steel, and aluminum in the 3F position using the GTAW process.

EQUIPMENT AND MATERIALS NEEDED FOR THIS PRACTICE

1. A properly set-up and adjusted GTA welding machine.

2. Proper safety protection.

3. Filler rods, 36-in. (0.9-m) long by 1/16-in. (1.6-mm), 3/32-in. (2.4-mm), and 1/8-in. (3-mm) diameter.

4. Two or more pieces of mild steel, stainless steel, and aluminum, 1 1/2-in. (38-mm) wide by 6-in. (152-mm) long by 1/16-in. (2-mm) and 1/8-in. (3-mm) thick, and aluminum plate 1/4-in. (6-mm) thick.

INSTRUCTIONS

1. Tack weld the plates together in the lap joint configuration with an overlap of 1/4 in. (6 mm) to 3/8 in. (10 mm). Then place them on the welding table at a 45° vertical angle. Start the weld at the bottom and weld in an upward direction.

2. Maintain a uniform rhythm so that a nice looking weld bead is formed.

3. It may be necessary to move the torch in and around the base of the weld pool to ensure adequate root fusion. The filler metal should be added along the top edge of the weld pool to ensure adequate fusion.

4. As you develop skill, increase the angle until the weld is being made in the vertical up position.

5. Repeat the process using all thicknesses and types of metal until you can consistently make the weld visually defect free.

6. Turn off the welding machine and shielding gas, and clean up your work area when you are finished.

INSTRUCTOR'S COMMENTS _____

_____

_____

■ PRACTICE 17-16

Name _____ Date _____

Base Metal _____ Thickness _____ Filler Dia. _____ Joint _____

Class _____ Instructor _____ Grade _____

OBJECTIVE: After completing this practice you should be able to make a tee joint on carbon steel, stainless steel, and aluminum at a 45° vertical angle using the GTAW process.

EQUIPMENT AND MATERIALS NEEDED FOR THIS PRACTICE

1. A properly set-up and adjusted GTA welding machine.

2. Proper safety protection.

3. Filler rods, 36-in. (0.9-m) long by 1/16-in. (1.6-mm), 3/32-in. (2.4-mm), and 1/8-in. (3-mm) diameter.

4. Two or more pieces of mild steel, stainless steel, and aluminum, 1 1/2-in. (38-mm) wide by 6-in. (152-mm) long by 1/16-in. (2-mm) and 1/8-in. (3-mm) thick, and aluminum plate 1/4-in. (6-mm) thick.

INSTRUCTIONS

1. Tack weld the plates together in the tee joint configuration then place them on the welding table at a 45° vertical angle. Start the weld at the bottom and weld in an upward direction.

2. The edge of the side plate will heat up more quickly than the back plate. This may lead to undercutting along the edge of the weld.

3. To control undercutting, keep the arc on the back plate and add the filler metal to the weld pool near the side plate.

4. Repeat the process using all thicknesses and types of metal until you can consistently make the weld with 100% penetration.

5. Turn off the welding machine and shielding gas, and clean up your work area when you are finished.

INSTRUCTOR'S COMMENTS _____

_____

_____

■ PRACTICE 17-17

Name _____ Date _____

Base Metal _____ Thickness _____ Filler Dia. _____ Joint _____

Class _____ Instructor _____ Grade _____

OBJECTIVE: After completing this practice you should be able to make a tee joint on carbon steel, stainless steel, and aluminum in the 3F position using the GTAW process.

EQUIPMENT AND MATERIALS NEEDED FOR THIS PRACTICE

1. A properly set-up and adjusted GTA welding machine.

2. Proper safety protection.

3. Filler rods, 36-in. (0.9-m) long by 1/16-in. (1.6-mm), 3/32-in. (2.4-mm), and 1/8-in. (3-mm) diameter.

4. Two or more pieces of mild steel, stainless steel, and aluminum, 1 1/2-in. (38-mm) wide by 6-in. (152-mm) long by 1/16-in. (2-mm) and 1/8-in. (3-mm) thick, and aluminum plate 1/4-in. (6-mm) thick.

INSTRUCTIONS

1. Tack weld the plates together in the tee joint configuration then place them on the welding table at a 45° vertical angle. Start the weld at the bottom and weld in an upward direction.

2. The edge of the side plate will heat up more quickly than the back plate. This may lead to undercutting along the edge of the weld.

3. To control undercutting, keep the arc on the back plate and add the filler metal to the weld pool near the side plate.

4. Gradually increase the plate angle as your skill develops until the weld is being made in the vertical up position.

5. Repeat the process using all thicknesses and types of metal until you can consistently make the weld visually defect free.

6. Turn off the welding machine and shielding gas, and clean up your work area when you are finished.

INSTRUCTOR'S COMMENTS _____

_____

_____

■ PRACTICE 17-18

Name _____ Date _____

Base Metal _____ Thickness _____ Filler Dia. _____ Joint _____

Class _____ Instructor _____ Grade _____

OBJECTIVE: After completing this practice you should be able to make a stringer bead on carbon steel, stainless steel, and aluminum at a 45° reclining angle using the GTAW process.

EQUIPMENT AND MATERIALS NEEDED FOR THIS PRACTICE

1. A properly set-up and adjusted GTA welding machine.

2. Proper safety protection.

3. Filler rods, 36-in. (0.9-m) long by 1/16-in. (1.6-mm), 3/32-in. (2.4-mm), and 1/8-in. (3-mm) diameter.

4. One or more pieces of mild steel, stainless steel, and aluminum, 1 1/2-in. (38-mm) wide by 6-in. (152-mm) long by 1/16-in. (2-mm) and 1/8-in. (3-mm) thick, and aluminum plate 1/4-in. (6-mm) thick.

INSTRUCTIONS

1. Place the plate on the welding table at a 45° reclining angle. Add the filler metal along the top leading edge of the weld pool.

2. Do not let the weld pool become too large so that surface tension will hold it in place.

3. Use a rhythmic movement of the torch and filler rod so that the bead will be uniform in width and contour.

4. Repeat the process using all thicknesses and types of metal until you can consistently make the weld visually defect free.

5. Turn off the welding machine and shielding gas, and clean up your work area when you are finished.

INSTRUCTOR'S COMMENTS _____

_____

_____

■ PRACTICE 17-19

Name _____ Date _____

Base Metal _____ Thickness _____ Filler Dia. _____ Joint _____

Class _____ Instructor _____ Grade _____

OBJECTIVE: After completing this practice you should be able to make a stringer bead on carbon steel, stainless steel, and aluminum in the horizontal position using the GTAW process.

EQUIPMENT AND MATERIALS NEEDED FOR THIS PRACTICE

1. A properly set-up and adjusted GTA welding machine.

2. Proper safety protection.

3. Filler rods, 36-in. (0.9-m) long by 1/16-in. (1.6-mm), 3/32-in. (2.4-mm), and 1/8-in. (3-mm) diameter.

4. One or more pieces of mild steel, stainless steel, and aluminum, 1 1/2-in. (38-mm) wide by 6-in. (152-mm) long by 1/16-in. (2-mm) and 1/8-in. (3-mm) thick, and aluminum plate 1/4-in. (6-mm) thick.

INSTRUCTIONS

1. Place the plate on the welding table at a 45° reclining angle. Add the filler metal along the top leading edge of the weld pool.

2. Do not let the weld pool become too large so that surface tension will hold it in place.

3. Use a rhythmic movement of the torch and filler rod so that the bead will be uniform in width and contour.

4. Gradually increase the plate angle as your skill develops until the weld is being made in the horizontal position on a vertical plate.

5. Repeat the process using all thicknesses and types of metal until you can consistently make the weld visually defect free.

6. Turn off the welding machine and shielding gas, and clean up your work area when you are finished.

INSTRUCTOR'S COMMENTS _____

_____

_____

■ PRACTICE 17-20

Name _____ Date _____

Base Metal _____ Thickness _____ Filler Dia. _____ Joint _____

Class _____ Instructor _____ Grade _____

OBJECTIVE: After completing this practice you should be able to make a butt joint on carbon steel, stainless steel, and aluminum in the horizontal 2G position using the GTAW process.

EQUIPMENT AND MATERIALS NEEDED FOR THIS PRACTICE

1. A properly set-up and adjusted GTA welding machine.

2. Proper safety protection.

3. Filler rods, 36-in. (0.9-m) long by 1/16-in. (1.6-mm), 3/32-in. (2.4-mm), and 1/8-in. (3-mm) diameter.

4. Two or more pieces of mild steel, stainless steel, and aluminum, 1 1/2-in. (38-mm) wide by 6-in. (152-mm) long by 1/16-in. (2-mm) and 1/8-in. (3-mm) thick, and aluminum plate 1/4-in. (6-mm) thick.

INSTRUCTIONS

1. Tack the plates together in the butt joint configuration and place them on the weld table at a 45° reclining angle.

2. Add the filler metal to the top plate and keep the bead small in size.

3. Use a rhythmic movement of the torch and filler rod so that the bead will be uniform in width and contour.

4. Gradually increase the plate angle as your skill develops until the weld is being made in the horizontal position on a vertical plate.

5. Repeat the process using all thicknesses and types of metal until you can consistently make the weld visually defect free.

6. Turn off the welding machine and shielding gas, and clean up your work area when you are finished.

INSTRUCTOR'S COMMENTS _____

_____

_____

■ PRACTICE 17-21

Name _____ Date _____

Base Metal _____ Thickness _____ Filler Dia. _____ Joint _____

Class _____ Instructor _____ Grade _____

OBJECTIVE: After completing this practice you should be able to make a lap joint on carbon steel, stainless steel, and aluminum in the horizontal 2F position using the GTAW process.

EQUIPMENT AND MATERIALS NEEDED FOR THIS PRACTICE

1. A properly set-up and adjusted GTA welding machine.

2. Proper safety protection.

3. Filler rods, 36-in. (0.9-m) long by 1/16-in. (1.6-mm), 3/32 in. (2.4-mm), and 1/8-in. (3-mm) diameter.

4. Two or more pieces of mild steel, stainless steel, and aluminum, 1 1/2 in. (38-mm) wide by 6 in. (152-mm) long by 1/16-in. (2-mm) and 1/8-in. (3-mm) thick, and aluminum plate 1/4-in. (6-mm) thick.

INSTRUCTIONS

1. Tack the plates together in the lap joint configuration with an overlap of 1/4 in. (6 mm) to 3/8 in. (10 mm). Place them on the weld table in the horizontal position.

2. Add the filler metal along the top edge of the weld pool and keep the bead small in size. Avoid heating the top plate too much which will result in undercutting.

3. The bottom plate will act as a shelf to support the molten weld pool.

4. Repeat the process using all thicknesses and types of metal until you can consistently make the weld visually defect free.

5. Turn off the welding machine and shielding gas, and clean up your work area when you are finished.

INSTRUCTOR'S COMMENTS _____

_____

_____

■ PRACTICE 17-22

Name _____ Date _____

Base Metal _____ Thickness _____ Filler Dia. _____ Joint _____

Class _____ Instructor _____ Grade _____

OBJECTIVE: After completing this practice you should be able to make a tee joint on carbon steel, stainless steel, and aluminum in the horizontal 2F position using the GTAW process.

EQUIPMENT AND MATERIALS NEEDED FOR THIS PRACTICE

1. A properly set-up and adjusted GTA welding machine.

2. Proper safety protection.

3. Filler rods, 36-in. (0.9-m) long by 1/16-in. (1.6-mm), 3/32-in. (2.4 mm), and 1/8-in. (3-mm) diameter.

4. Two or more pieces of mild steel, stainless steel, and aluminum, 1 1/2 in. (38 mm) wide by 6 in. (152 mm) long by 1/16-in. (2-mm) and 1/8-in. (3-mm) thick, and aluminum plate 1/4-in. (6-mm) thick.

INSTRUCTIONS

1. Tack the plates together in the tee joint configuration and place them on the weld table in the horizontal position.

2. Add the filler metal along the top leading edge of the weld pool. Beware of undercutting.

3. The bottom plate will act as a shelf to support the molten weld pool.

4. Repeat the process using all thicknesses and types of metal until you can consistently make the weld visually defect free.

5. Turn off the welding machine and shielding gas, and clean up your work area when you are finished.

INSTRUCTOR'S COMMENTS _____

_____

_____

■ PRACTICE 17-23

Name _____ Date _____

Base Metal _____ Thickness _____ Filler Dia. _____ Joint _____

Class _____ Instructor _____ Grade _____

OBJECTIVE: After completing this practice you should be able to make a stringer bead on carbon steel, stainless steel, and aluminum in the overhead position using the GTAW process.

EQUIPMENT AND MATERIALS NEEDED FOR THIS PRACTICE

1. A properly set-up and adjusted GTA welding machine.

2. Proper safety protection.

3. Filler rods, 36-in. (0.9-m) long by 1/16-in. (1.6-mm), 3/32-in. (2.4-mm), and 1/8-in. (3-mm) diameter.

4. One or more pieces of mild steel, stainless steel, and aluminum, 1 1/2-in. (38-mm) wide by 6 in. (152 mm) long by 1/16-in. (2-mm) and 1/8-in. (3-mm) thick, and aluminum plate 1/4-in. (6-mm) thick.

INSTRUCTIONS

1. Place the plate in the overhead position.

2. Keep the bead small and add the filler metal to the leading edge of the weld pool. Surface tension will hold the weld pool in place unless it gets too large.

3. A wide bead with little buildup will be easier to control and less likely to undercut along the edges.

4. Repeat the process using all thicknesses and types of metal until you can consistently make the weld visually defect free.

5. Turn off the welding machine and shielding gas, and clean up your work area when you are finished.

INSTRUCTOR'S COMMENTS _____

_____

_____

■ PRACTICE 17-24

Name _____ Date _____

Base Metal _____ Thickness _____ Filler Dia. _____ Joint _____

Class _____ Instructor _____ Grade _____

OBJECTIVE: After completing this practice you should be able to make a butt joint on carbon steel, stainless steel, and aluminum in the overhead 4G position using the GTAW process.

EQUIPMENT AND MATERIALS NEEDED FOR THIS PRACTICE

1. A properly set-up and adjusted GTA welding machine.

2. Proper safety protection.

3. Filler rods, 36-in. (0.9-m) long by 1/16-in. (1.6-mm), 3/32-in. (2.4-mm), and 1/8-in. (3-mm) diameter.

4. One or more pieces of mild steel, stainless steel, and aluminum, 1 1/2-in. (38-mm) wide by 6-in. (152-mm) long by 1/16-in. (2-mm) and 1/8-in. (3-mm) thick, and aluminum plate 1/4-in. (6-mm) thick.

INSTRUCTIONS

1. Tack weld the pieces together in the butt joint configuration and place them in the overhead position.

2. Keep the bead small and add the filler metal to the leading edge of the weld pool. Surface tension will hold the weld pool in place unless it gets too large.

3. The complete bead should have a uniform width.

4. Repeat the process using all thicknesses and types of metal until you can consistently make the weld visually defect free.

5. Turn off the welding machine and shielding gas, and clean up your work area when you are finished.

INSTRUCTOR'S COMMENTS _____

_____

_____

■ PRACTICE 17-25

Name _____ Date _____

Base Metal _____ Thickness _____ Filler Dia. _____ Joint _____

Class _____ Instructor _____ Grade _____

OBJECTIVE: After completing this practice you should be able to make a fillet weld on a lap joint on carbon steel, stainless steel, and aluminum in the overhead 4F position using the GTAW process.

EQUIPMENT AND MATERIALS NEEDED FOR THIS PRACTICE

1. A properly set-up and adjusted GTA welding machine.

2. Proper safety protection.

3. Filler rods, 36-in. (0.9-m) long by 1/16-in. (1.6-mm), 3/32-in. (2.4-mm), and 1/8-in. (3-mm) diameter.

4. Two or more pieces of mild steel, stainless steel, and aluminum, 1 1/2-in. (38-mm) wide by 6-in. (152-mm) long by 1/16-in. (2-mm) and 1/8-in. (3-mm) thick, and aluminum plate 1/4-in. (6-mm) thick.

INSTRUCTIONS

1. Tack weld the plates together in the lap joint configuration with an overlap of 1/4 in. (6 mm) to 3/8 in. (10 mm). Place them in the overhead position.

2. Concentrate the heat and filler metal on the top plate.

3. Gravity and an occasional sweep of the torch along the bottom plate will pull the weld pool down.

4. Control undercutting along the top edge of the weld by putting most of the filler metal along the top edge.

5. Repeat the process using all thicknesses and types of metal until you can consistently make the weld visually defect free.

6. Turn off the welding machine and shielding gas, and clean up your work area when you are finished.

INSTRUCTOR'S COMMENTS _____

_____

_____

■ PRACTICE 17-26

Name _____ Date _____

Base Metal _____ Thickness _____ Filler Dia. _____ Joint _____

Class _____ Instructor _____ Grade _____

OBJECTIVE: After completing this practice you should be able to make a fillet weld on a tee joint on carbon steel, stainless steel, and aluminum in the overhead 4F position using the GTAW process.

EQUIPMENT AND MATERIALS NEEDED FOR THIS PRACTICE

1. A properly set-up and adjusted GTA welding machine.

2. Proper safety protection.

3. Filler rods, 36-in. (0.9-m) long by 1/16-in. (1.6-mm), 3/32-in. (2.4-mm), and 1/8-in. (3-mm) diameter.

4. Two or more pieces of mild steel, stainless steel, and aluminum, 1 1/2-in. (38-mm) wide by 6-in. (152-mm) long by 1/16-in. (2-mm) and 1/8-in. (3-mm) thick, and aluminum plate 1/4-in. (6-mm) thick.

INSTRUCTIONS

1. Tack weld the plates together in the tee joint configuration and place them in the overhead position.

2. Concentrate the heat and filler metal on the top plate. Beware of undercutting.

3. A "J" weave pattern will help pull down any needed metal to the side plate.

4. Repeat the process using all thicknesses and types of metal until you can consistently make the weld visually defect free.

5. Turn off the welding machine and shielding gas, and clean up your work area when you are finished.

INSTRUCTOR'S COMMENTS _____

_____

_____

CHAPTER 17: GAS TUNGSTEN ARC WELDING OF PLATE QUIZ

Name _____ Date _____

Class _____ Instructor _____ Grade _____

Instructions: Carefully read Chapter 17 in the text and answer the following questions.

A. MATCHING

In the space provided to the left, write the letter from Column B that best answers the question or completes the statement in Column A.

1. Match the following filler metals to the correct base metals.

Column A	Column B
BASE METAL	FILLER ROD
_____ a. 316	1. ER309
_____ b. 1030	2. ER316L
_____ c. 304	3. RG60
_____ d. 1010	4. ER70S-3
_____ e. 309	5. ER308

B. SHORT ANSWER

Write a brief answer in the space provided that will answer the question or complete the statement.

2. List seven factors that determine the proper preflow and postflow times required to protect the tungsten and the weld.

a. _____

b. _____

c. _____

d. _____

e. _____

f. _____

g. _____

3. Name four chemicals that can be used for chemical cleaning of metal for the GTAW process.

   a. _____

   b. _____

   c. _____

   d. _____

4. List five methods of mechanically cleaning metal for the GTAW process.

   a. _____

   b. _____

   c. _____

   d. _____

   e. _____

C. FILL IN THE BLANK

Fill in the blank with the correct word. Answers may be more than one word.

5. Welders can have a clear unobstructed view of the molten weld pool because GTA welding is

   _____, _____, and _____.

6. The torch may be angled from _____° to _____° from perpendicular for better visibility and still have the proper shielding gas coverage.

7. The GTAW rod should enter the shielding gas as _____ to the base metal as possible.

8. The tungsten becomes _____ when it touches the molten weld pool or when it is touched by the filler metal.

9. The _____ possible gas flow rates and the _____ preflow or postflow time can help reduce the cost of welding by saving the expensive shielding gas.

10. The _____ flow rates and times must be increased to weld in drafty areas or for out-of-position welds.

11. Most GTA filler metals have some alloys, called _____, that can help prevent porosity caused by gases trapped in the base metal.

12. Using a low arc current setting with faster travel speeds is important when welding stainless steel because some stainless steels are subject to _____.

13. Black crusty spots may appear on weld beads. These spots are often caused by improper cleaning of the filler rod or failure to keep the end of the rod inside the shielding gas.

14. The molten aluminum weld pool has _____, which allows large weld beads to be controlled easily.

15. The _____ of the metal may make starting a weld on thick sections difficult without first preheating the base metal.

16. _____, for metal preparation, may be done by using acids, alkalis, solvents, or detergents.

D. ESSAY

Provide complete answers for all of the following questions.

17. Why is it necessary for the welder to experiment, by trial and error, in order to find the minimum machine amperage settings?

   _____

   _____

18. What effect does torch angle have on the shielding gas coverage?

   _____

   _____

19. When should the tungsten be cleaned?

   _____

   _____

20. From the experiments performed in this chapter, what effect did the cup size have on gas coverage?

   _____

   _____

21. How are mild steel, stainless steel, and aluminum cleaned before GTA welding?

   _____

   _____

22. Why must metals be cleaned before GTA welding?

   _____

   _____

■ PRACTICE 18-4

Name _____ Date _____

Pipe Diameter_____ Filler Metal Type _____ Diameter _____

Class _____ Instructor _____ Grade _____

OBJECTIVE: After completing this practice you should be able to make straight weave and lace beads on mild steel pipe in the horizontal rolled 1G position using the GTAW process.

EQUIPMENT AND MATERIALS NEEDED FOR THIS PRACTICE

1. A properly set-up and adjusted GTA welding machine.

2. Proper safety protection.

3. Filler rods, 36-in. (0.9-m) long by 1/16-in. (1.6-mm), 3/32-in. (2.4-mm), and 1/8-in. (3-mm) diameter.

4. One or more pieces of mild steel pipe, 3 in. (76 mm) to 10 in. (254 mm) in diameter.

INSTRUCTIONS

1. Clean a strip 1-in. (25-mm) wide around the pipe. Place the pipe securely in an angle iron vee block in the flat position.

2. Ensure that you have enough freedom of movement weld from the 3 o'clock to the 12 o'clock position. Hold the torch with a 5° to 10° upward angle. Start the weld in the 3 o'clock position.

3. To make a weave bead move the torch in a "C," "U," or zigzag pattern across the weld pool. Keep the pattern no wider than one-half the diameter of the cup.

4. To make a lace bead move the torch in a zigzag pattern across the pipe. Keep the weld pool about the same size as for a stringer bead. The distance of the zigzag can be anything desired to produce a lace bead of any width.

5. Continue the weave or lace bead up the pipe to the top. Lower the current and fill the crater.

6. Rotate the pipe 90° and, beginning where you left off, repeat the steps above. Do this until you have welded 360° around the pipe.

7. After you have welded all the way around the pipe, visually inspect the bead. It must be straight, have good contour, and be visually defect free.

8. Turn off the welding machine and shielding gas, and clean up your work area when you are finished.

INSTRUCTOR'S COMMENTS _____

_____

_____

■ PRACTICE 18-5

Name _____ Date _____

Pipe Diameter_____ Filler Metal Type _____ Diameter _____

Class _____ Instructor _____ Grade _____

OBJECTIVE: After completing this practice you should be able to make a filler pass on mild steel pipe in the horizontal rolled 1G position using the GTAW process.

EQUIPMENT AND MATERIALS NEEDED FOR THIS PRACTICE

1. A properly set-up and adjusted GTA welding machine.

2. Proper safety protection.

3. Filler rods, 36-in. (0.9-m) long by 1/16-in. (1.6-mm), 3/32-in. (2.4-mm), and 1/8-in. (3-mm) diameter.

4. Two or more pieces of schedule 40 mild steel pipe, 3 in. (76 mm) to 10 in. (254 mm) in diameter with single V-groove prepared ends and already having a root pass in place.

INSTRUCTIONS

1. Place the pipe securely in an angle iron vee block in the flat position.

2. Place the cup against the beveled sides of the joint with a 5° to 10° upward angle. Be sure that you have full freedom of movement.

3. Establish the weld pool and use a forward and backward motion as you did in Practice 17-3. Start the weld near the 1 o'clock position and end just beyond the 12 o'clock position.

4. Add the filler metal at the top center of the molten weld pool until the bead surface is flat or slightly convex. Filler passes should have very little penetration so that maximum reinforcement can be added with each pass. Deep penetration will slow the joint fill-up rate.

5. Lower the current and fill the crater. Another method of ending the bead is to slowly decrease the current and pull the bead up on the beveled side of the joint.

6. Turn the pipe and, beginning where you left off, repeat the steps above. Do this until you have welded 360° around the pipe.

7. After you have welded all the way around the pipe, visually inspect the bead.

8. Turn off the welding machine and shielding gas, and clean up your work area when you are finished.

INSTRUCTOR'S COMMENTS _____

_____

_____

■ PRACTICE 18-6

Name _____ Date _____

Pipe Diameter_____ Filler Metal Type _____ Diameter _____

Class _____ Instructor _____ Grade _____

OBJECTIVE: After completing this practice you should be able to make a cover pass on mild steel pipe in the horizontal rolled 1G position using the GTAW process.

EQUIPMENT AND MATERIALS NEEDED FOR THIS PRACTICE

1. A properly set-up and adjusted GTA welding machine.

2. Proper safety protection.

3. Filler rods, 36-in. (0.9-m) long by 1/16 in. (1.6-mm), 3/32-in. (2.4-mm), and 1/8-in. (3-mm) diameter.

4. Two or more pieces of schedule 40 mild steel pipe, 3 in. (76 mm) to 10 in. (254 mm) in diameter with single V-groove prepared ends and already having a root pass and flush filler passes in place.

INSTRUCTIONS

1. Place the pipe securely in an angle iron vee block in the flat position.

2. Place the cup against the beveled sides of the joint with a 5° to 10° upward angle. Be sure that you have full freedom of movement. Start the weld near the 1 o'clock position and end just beyond the 12 o'clock position.

3. Using the same techniques and skill developed in Practice 17-5, make a cover pass.

4. The weld should not be more than 3/32 in. (2.4 mm) wider than the groove and should have a buildup of no more than 3/32 in. (2.4 mm). Fill the weld craters as you go.

5. Turn the pipe and, beginning where you left off, repeat the steps above. Do this until you have welded 360° around the pipe.

6. After you have welded all the way around the pipe, visually inspect the bead.

7. Turn off the welding machine and shielding gas, and clean up your work area when you are finished.

INSTRUCTOR'S COMMENTS _____

_____

_____

■ PRACTICE 18-7

Name _____ Date _____

Pipe Diameter_____ Filler Metal Type _____ Diameter _____

Class _____ Instructor _____ Grade _____

OBJECTIVE: After completing this practice you should be able to make a stringer bead on mild steel pipe in the horizontal fixed 5G position using the GTAW process.

EQUIPMENT AND MATERIALS NEEDED FOR THIS PRACTICE

1. A properly set-up and adjusted GTA welding machine.

2. Proper safety protection.

3. Filler rods, 36-in. (0.9-m) long by 1/16-in. (1.6-mm), 3/32-in. (2.4-mm), and 1/8-in. (3-mm) diameter.

4. One or more pieces of schedule 40 mild steel pipe, 3 in. (76 mm) to 10 in. (254 mm) in diameter.

INSTRUCTIONS

1. Clean a strip 1-in. (25-mm) wide around the pipe.

2. Mark the top for future reference and clamp the pipe at a comfortable work height. Be sure that you have complete freedom of movement.

3. Establish a weld pool at the 6 o'clock position. Keep the weld pool small for controllability and uniformity. Add filler metal at the front leading edge of the weld pool.

4. Move the torch forward and backward as the filler is added. The frequency of movement will increase as the weld becomes more vertical.

5. Continue the weld without stopping, if possible, to the 12 o'clock position.

6. Repeat the weld up the other side of the pipe in the same manner.

7. Inspect the weld for straightness, uniformity, and visual defects.

8. Repeat this practice until you can consistently make welds that are visually defect free.

9. Turn off the welding machine and shielding gas, and clean up your work area when you are finished.

INSTRUCTOR'S COMMENTS _____

_____

_____

■ PRACTICE 18-8

Name _____ Date _____

Pipe Diameter_____ Filler Metal Type _____ Diameter _____

Class _____ Instructor _____ Grade _____

OBJECTIVE: After completing this practice you should be able to make a stringer bead on mild steel pipe in the vertical fixed 2G position using the GTAW process.

EQUIPMENT AND MATERIALS NEEDED FOR THIS PRACTICE

1. A properly set-up and adjusted GTA welding machine.

2. Proper safety protection.

3. Filler rods, 36-in. (0.9-m) long by 1/16-in. (1.6-mm), 3/32-in. (2.4-mm), and 1/8-in. (3-mm) diameter.

4. One or more pieces of schedule 40 mild steel pipe, 3 in. (76 mm) to 10 in. (254 mm) in diameter.

INSTRUCTIONS

1. Clean a strip 1-in. (25-mm) wide around the pipe.

2. Hold the torch at a 5° angle from horizontal and 5° to 10° from perpendicular to the pipe. (Refer to Figure 17-46 in the textbook.)

3. Establish a small weld pool and add filler along the top front edge. Keep the weld pool small for controllability and uniformity.

4. A slight "J" pattern may help to control weld bead sag.

5. Continue the weld without stopping, if possible, all the way around the pipe.

6. Inspect the weld for straightness, uniformity, and visual defects.

7. Repeat this practice until you can consistently make welds that are visually defect free.

8. Turn off the welding machine and shielding gas, and clean up your work area when you are finished.

INSTRUCTOR'S COMMENTS _____

_____

_____

■ PRACTICE 18-9

Name _____ Date _____

Pipe Diameter_____ Filler Metal Type _____ Diameter _____

Class _____ Instructor _____ Grade _____

OBJECTIVE: After completing this practice you should be able to make a stringer bead on mild steel pipe at a 45° inclined angle (6G position) using the GTAW process.

EQUIPMENT AND MATERIALS NEEDED FOR THIS PRACTICE

1. A properly set-up and adjusted GTA welding machine.

2. Proper safety protection.

3. Filler rods, 36-in. (0.9-m) long by 1/16-in. (1.6-mm), 3/32-in. (2.4-mm), and 1/8-in. (3-mm) diameter.

4. One or more pieces of schedule 40 mild steel pipe, 3 in. (76 mm) to 10 in. (254 mm) in diameter.

INSTRUCTIONS

1. Clean a strip 1-in. (25-mm) wide around the pipe.

2. Clamp the pipe at a 45° inclined angle (6G position) to the table.

3. Starting at the 6:30 o'clock position with the torch at a slight downward angle, establish a small molten weld pool.

4. Add the filler metal at the upper leading edge of the molten weld pool. Keep the molten weld pool small. A slight "J" pattern may help to control weld bead shape. As the weld progresses around the pipe, the angle becomes more vertical and the rate of movement should increase.

5. Continue the weld without stopping, if possible, all the way around the pipe.

6. Inspect the weld for straightness, uniformity, and visual defects.

7. Repeat this practice until you can consistently make welds that are visually defect free.

8. Turn off the welding machine and shielding gas, and clean up your work area when you are finished.

INSTRUCTOR'S COMMENTS _____

_____

_____

CHAPTER 18: GAS TUNGSTEN ARC WELDING OF PIPE QUIZ

Name _____ Date _____

Class _____ Instructor _____ Grade _____

Instructions: Carefully read Chapter 18 in the text and answer the following questions.

A. IDENTIFICATION

In the space provided, identify the items shown in the illustration.

1. From the illustrations below, which of them is testing the root of the weld? Circle the best answer.

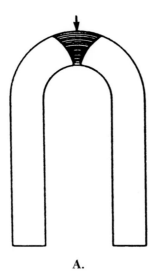

A.

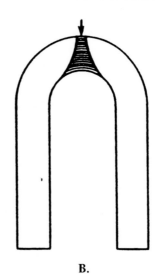

B.

A. is testing the root of the weld.

B. is testing the root of the weld.

2. Identify each of the following parts of a single-V pipe groove.

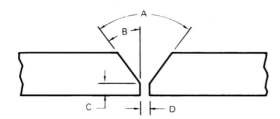

A. _____

B. _____

C. _____

D. _____

B. MATCHING

In the space provided to the left, write the letter from Column B that best answers the question or completes the statement in Column A.

Column A   Column B

_____ 3. What is another name for excessive reinforcement, the   a. class 2
excessive buildup of metal on the back side of a weld?

b. class 5

_____ 4. What occurs when the back side of the root weld is
concave in shape?   c. class 4

_____ 5. What are preplaced filler metals which are used for the root   d. class 1
pass when consistent, high-quality welds are required?

e. root reinforcement

_____ 6. What amount of metal is deposited on the back side of a
welded joint?   f. consumable inserts

_____ 7. What class is a Y-shaped consumable insert?   g. grapes

_____ 8. What class is an A-shaped consumable insert?   h. suck back

_____ 9. What class is a rectangular-shaped consumable insert
(contoured edges)?

_____ 10. What class is a J-shaped consumable insert?

C. SHORT ANSWER

Write a brief answer in the space provided that will answer the question or complete the statement.

11. List three ways of determining excessive reinforcement of a weld.

a. _____

b. _____

c. _____

12. List the four most common root defects.

a. _____

b. _____

c. _____

d. _____

13. List three ways of determining the concavity of a weld.

a. _____

b. _____

c. _____

D. ESSAY

Provide complete answers for all of the following questions.

14. Why does a pipe bevel have a root face?

    _____

    _____

15. What is incomplete fusion?

    _____

    _____

16. Explain how to use a backing gas on both large- and small-diameter pipe.

    _____

    _____

17. Explain the technique of preplacing filler metal.

    _____

    _____

18. For what purpose are consumable inserts used?

    _____

    _____

19. How should a weld bead be ended to prevent problems with weld craters?

    _____

    _____

20. Why should a filler pass not penetrate deeply?

    _____

    _____

E. DEFINITIONS

21. Define the following terms:

    consumable inserts _____

    backing gas _____

    stress points _____

    root penetration _____

INSTRUCTOR'S COMMENTS _____

_____

_____

Chapter 19

Gas Tungsten Arc Welding Plate and Pipe AWS SENSE Certification

AWS SENSE CERTIFICATION

Metal Specifications		Gas Flow			Nozzle Size in. (mm)	Amperage Min. to Max.
Thickness	Diameter of E70S-3*	Rates cfm (L/min)	Preflow Times	Postflow Times		
18 ga	1/16 in. (2 mm)	15 to 20 (7 to 9)	10 to 15 sec	10 to 25 sec	1/4 to 3/8 (6 to 10)	45 to 65
17 ga	1/16 in. (2 mm)	15 to 20 (7 to 9)	10 to 15 sec	10 to 25 sec	1/4 to 3/8 (6 to 10)	45 to 70
16 ga	1/16 in. (2 mm)	15 to 20 (7 to 9)	10 to 15 sec	10 to 25 sec	1/4 to 3/8 (6 to 10)	50 to 75
15 ga	1/16 in. (2 mm)	15 to 20 (7 to 9)	10 to 15 sec	10 to 25 sec	1/4 to 3/8 (6 to 10)	55 to 80
14 ga	3.32 in. (2.4 mm)	20 to 25 (9 to 12)	10 to 20 sec	10 to 30 sec	3/8 to 5/8 (10 to 16)	60 to 90
13 ga	3.32 in. (2.4 mm)	20 to 25 (9 to 12)	10 to 20 sec	10 to 30 sec	3/8 to 5/8 (10 to 16)	60 to 100
12 ga	3.32 in. (2.4 mm)	20 to 25 (9 to 12)	10 to 20 sec	10 to 30 sec	3/8 to 5/8 (10 to 16)	60 to 110
11 ga	3.32 in. (2.4 mm)	20 to 25 (9 to 12)	10 to 20 sec	10 to 30 sec	3/8 to 5/8 (10 to 16)	65 to 120
10 ga	3.32 in. (2.4 mm)	20 to 25 (9 to 12)	10 to 20 sec	10 to 30 sec	3/8 to 5/8 (10 to 16)	70 to 130

*Other E70S-X filler metal may be used.

Metal Specifications		Gas Flow			Nozzle Size in. (mm)	Amperage Min. to Max.
Thickness	Diameter of ER3XX* (in.)	Rates cfm (L/min)	Preflow Times	Postflow Times		
18 ga	1/16 in. (2 mm)	15 to 20 (7 to 9)	10 to 15 sec	10 to 25 sec	1/4 to 3/8 (6 to 10)	35 to 60
17 ga	1/16 in. (2 mm)	15 to 20 (7 to 9)	10 to 15 sec	10 to 25 sec	1/4 to 3/8 (6 to 10)	40 to 65
16 ga	1/16 in. (2 mm)	15 to 20 (7 to 9)	10 to 15 sec	10 to 25 sec	1/4 to 3/8 (6 to 10)	40 to 75
15 ga	1/16 in. (2 mm)	15 to 20 (7 to 9)	10 to 15 sec	10 to 25 sec	1/4 to 3/8 (6 to 10)	50 to 80
14 ga	3.32 in. (2.4 mm)	20 to 25 (9 to 12)	10 to 20 sec	10 to 30 sec	3/8 to 5/8 (10 to 16)	50 to 90
13 ga	3.32 in. (2.4 mm)	20 to 25 (9 to 12)	10 to 20 sec	10 to 30 sec	3/8 to 5/8 (10 to 16)	55 to 100
12 ga	3.32 in. (2.4 mm)	20 to 25 (9 to 12)	10 to 20 sec	10 to 30 sec	3/8 to 5/8 (10 to 16)	60 to 110
11 ga	3.32 in. (2.4 mm)	20 to 25 (9 to 12)	10 to 20 sec	10 to 30 sec	3/8 to 5/8 (10 to 16)	65 to 120
10 ga	3.32 in. (2.4 mm)	20 to 25 (9 to 12)	10 to 20 sec	10 to 30 sec	3/8 to 5/8 (10 to 16)	70 to 130

*Other ER3XX stainless steel A5.9 filler metal may be used.

Metal Specifications		Gas Flow			Nozzle Size in. (mm)	Amperage Min. to Max.
Thickness	Diameter of ER4043*	Rates cfm (L/min)	Preflow Times	Postflow Times		
18 ga	3/32 in. (2.4 mm)	20 to 30 (9 to 14)	10 to 15 sec	10 to 25 sec	1/4 to 3/8 (6 to 10)	40 to 60
17 ga	3/32 in. (2.4 mm)	20 to 30 (9 to 14)	10 to 15 sec	10 to 25 sec	1/4 to 3/8 (6 to 10)	50 to 70
16 ga	3/32 in. (2.4 mm)	20 to 30 (9 to 14)	10 to 15 sec	10 to 25 sec	1/4 to 3/8 (6 to 10)	60 to 75
15 ga	3/32 in. (2.4 mm)	20 to 30 (9 to 14)	10 to 15 sec	10 to 25 sec	1/4 to 3/8 (6 to 10)	65 to 85
14 ga	3/32 in. (2.4 mm)	20 to 30 (9 to 14)	10 to 15 sec	10 to 25 sec	1/4 to 3/8 (6 to 10)	75 to 90
13 ga	1/8 in. (3 mm)	25 to 40 (12 to 19)	10 to 20 sec	10 to 30 sec	3/8 to 5/8 (10 to 16)	85 to 100
12 ga	1/8 in. (3 mm)	25 to 40 (12 to 19)	10 to 20 sec	10 to 30 sec	3/8 to 5/8 (10 to 16)	90 to 110
11 ga	1/8 in. (3 mm)	25 to 40 (12 to 19)	10 to 20 sec	10 to 30 sec	3/8 to 5/8 (10 to 16)	100 to 115
10 ga	1/8 in. (3 mm)	25 to 40 (12 to 19)	10 to 20 sec	10 to 30 sec	3/8 to 5/8 (10 to 16)	100 to 125

*Other aluminum AWS A5.10 filler metal may be used if needed.

■ PRACTICE 19-1

Name _____ Date _____

Class _____ Instructor _____ Grade _____

OBJECTIVE: After completing this practice, you will make GTA welded butt joints, all four positions on 11 gauge to 14 gauge thickness mild steel with 100% joint penetration to be bend tested.

EQUIPMENT AND MATERIALS NEEDED FOR THIS PRACTICE

1. A properly set up and adjusted GTA welding machine.

2. Assorted hand tools, vice, ball-peen hammer, saw or PAC torch, spare parts, and any other required materials.

3. Proper PPE.

4. ER70S-2 and/or ER70S-3 filler rod with 1/16 in. and/or 3/32 in. (1.6 mm and/or 2.4 mm) in diameter.

5. 100% Ar shielding gas.

6. Tungsten electrode ranging from 1/16 in. to 1/8 in. (1.6 mm to 3 mm) in diameter.

7. Two or more pieces of 6 in. long 1-1/2 in. wide (150 mm by 38 mm) 11 to 14 gauge mild steel.

INSTRUCTIONS

1. One at a time tack weld the sheets together to form the joints and place them in the desired position on a welding stand.

2. Starting at one end, run a bead along the joint. Watch the molten weld pool and bead for signs it is increasing or decreasing in size. Make any necessary changes in your technique to keep the weld consistent.

3. When the welds have cooled use a hack saw or thermal cutting torch to cut out 1-in. (25-mm) wide strips as shown in Figure 19-1.

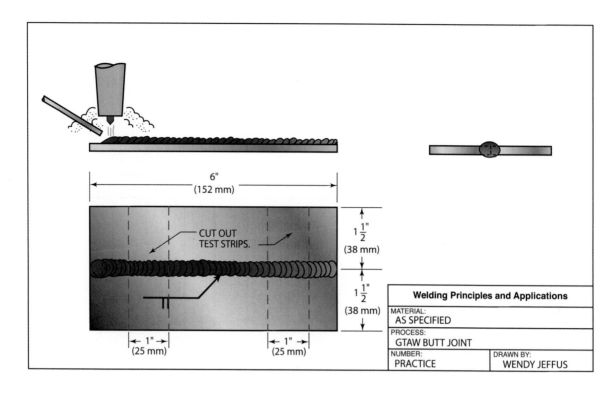

Welding Principles and Applications

MATERIAL:	
AS SPECIFIED	
PROCESS:	
GTAW BUTT JOINT	
NUMBER:	DRAWN BY:
PRACTICE	WENDY JEFFUS

4. Using a vice and a hammer bend the strip of welded joint as shown in Figure 19-2.

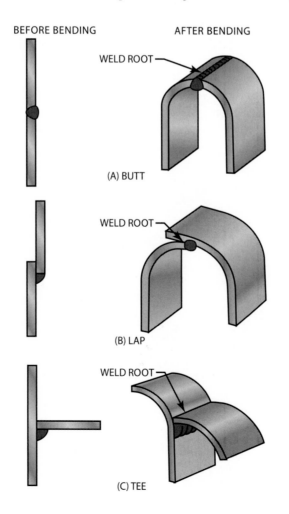

BEFORE BENDING   AFTER BENDING

WELD ROOT

(A) BUTT

WELD ROOT

(B) LAP

WELD ROOT

(C) TEE

5. Check for complete root fusion on the lap and tee joints and 100% penetration on the butt joint.

6. Repeat each type of joint in each position as needed until consistently good beads are obtained. Turn off the welding machine and shielding gas and clean up your work area when you are finished welding.

INSTRUCTOR'S COMMENTS _____

_____

_____

■ PRACTICE 19-2

Name _____ Date _____

Class _____ Instructor _____ Grade _____

OBJECTIVE: After completing this practice, you will make GTA welded lap and tee joints, all four positions on 11 gauge to 14 gauge thickness mild steel with 100% joint penetration to be bend tested.

EQUIPMENT AND MATERIALS NEEDED FOR THIS PRACTICE

1. A properly set up and adjusted GTA welding machine.

2. Assorted hand tools, vice, ball-peen hammer, saw or PAC torch, spare parts, and any other required materials.

3. Proper PPE.

4. ER70S-2 and/or ER70S-3 filler rod with 1/16 in. and/or 3/32 in. (1.6 mm and/or 2.4 mm) in diameter.

5. 100% Ar shielding gas.

6. Tungsten electrode ranging from 1/16 in. to 1/8 in. (1.6 mm to 3 mm) in diameter.

7. Two or more pieces of 6 in. long 1-1/2 in. wide (150 mm by 38 mm) 11 to 14 gauge mild steel.

INSTRUCTIONS

1. One at a time tack weld the sheets together to form the joints and place them in the desired position on a welding stand.

2. Starting at one end, run a bead along the joint. Watch the molten weld pool and bead for signs it is increasing or decreasing in size. Make any necessary changes in your technique to keep the weld consistent.

3. When the welds have cooled use a hack saw or thermal cutting torch to cut out 1-in. (25-mm) wide strips as shown in Figure 19-3 and Figure 19-4.

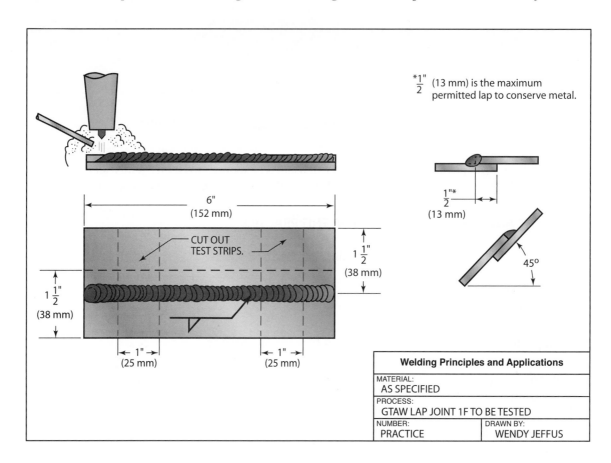

$*\frac{1}{2}"$ (13 mm) is the maximum permitted lap to conserve metal.

6" (152 mm)

CUT OUT TEST STRIPS.

$1\frac{1}{2}"$ (38 mm)

$1\frac{1}{2}"$ (38 mm)

1" (25 mm)

1" (25 mm)

$\frac{1}{2}"^*$ (13 mm)

45°

Welding Principles and Applications

MATERIAL:
AS SPECIFIED

PROCESS:
GTAW LAP JOINT 1F TO BE TESTED

NUMBER:
PRACTICE

DRAWN BY:
WENDY JEFFUS

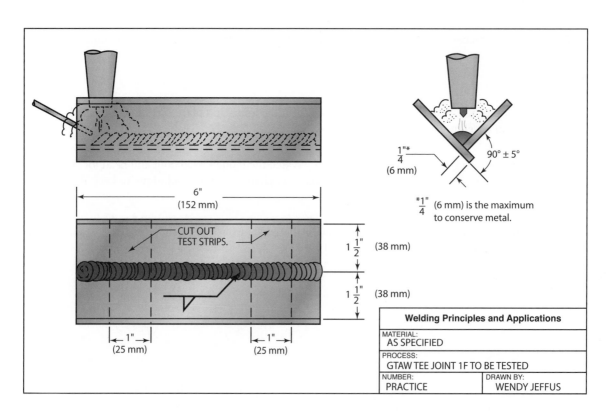

6" (152 mm)

CUT OUT TEST STRIPS.

$1\frac{1}{2}"$ (38 mm)

$1\frac{1}{2}"$ (38 mm)

1" (25 mm)

1" (25 mm)

$\frac{1}{4}"^*$ (6 mm)

90° ± 5°

$*\frac{1}{4}"$ (6 mm) is the maximum to conserve metal.

Welding Principles and Applications

MATERIAL:
AS SPECIFIED

PROCESS:
GTAW TEE JOINT 1F TO BE TESTED

NUMBER:
PRACTICE

DRAWN BY:
WENDY JEFFUS

4. Using a vice and a hammer bend the strip of welded joint as shown in Figure 19-2.

5. Check for complete root fusion on the lap and tee joints and 100% penetration on the butt joint.

6. Repeat each type of joint in each position as needed until consistently good beads are obtained. Turn off the welding machine and shielding gas and clean up your work area when you are finished welding.

INSTRUCTOR'S COMMENTS _____

_____

_____

■ PRACTICE 19-3

Name _____ Date _____

Class _____ Instructor _____ Grade _____

OBJECTIVE: After completing this practice, you will make GTA welded butt joints, all four positions on 11 gauge to 14 gauge thickness stainless steel with 100% joint penetration to be bend tested.

EQUIPMENT AND MATERIALS NEEDED FOR THIS PRACTICE

1. A properly set up and adjusted GTA welding machine.

2. Assorted hand tools, vice, ball-peen hammer, saw or PAC torch, spare parts, and any other required materials.

3. Proper PPE.

4. ER3XX stainless steel filler rod with 1/16 in. and/or 3/32 in. (1.6 mm and/or 2.4 mm) in diameter.

5. 100% Ar shielding gas.

6. Tungsten electrode ranging from 1/16 in. to 1/8 in. (1.6 mm to 3 mm) in diameter.

7. Two or more pieces of 6 in. long 1-1/2 in. wide (150 mm by 38 mm) 11 to 14 gauge stainless steel.

> **NOTE: Mild steel can be used in place of stainless steel as long as stainless steel filler metal is used.**

INSTRUCTIONS

1. One at a time tack weld the sheets together to form the joints and place them in the desired position on a welding stand.

2. Starting at one end, run a bead along the joint. Watch the molten weld pool and bead for signs it is increasing or decreasing in size. Make any necessary changes in your technique to keep the weld consistent.

3. When the welds have cooled use a hack saw or thermal cutting torch to cut out 1-in. (25-mm) wide strips.

4. Using a vice and a hammer bend the strip of welded joint.

5. Check for complete root fusion on the lap and tee joints and 100% penetration on the butt joint.

6. Repeat each type of joint in each position as needed until consistently good beads are obtained. Turn off the welding machine and shielding gas and clean up your work area when you are finished welding.

INSTRUCTOR'S COMMENTS _____

_____

_____

■ PRACTICE 19-4

Name _____ Date _____

Class _____ Instructor _____ Grade _____

OBJECTIVE: After completing this practice, you will make GTA welded lap and tee joints, all four positions on 11 gauge to 14 gauge thickness stainless steel with 100% joint penetration to be bend tested.

EQUIPMENT AND MATERIALS NEEDED FOR THIS PRACTICE

1. A properly set up and adjusted GTA welding machine.

2. Assorted hand tools, vice, ball-peen hammer, saw or PAC torch, spare parts, and any other required materials.

3. Proper PPE.

4. ER3XX stainless steel filler rod with 1/16 in. and/or 3/32 in. (1.6 mm and/or 2.4 mm) in diameter.

5. 100% Ar shielding gas.

6. Tungsten electrode ranging from 1/16 in. to 1/8 in. (1.6 mm to 3 mm) in diameter.

7. Two or more pieces of 6 in. long 1-1/2 in. wide (150 mm by 38 mm) 11 to 14 gauge stainless steel.

> **NOTE: Mild steel can be used in place of stainless steel as long as stainless steel filler metal is used.**

INSTRUCTIONS

1. One at a time tack weld the sheets together to form the joints and place them in the desired position on a welding stand.

2. Starting at one end, run a bead along the joint. Watch the molten weld pool and bead for signs it is increasing or decreasing in size. Make any necessary changes in your technique to keep the weld consistent.

3. When the welds have cooled use a hack saw or thermal cutting torch to cut out 1-in. (25-mm) wide strips.

4. Using a vice and a hammer bend the strip of welded joint.

5. Check for complete root fusion on the lap and tee joints and 100% penetration on the butt joint.

6. Repeat each type of joint in each position as needed until consistently good beads are obtained. Turn off the welding machine and shielding gas and clean up your work area when you are finished welding.

INSTRUCTOR'S COMMENTS _____

_____

_____

■ PRACTICE 19-5

Name _____ Date _____

Class _____ Instructor _____ Grade _____

OBJECTIVE: After completing this practice, you will make GTA welded butt joints, all four positions on 11 gauge to 14 gauge thickness aluminum with 100% joint penetration to be bend tested.

EQUIPMENT AND MATERIALS NEEDED FOR THIS PRACTICE

1. A properly set up and adjusted GTA welding machine.

2. Assorted hand tools, vice, ball-peen hammer, saw or PAC torch, spare parts, and any other required materials.

3. Proper PPE.

4. ER4043 filler rod with 1/16 in. and/or 3/32 in. (1.6 mm and/or 2.4 mm) in diameter.

5. 100% Ar shielding gas.

6. Tungsten electrode ranging from 1/16 in. to 1/8 in. (1.6 mm to 3 mm) in diameter.

7. Two or more pieces of 6 in. long 1-1/2 in. wide (150 mm by 38 mm) 11 to 14 aluminum.

INSTRUCTIONS

1. One at a time tack weld the sheets together to form the joints and place them in the desired position on a welding stand.

2. Starting at one end, run a bead along the joint. Watch the molten weld pool and bead for signs it is increasing or decreasing in size. Make any necessary changes in your technique to keep the weld consistent.

3. When the welds have cooled use a hack saw or thermal cutting torch to cut out 1-in. (25-mm) wide strips.

4. Using a vice and a hammer bend the strip of welded joint.

5. Check for complete root fusion on the lap and tee joints and 100% penetration on the butt joint.

6. Repeat each type of joint in each position as needed until consistently good beads are obtained. Turn off the welding machine and shielding gas and clean up your work area when you are finished welding.

INSTRUCTOR'S COMMENTS _____

_____

_____

■ PRACTICE 19-6

Name _____ Date _____

Class _____ Instructor _____ Grade _____

OBJECTIVE: After completing this practice, you will make GTA welded lap and tee joints, all four positions on 11 gauge to 14 gauge thickness aluminum with 100% joint penetration to be bend tested.

EQUIPMENT AND MATERIALS NEEDED FOR THIS PRACTICE

1. A properly set up and adjusted GTA welding machine.

2. Assorted hand tools, vice, ball-peen hammer, saw or PAC torch, spare parts, and any other required materials.

3. Proper PPE.

4. ER4043 filler rod with 1/16 in. and/or 3/32 in. (1.6 mm and/or 2.4 mm) diameter.

5. 100% Ar shielding gas.

6. Tungsten electrode ranging from 1/16 in. to 1/8 in. (1.6 mm to 3 mm) in diameter.

7. Two or more pieces of 6 in. long 1-1/2 in. wide (150 mm by 38 mm) 11 to 14 aluminum.

INSTRUCTIONS

1. One at a time tack weld the sheets together to form the joints and place them in the desired position on a welding stand.

2. Starting at one end, run a bead along the joint. Watch the molten weld pool and bead for signs it is increasing or decreasing in size. Make any necessary changes in your technique to keep the weld consistent.

3. When the welds have cooled use a hack saw or thermal cutting torch to cut out 1-in. (25-mm) wide strips.

4. Using a vice and a hammer bend the strip of welded joint.

5. Check for complete root fusion on the lap and tee joints and 100% penetration on the butt joint.

6. Repeat each type of joint in each position as needed until consistently good beads are obtained. Turn off the welding machine and shielding gas and clean up your work area when you are finished welding.

INSTRUCTOR'S COMMENTS _____

_____

_____

■ PRACTICE 19-7 AWS SENSE

GAS TUNGSTEN ARC WELDING (GTAW) ON PLAIN CARBON STEEL WORKMANSHIP SAMPLE

Name _____ Date _____

Class _____ Instructor _____ Grade _____

Welding Procedure Specification (WPS) No.: Practice 19-7.

TITLE:
Welding GTAW of sheet to sheet.

SCOPE:
This procedure is applicable for square groove and fillet welds within the range of 18 gauge through 10 gauge.

WELDING MAY BE PERFORMED IN THE FOLLOWING POSITIONS:
1G and 2F.

BASE METAL:
The base metal shall conform to carbon steel M-1, Group 1.

BACKING MATERIAL SPECIFICATION:
None.

FILLER METAL:
The filler metal shall conform to AWS specification no. #70S-3 for 1/16 in. (1.6 mm) to 3/32 in. (2.4 mm) diameter as listed in AWS specification A5.18. This filler metal falls into F-number F-6 and A-number A-1.

ELECTRODE:
The tungsten electrode shall conform to AWS specification no. EWTh-2, EWCe-2, or EWLa from AWS specification A5.12. The tungsten diameter shall be 1/8 in. (3.2 mm) maximum.

The tungsten end shape shall be tapered at two to three times its length to its diameter.

SHIELDING GAS:
The shielding gas, or gases, shall conform to the following compositions and purity: welding grade argon.

JOINT DESIGN AND TOLERANCES:
Refer to the drawing and specifications in Figure 19-5 for the workmanship sample layout.

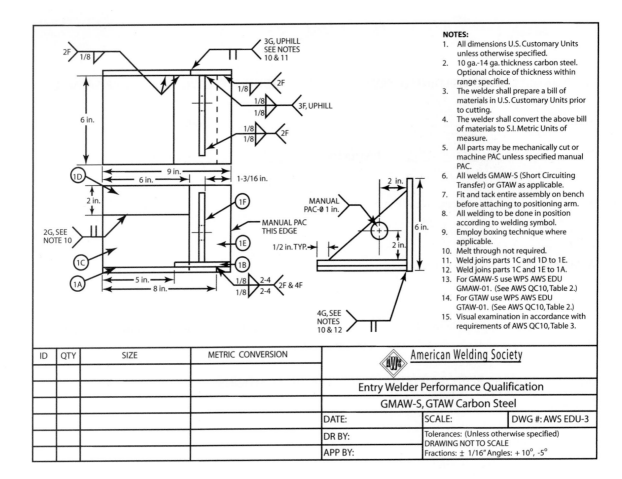

NOTES:
1. All dimensions U.S. Customary Units unless otherwise specified.
2. 10 ga.-14 ga. thickness carbon steel. Optional choice of thickness within range specified.
3. The welder shall prepare a bill of materials in U.S. Customary Units prior to cutting.
4. The welder shall convert the above bill of materials to S.I. Metric Units of measure.
5. All parts may be mechanically cut or machine PAC unless specified manual PAC.
6. All welds GMAW-S (Short Circuiting Transfer) or GTAW as applicable.
7. Fit and tack entire assembly on bench before attaching to positioning arm.
8. All welding to be done in position according to welding symbol.
9. Employ boxing technique where applicable.
10. Melt through not required.
11. Weld joins parts 1C and 1D to 1E.
12. Weld joins parts 1C and 1E to 1A.
13. For GMAW-S use WPS AWS EDU GMAW-01. (See AWS QC10, Table 2.)
14. For GTAW use WPS AWS EDU GTAW-01. (See AWS QC10, Table 2.)
15. Visual examination in accordance with requirements of AWS QC10, Table 3.

ID	QTY	SIZE	METRIC CONVERSION	American Welding Society
				Entry Welder Performance Qualification
				GMAW-S, GTAW Carbon Steel

DATE:	SCALE:	DWG #: AWS EDU-3
DR BY:	Tolerances: (Unless otherwise specified) DRAWING NOT TO SCALE	
APP BY:	Fractions: ± 1/16" Angles: + 10°, -5°	

PREPARATION OF BASE METAL:

All hydrocarbons and other contaminants, such as cutting fluids, grease, oil, and primers, must be cleaned off all parts and filler metals before welding. This cleaning can be done with any suitable solvents or detergents. The joint face and inside and outside plate surface within 1 in. (25 mm) of the joint must be mechanically cleaned of slag, rust, and mill scale. Cleaning must be done with a wire brush or grinder down to bright metal.

ELECTRICAL CHARACTERISTICS:

Set the welding current to DCEN and the amperage and shielding gas flow according to Table 19-1.

PREHEAT:

The parts must be heated to a temperature higher than 50°F (10°C) before any welding is started.

BACKING GAS:

None.

SAFETY:

Proper protective clothing and equipment must be used. The area must be free of all hazards that may affect the welder or others in the area. The welding machine, welding leads, work clamp, electrode holder, and other equipment must be in safe working order.

WELDING TECHNIQUE:

Tack Welds: With the parts securely clamped in place with the correct root gap, the tack welds are to be performed. Holding the electrode so that it is very close to the root face but not touching, slowly increase the current until the arc starts and a molten weld pool is formed. Add filler metal as required to maintain a slightly convex weld face and a flat or slightly concave root face. When it is time to end the tack weld, lower the current slowly so that the molten weld pool can be tapered down in size. When all tack welds are complete, allow the parts to cool as needed before assembling the remaining parts. Repeat the tack welding procedure until the entire part is assembled.

Square Groove and Fillet Welds: Holding the electrode so that it is very close to the metal surface but not touching, slowly increase the current until the arc starts and a molten weld pool is formed. As the weld progresses, add filler metal as required to maintain a flat or slightly convex weld face. If it is necessary to stop the weld or to reposition yourself or if the weld is completed, the current must be lowered slowly so that the molten weld pool can be tapered down in size.

INTERPASS TEMPERATURE:

The plate should not be heated to a temperature higher than 120°F (49°C) during the welding process. After each weld pass is completed, allow it to cool but never to a temperature below 50°F (10°C). The weldment must not be quenched in water.

CLEANING:

Recleaning may be required if the parts or filler metal becomes contaminated or reoxidized to a degree that the weld quality will be affected. Reclean using the same procedure used for the original metal preparation.

VISUAL INSPECTION:

Visual inspection criteria for entry welders:

- There shall be no cracks, no incomplete fusion.
- There shall be no incomplete joint penetration in groove welds except as permitted for partial joint penetration groove welds.
- The Test Supervisor shall examine the weld for acceptable appearance and shall be satisfied that the welder is skilled in using the process and procedure specified for the test.
- Undercut shall not exceed the lesser of 10% of the base metal thickness or 1/32 in. (0.8 mm).
- Where visual examination is the only criterion for acceptance, all weld passes are subject to visual examination at the discretion of the Test Supervisor.
- The frequency of porosity shall not exceed one in each 4 in. (100 mm) of weld length, and the maximum diameter shall not exceed 3/32 in. (2.4 mm).
- Welds shall be free from overlap.

SKETCHES:

Gas Tungsten Arc Welding (GTAW) Workmanship Sample drawing for Carbon Steel.

INSTRUCTOR'S COMMENTS _____

_____

_____

■ PRACTICE 19-8 AWS SENSE

GAS TUNGSTEN ARC WELDING (GTAW) ON STAINLESS STEEL WORKMANSHIP SAMPLE

Name _____ Date _____

Class _____ Instructor _____ Grade _____

Welding Procedure Specification (WPS) No.: Practice 19-8.

TITLE:

Welding GTAW of sheet to sheet.

SCOPE:

This procedure is applicable for square groove and fillet welds within the range of 18 gauge through 10 gauge

WELDING MAY BE PERFORMED IN THE FOLLOWING POSITIONS:

1G and 2F.

BASE METAL:

The base metal shall conform to austenitic stainless steel M-8 or P-8.

BACKING MATERIAL SPECIFICATION:

None.

FILLER METAL:

The filler metal shall conform to AWS specification no. ER3XX from AWS specification A5.9. This filler metal falls into F-number F-6 and A-number A-8.

ELECTRODE:

The tungsten electrode shall conform to AWS specification no. EWTh-2, EWCe-2, or EWLa from AWS specification A5.12. The tungsten diameter shall be 1/8 in. (3.2 mm) maximum. The tungsten end shape shall be tapered at two to three times its length to its diameter.

SHIELDING GAS:

The shielding gas, or gases, shall conform to the following compositions and purity: welding grade argon.

JOINT DESIGN AND TOLERANCES:

Refer to the drawing and specifications in Figure 19-6 for the workmanship sample layout.

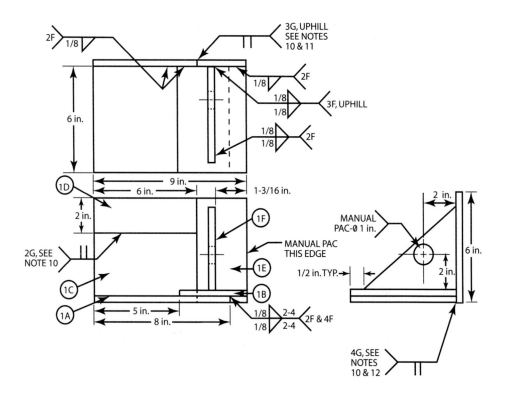

PREPARATION OF BASE METAL:

All hydrocarbons and other contaminants, such as cutting fluids, grease, oil, and primers, must be cleaned off all parts and filler metals before welding. This cleaning can be done with any suitable solvents or detergents. The joint face and inside and outside plate surface within 1 in. (25 mm) of the joint must be cleaned of slag, oxide, and scale. Cleaning can be mechanical or chemical. Mechanical metal cleaning can be done by grinding, stainless steel wire brushing, scraping, machining, or filing. Chemical cleaning can be done by using acids, alkalis, solvents, or detergents. Cleaning must be done down to bright metal.

ELECTRICAL CHARACTERISTICS:

Set the welding current to DCEN and the amperage and shielding gas flow according to Table 19-2.

PREHEAT:

The parts must be heated to a temperature higher than 50°F (10°C) before any welding is started.

BACKING GAS:

None.

SAFETY:

Proper protective clothing and equipment must be used. The area must be free of all hazards that may affect the welder or others in the area. The welding machine, welding leads, work clamp, electrode holder, and other equipment must be in safe working order.

WELDING TECHNIQUE:

Tack Welds: With the parts securely clamped in place with the correct root gap, the tack welds are to be performed. Holding the electrode so that it is very close to the root face but not touching, slowly increase the current until the arc starts and a molten weld pool is formed. Add filler metal as required to maintain a slightly convex weld face and a flat or slightly concave root face. When it is time to end the tack weld, lower the current slowly so that the molten weld pool can be tapered down in size. When all tack welds are complete, allow the parts to cool as needed before assembling the remaining parts. Repeat the tack welding procedure until the entire part is assembled.

Square Groove and Fillet Welds: Holding the electrode so that it is very close to the metal surface but not touching, slowly increase the current until the arc starts and a molten weld pool is formed. As the weld progresses, add filler metal as required to maintain a flat or slightly convex weld face. If it is necessary to stop the weld or to reposition yourself or if the weld is completed, the current must be lowered slowly so that the molten weld pool can be tapered down in size.

INTERPASS TEMPERATURE:

The plate should not be heated to a temperature higher than 350°F (180°C) during the welding process. After each weld pass is completed, allow it to cool but never to a temperature below 50°F (10°C). The weldment must not be quenched in water.

CLEANING:

Recleaning may be required if the parts or filler metal become contaminated or oxidized to a degree that the weld quality will be affected. Reclean using the same procedure used for the original metal preparation.

VISUAL INSPECTION:

Visual inspection criteria for entry welders:

- There shall be no cracks, no incomplete fusion.
- There shall be no incomplete joint penetration in groove welds except as permitted for partial joint penetration groove welds.
- The Test Supervisor shall examine the weld for acceptable appearance and shall be satisfied that the welder is skilled in using the process and procedure specified for the test.
- Undercut shall not exceed the lesser of 10% of the base metal thickness or 1/32 in. (0.8 mm).
- Where visual examination is the only criterion for acceptance, all weld passes are subject to visual examination at the discretion of the Test Supervisor.
- The frequency of porosity shall not exceed one in each 4 in. (100 mm) of weld length and the maximum diameter shall not exceed 3/32 in. (2.4 mm).
- Welds shall be free from overlap.

SKETCHES:

Gas Tungsten Arc Welding (GTAW) Workmanship Sample drawing for Stainless Steel.

INSTRUCTOR'S COMMENTS _____

_____

_____

■ PRACTICE 19-9 AWS SENSE

GAS TUNGSTEN ARC WELDING (GTAW) ON ALUMINUM WORKMANSHIP SAMPLE

Name _____ Date _____

Class _____ Instructor _____ Grade _____

Welding Procedure Specification (WPS) No.: Practice 19-9.

TITLE:

Welding GTAW of sheet to sheet.

SCOPE:

This procedure is applicable for square groove and fillet welds within the range of 18 gauge through 10 gauge.

WELDING MAY BE PERFORMED IN THE FOLLOWING POSITIONS:

1G and 2F.

BASE METAL:

The base metal shall conform to aluminum M-22 or P-22.

Backing Material Specification: None.

FILLER METAL:

The filler metal shall conform to AWS specification no. ER4043 from AWS specification A5.10. This filler metal falls into F-number F-22 and A-number A-5.10.

ELECTRODE:

The tungsten electrode shall conform to AWS specification no. EWCe-2, EWZr, EWLa, or EWP from AWS specification A5.12. The tungsten diameter shall be 1/8 in. (3.2 mm) maximum. The tungsten end shape shall be rounded.

SHIELDING GAS:

The shielding gas, or gases, shall conform to the following compositions and purity: welding grade argon.

JOINT DESIGN AND TOLERANCES:

Refer to the drawing and specifications in Figure 19-7 for the workmanship sample layout.

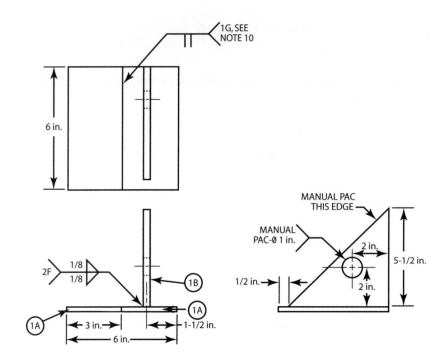

PREPARATION OF BASE METAL:

All hydrocarbons and other contaminants, such as cutting fluids, grease, oil, and primers, must be cleaned off all parts and filler metals before welding. This cleaning can be done with any suitable solvents or detergents. The joint face and inside and outside plate surface within 1 in. (25 mm) of the joint must be mechanically or chemically cleaned of oxides. Mechanical cleaning may be done by stainless steel wire brushing, scraping, machining, or filing. Chemical cleaning may be done by using acids, alkalis, solvents, or detergents. Because the oxide layer may reform quickly and affect the weld, welding should be started within 10 minutes of cleaning.

ELECTRICAL CHARACTERISTICS:

Set the welding current to AC high-frequency stabilized and the amperage and shielding gas flow according to Table 19-3.

PREHEAT:

The parts must be heated to a temperature higher than 50°F (10°C) before any welding is started.

BACKING GAS:

N/A.

SAFETY:

Proper protective clothing and equipment must be used. The area must be free of all hazards that may affect the welder or others in the area. The welding machine, welding leads, work clamp, electrode holder, and other equipment must be in safe working order.

WELDING TECHNIQUE:

The welder's hands or gloves must be clean and oil free to prevent contamination of the metal or filler rods.

Tack Welds: With the parts securely clamped in place with the correct root gap, the tack welds are to be performed. Holding the electrode so that it is very close to the root face but not touching, slowly increase the current until the arc starts and a molten weld pool is formed. Add filler metal as required to maintain a slightly convex weld face and a flat or slightly concave root face. When it is time to end the tack weld, lower the current slowly so that the molten weld pool can be tapered down in size. When all tack welds are complete, allow the parts to cool as needed before assembling the remaining parts. Repeat the tack welding procedure until the entire part is assembled.

Square Groove and Fillet Welds: Holding the electrode so that it is very close to the metal surface but not touching, slowly increase the current until the arc starts and a molten weld pool is formed. As the weld progresses, add filler metal as required to maintain a flat or slightly convex weld face. If it is necessary to stop the weld or to reposition yourself or the weld is completed, the current must be lowered slowly so that the molten weld pool can be tapered down in size.

INTERPASS TEMPERATURE:

The plate should not be heated to a temperature higher than 120°F (49°C) during the welding process. After each weld pass is completed, allow it to cool but never to a temperature below 50°F (10°C). The weldment must not be quenched in water.

CLEANING:

Recleaning may be required if the parts or filler metal become contaminated or oxidized to a degree that the weld quality will be affected. Reclean using the same procedure used for the original metal preparation.

VISUAL INSPECTION:

Visual inspection criteria for entry welders:

- There shall be no cracks, no incomplete fusion.
- There shall be no incomplete joint penetration in groove welds except as permitted for partial joint penetration groove welds.
- The Test Supervisor shall examine the weld for acceptable appearance and shall be satisfied that the welder is skilled in using the process and procedure specified for the test.
- Undercut shall not exceed the lesser of 10% of the base metal thickness or 1/32 in. (0.8 mm).
- Where visual examination is the only criterion for acceptance, all weld passes are subject to visual examination at the discretion of the Test Supervisor.
- The frequency of porosity shall not exceed one in each 4 in. (100 mm) of weld length, and the maximum diameter shall not exceed 3/32 in. (2.4 mm).
- Welds shall be free from overlap.

SKETCHES:

Gas Tungsten Arc Welding (GTAW) Workmanship Sample drawing for Aluminum.

INSTRUCTOR'S COMMENTS _____

_____

_____

■ PRACTICE 19-10 AWS SENSE

Name _____ Date _____

Class _____ Instructor _____ Grade _____

OBJECTIVE: After completing this practice, you will make GTA welded butt pipe joints, in 2G and 5G positions on 1 in. to 2-7/8 in. (25 mm to 70 mm) 10 gauge to 18 gauge thickness mild steel tubing with 100% joint penetration.

EQUIPMENT AND MATERIALS NEEDED FOR THIS PRACTICE

1. A properly set up and adjusted GTA welding machine.

2. Assorted hand tools, spare parts, and any other required materials.

3. Proper PPE.

4. ER70S-2 and/or ER70S-3 filler rod with 1/16 in. and/or 3/32 in. (1.6 mm and/or 2.4 mm) in diameter.

5. 100% Ar shielding gas.

6. Tungsten electrode ranging from 1/16 in. to 1/8 in. (1.6 mm to 3 mm) in diameter.

7. Two or more pieces of 3 in. long 1 in. to 2-7/8 in. (25 mm to 70 mm) 10 gauge to 18 gauge thickness mild steel tubing.

8. One or more backing rings for the size of tubing being welded.

INSTRUCTIONS

1. Tack weld the tubing together with the proper root opening, Figure 19-8. Make the tack welds at the 12:00, 3:00, 6:00, and 9:00 o'clock positions.

2. Grind the ends of the tack welds to a featheredge.

3. Attach the pipe section in the proper position to the welding stand at a comfortable height.

   NOTE: You may use the cup walking or freehand technique for making these welds.

2G Position (Pipe is Vertical and the Weld is Horizontal)

4. Start welding on one of the tack welds by resting your gloved hand with the torch on the pipe or pipe stand.

5. Once the tack weld is melted, slowly add filler metal to the leading edge so that the keyhole stays about the same size as the root opening. If the keyhole increases in size reduce the power setting or add filler metal faster.

6. Stop your weld when you reach the next tack weld and reposition to continue the weld.

5G Position (Pipe is Horizontal and the Weld is Vertical)

7. Starting your weld at the 3:00 or 9:00 o'clock positions and welding upward is an easier way of developing the skills of watching the keyhole and adding filler metal than starting at the 6:00 o'clock position.

8. Start welding on one of the tack welds by resting your gloved hand with the torch on the pipe or pipe stand.

9. Once the tack weld is melted, slowly add filler metal to the leading edge so that the keyhole stays about the same size as the root opening. If the keyhole increases in size reduce the power setting or add filler metal faster.

10. Stop your weld when you reach the next tack weld and reposition to continue the weld.

11. Repeat each of joint in each position as needed until consistently good beads are obtained. Turn off the welding machine and shielding gas and clean up your work area when you are finished welding.

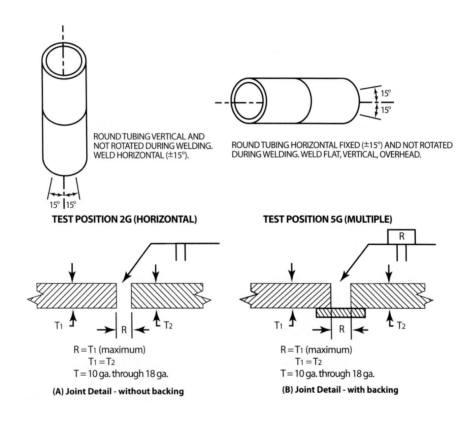

ROUND TUBING VERTICAL AND NOT ROTATED DURING WELDING. WELD HORIZONTAL (±15°).

ROUND TUBING HORIZONTAL FIXED (±15°) AND NOT ROTATED DURING WELDING. WELD FLAT, VERTICAL, OVERHEAD.

TEST POSITION 2G (HORIZONTAL)

TEST POSITION 5G (MULTIPLE)

$R = T_1$ (maximum)
$T_1 = T_2$
$T = 10$ ga. through 18 ga.

(A) Joint Detail - without backing

$R = T_1$ (maximum)
$T_1 = T_2$
$T = 10$ ga. through 18 ga.

(B) Joint Detail - with backing

INSTRUCTOR'S COMMENTS _____

_____

_____

■ PRACTICE 19-11 AWS SENSE

Name _____ Date _____

Class _____ Instructor _____ Grade _____

OBJECTIVE: After completing this practice, you will make GTA welded butt pipe joints, in 2G and 5G positions on 1 in. to 2-7/8 in. (25 mm to 70 mm) 10 gauge to 18 gauge thickness stainless steel tubing with 100% joint penetration.

EQUIPMENT AND MATERIALS NEEDED FOR THIS PRACTICE

1. A properly set up and adjusted GTA welding machine.

2. Assorted hand tools, spare parts, and any other required materials.

3. Proper PPE.

4. ER3XX filler rod with 1/16 in. and/or 3/32 in. (1.6 mm and/or 2.4 mm) in diameter.

5. 100% Ar shielding gas.

6. Tungsten electrode ranging from 1/16 in. to 1/8 in. (1.6 mm to 3 mm) in diameter.

7. Two or more pieces of 3 in. long 1 in. to 2-7/8 in. (25 mm to 70 mm) 10 gauge to 18 gauge thickness stainless steel tubing.

8. One or more backing rings for the size of tubing being welded.

INSTRUCTIONS

1. Tack weld the tubing together with the proper root opening, Figure 19-3. Make the tack welds at the 12:00, 3:00, 6:00, and 9:00 o'clock positions.

2. Grind the ends of the tack welds to a featheredge.

3. Attach the pipe section in the proper position to the welding stand at a comfortable height.

   NOTE: You may use the cup walking or freehand technique for making these welds.

2G Position (Pipe is Vertical and the Weld is Horizontal)

4. Start welding on one of the tack welds by resting your gloved hand with the torch on the pipe or pipe stand.

5. Once the tack weld is melted, slowly add filler metal to the leading edge so that the keyhole stays about the same size as the root opening. If the keyhole increases in size reduce the power setting or add filler metal faster.

6. Stop your weld when you reach the next tack weld and reposition to continue the weld.

5G Position (Pipe is Horizontal and the Weld is Vertical)

7. Starting your weld at the 3:00 or 9:00 o'clock positions and welding upward is an easier way of developing the skills of watching the keyhole and adding filler metal than starting at the 6:00 o'clock position.

8. Start welding on one of the tack welds by resting your gloved hand with the torch on the pipe or pipe stand.

9. Once the tack weld is melted, slowly add filler metal to the leading edge so that the keyhole stays about the same size as the root opening. If the keyhole increases in size reduce the power setting or add filler metal faster.

10. Stop your weld when you reach the next tack weld and reposition to continue the weld.

11. Repeat each of joint in each position as needed until consistently good beads are obtained. Turn off the welding machine and shielding gas and clean up your work area when you are finished welding.

INSTRUCTOR'S COMMENTS _____

_____

_____

■ PRACTICE 19-12

Name _____ Date _____

Class _____ Instructor _____ Grade _____

OBJECTIVE: After completing this practice, you will make GTA welded butt pipe joints, in 2G and 5G positions on 1 in. to 2-7/8 in. (25 mm to 70 mm) 10 gauge to 18 gauge thickness aluminum tubing with 100% joint penetration.

EQUIPMENT AND MATERIALS NEEDED FOR THIS PRACTICE

1. A properly set up and adjusted GTA welding machine.

2. Assorted hand tools, spare parts, and any other required materials.

3. Proper PPE.

4. ER4043 or ER5XXX filler rod with 1/16 in. and/or 3/32 in. (1.6 mm and/or 2.4 mm) in diameter.

5. 100% Ar shielding gas.

6. Tungsten electrode ranging from 1/16 in. to 1/8 in. (1.6 mm to 3 mm) in diameter.

7. Two or more pieces of 3 in. long 1 in. to 2-7/8 in. (25 mm to 70 mm) 10 gauge to 18 gauge thickness aluminum tubing.

8. One or more backing rings for the size of tubing being welded.

INSTRUCTIONS

1. Tack weld the tubing together with the proper root opening, Figure 19-3. Make the tack welds at the 12:00, 3:00, 6:00, and 9:00 o'clock positions.

2. Grind the ends of the tack welds to a featheredge.

3. Attach the pipe section in the proper position to the welding stand at a comfortable height.

   NOTE: You may use the cup walking or freehand technique for making these welds.

2G Position (Pipe is Vertical and the Weld is Horizontal)

4. Start welding on one of the tack welds by resting your gloved hand with the torch on the pipe or pipe stand.

5. Once the tack weld is melted, slowly add filler metal to the leading edge so that the keyhole stays about the same size as the root opening. If the keyhole increases in size reduce the power setting or add filler metal faster.

6. Stop your weld when you reach the next tack weld and reposition to continue the weld.

5G Position (Pipe is Horizontal and the Weld is Vertical)

7. Starting your weld at the 3:00 or 9:00 o'clock positions and welding upward is an easier way of developing the skills of watching the keyhole and adding filler metal than starting at the 6:00 o'clock position.

8. Start welding on one of the tack welds by resting your gloved hand with the torch on the pipe or pipe stand.

9. Once the tack weld is melted, slowly add filler metal to the leading edge so that the keyhole stays about the same size as the root opening. If the keyhole increases in size reduce the power setting or add filler metal faster.

10. Stop your weld when you reach the next tack weld and reposition to continue the weld.

11. Repeat each of joint in each position as needed until consistently good beads are obtained. Turn off the welding machine and shielding gas and clean up your work area when you are finished welding.

INSTRUCTOR'S COMMENTS _____

_____

_____

■ PRACTICE 19-13

Name _____ Date _____

Pipe Diameter_____ Filler Metal Type _____ Diameter _____

Class _____ Instructor _____ Grade _____

OBJECTIVE: After completing this practice you should be able to make a pipe weld on mild steel pipe in the horizontal rolled 1G position using the GTAW process. The weld is to be tested.

EQUIPMENT AND MATERIALS NEEDED FOR THIS PRACTICE

1. A properly set-up and adjusted GTA welding machine.

2. Proper safety protection.

3. Filler rods, 36-in. (0.9-m) long by 1/16-in. (1.6-mm), 3/32-in. (2.4-mm), and 1/8-in. (3-mm) diameter.

4. Two or more pieces of schedule 40 mild steel pipe, 3 in. (76 mm) to 10 in. (254 mm) in diameter with single V-groove prepared ends.

INSTRUCTIONS

1. Using the skills you have developed so far, tack weld two pieces of pipe together, then do a root pass, filler passes (until flush), and a cover pass.

2. After you have completed all passes all the way around the pipe, visually inspect the bead.

3. If it passes the visual inspection, cut out guided-bend test specimens.

4. Subject the specimens to the bend test.

5. Repeat this practice until all the specimens consistently pass the bend test.

6. Turn off the welding machine and shielding gas, and clean up your work area when you are finished.

INSTRUCTOR'S COMMENTS _____

_____

_____

■ PRACTICE 19-14

Name _____ Date _____

Pipe Diameter_____ Filler Metal Type _____ Diameter _____

Class _____ Instructor _____ Grade _____

OBJECTIVE: After completing this practice you should be able to make a single-V butt joint bead on mild steel pipe in the horizontal fixed 5G position using the GTAW process, 100% root penetration to be tested.

EQUIPMENT AND MATERIALS NEEDED FOR THIS PRACTICE

1. A properly set-up and adjusted GTA welding machine.

2. Proper safety protection.

3. Filler rods, 36-in. (0.9-m) long by 1/16-in. (1.6-mm), 3/32-in. (2.4-mm), and 1/8-in. (3-mm) diameter.

4. Two or more pieces of schedule 40 mild steel pipe, 3 in. (76 mm) to 10 in. (254 mm) in diameter having the ends prepared with a single V-groove.

INSTRUCTIONS

1. Tack weld the pipes together in the 12:00, 3:00, 6:00, and 9:00 o'clock positions. Clamp the pipe in the 5G position.

2. Start the weld at the 6:30 o'clock position and weld uphill around the pipe to the 12:30 o'clock position. Place the cup against the pipe bevels and add the filler rod at the leading edge of the molten weld pool.

3. Repeat this process up the other side of the pipe. The starts and stops should overlap slightly to ensure a good tie-in.

4. The filler passes should also start at the 6:30 o'clock position and go up to the 12:30 o'clock position. Stagger the bead locations to prevent defects arising from discontinuities at the start and stop points.

5. Use a lace or weave bead and put on a cover pass.

6. Inspect the weld for straightness, uniformity, and visual defects. Repeat this practice until you can consistently make welds that are visually defect free.

7. Cut out guided-bend test specimens and test them. Repeat this practice until you can consistently make welds that pass the bend test.

8. Turn off the welding machine and shielding gas, and clean up your work area when you are finished.

INSTRUCTOR'S COMMENTS _____

_____

_____

■ PRACTICE 19-15

Name _____ Date _____

Pipe Diameter_____ Filler Metal Type _____ Diameter _____

Class _____ Instructor _____ Grade _____

OBJECTIVE: After completing this practice you should be able to make a single-V butt joint bead on mild steel pipe in the vertical fixed 2G position using the GTAW process, 100% root penetration to be tested.

EQUIPMENT AND MATERIALS NEEDED FOR THIS PRACTICE

1. A properly set-up and adjusted GTA welding machine.

2. Proper safety protection.

3. Filler rods, 36-in. (0.9-m) long by 1/16-in. (1.6-mm), 3/32-in. (2.4-mm), and 1/8-in. (3-mm) diameter.

4. Two or more pieces of schedule 40 mild steel pipe, 3 in. (76 mm) to 10 in. (254 mm) in diameter having the ends prepared with a single V-groove.

INSTRUCTIONS

1. Tack weld the pipes together in the 12:00, 3:00, 6:00, and 9:00 o'clock positions. Clamp the pipe in the 2G position.

2. Start the root weld at a point between the tack welds. The root pass should be small enough that surface tension will hold it in place.

3. The filler passes can be larger if they are made along the lower beveled surface which will support the bead. The next pass goes on top side of the first, and so forth.

4. The cover pass is started around the lower side of the joint, overlapping the pipe surface by no more than 1/8 in. (3 mm). The next pass covers about one-half of the first pass and should be slightly larger. This process of making each pass larger continues until the center of the weld is reached, then each weld is made successively smaller.

5. Inspect the weld for uniformity and visual defects. Repeat this practice until you can consistently make welds that are visually defect free.

6. Cut out guided-bend test specimens and test them. Repeat this practice until you can consistently make welds that pass the bend test.

7. Turn off the welding machine and shielding gas, and clean up your work area when you are finished.

INSTRUCTOR'S COMMENTS _____

_____

_____

■ PRACTICE 19-16

Name _____ Date _____

Pipe Diameter_____ Filler Metal Type _____ Diameter _____

Class _____ Instructor _____ Grade _____

OBJECTIVE: After completing this practice you should be able to make a single-V butt joint bead on mild steel pipe at a 45° fixed inclined angle (6G position) using the GTAW process, 100% root penetration to be tested.

EQUIPMENT AND MATERIALS NEEDED FOR THIS PRACTICE

1. A properly set-up and adjusted GTA welding machine.

2. Proper safety protection.

3. Filler rods, 36-in. (0.9-m) long by 1/16-in. (1.6-mm), 3/32-in. (2.4-mm), and 1/8-in. (3-mm) diameter.

4. Two or more pieces of schedule 40 mild steel pipe, 3 in. (76 mm) to 10 in. (254 mm) in diameter having the ends prepared with a single V-groove.

INSTRUCTIONS

1. Tack weld the pipes together in the 12:00, 3:00, 6:00, and 9:00 o'clock positions. Clamp the pipe in the 6G position.

2. Start the root weld at the bottom and at a point between the tack welds. The root pass should be small enough that surface tension will hold it in place. The root pass may be off to one side.

3. The filler passes are applied to the downhill side first so that the upper side will be supported. Stagger the starts and stops to ensure against proximity of defects.

4. The cover pass is easy to control if it is a series of stringer beads. Start on the lower side and build up the cover pass as you did in the 2G position. Each weld should overlap the preceding weld for continuity.

5. Inspect the weld for uniformity and visual defects. Repeat this practice until you can consistently make welds that are visually defect free.

6. Cut out guided-bend test specimens and test them. Repeat this practice until you can consistently make welds that pass the bend test.

7. Turn off the welding machine and shielding gas, and clean up your work area when you are finished.

INSTRUCTOR'S COMMENTS _____

_____

_____

CHAPTER 19: GAS TUNGSTEN ARC WELDING PLATE AND PIPE AWS SENSE QUIZ

Name _____ Date _____

Class _____ Instructor _____ Grade _____

Instructions: Carefully read Chapter 19 in the text and answer the following questions.

A. SHORT ANSWER

Write a brief answer in the space provided that will answer the question or complete the statement.

1. List the visual inspection acceptance criteria for pipe butt joint welds.

    _____

    _____

    _____

    _____

    _____

    _____

2. List the visual inspection acceptance criteria for tubing fillet welds.

    _____

    _____

    _____

    _____

    _____

    _____

3. List the visual inspection acceptance criteria for tubing butt joint welds.

    _____

    _____

    _____

    _____

    _____

    _____

4. Describe how fillet welds are to be tested.

_____

_____

_____

_____

_____

_____

5. How much of the side of a weld groove needs to be cleaned?

_____

B. FILL IN THE BLANK

Fill in the blank with the correct word. Answers may be more than one word.

6. The free bend test is _____ designed to help you identify any root problems that might exist in your weld.

7. Often it is possible to carefully grind off the _____ as a way of conserving tungsten.

8. For Practice 19-3, you will need: properly set up and adjusted GTA welding machine, proper safety protection including all required PPE, ER _____ (stainless steel).

9. On Practice 19-7 you add filler metal as required to maintain a slightly _____ weld face and a flat or slightly concave root face.

10. Practice 19-12 is an AWS SENSE _____ GTA welding workmanship qualification practice.

INSTRUCTOR'S COMMENTS _____

_____

_____

Chapter 20

Shop Math and Weld Cost

■ PRACTICE 20-1

Name _____ Date _____

Class _____ Instructor _____ Grade _____

OBJECTIVE: After completing this practice, you will be able to determine the total volume of groove welds having various dimensions.

EQUIPMENT AND MATERIALS NEEDED FOR THIS PRACTICE

Piece of paper, pencil, calculator, and weld groove dimensions.

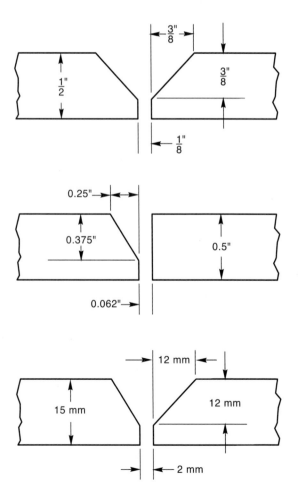

INSTRUCTIONS

1. Using Formulas 20-1, 20-2, 20-3, and 20-4 from the text, you will figure the total volume given the following groove weld dimensions.

2. Using Formula 20-1, determine the cross-sectional area for the given groove joints.

3. Using Formula 20-2, determine the cross-sectional area for the root opening.

4. Using Formula 20-3, determine the total cross-sectional area for each weld groove.

5. Plug the results from Formula 20-1 and Formula 20-2 into Formula 20-4. This will give you the total groove volume.

6. V-groove joint with the following dimensions:

   Width 3/8 in. _____

   Depth 3/8 in. _____

   Root opening 1/8 in. _____

   Thickness 1/2 in. _____

   Weld length 144 in. _____

7. Single bevel joint with the following dimensions:

   Width 0.25 in. _____

   Depth 0.375 in. _____

   Root opening 0.062 in. _____

   Thickness 0.5 in. _____

   Weld length 96 in. _____

8. V-groove joint with the following dimensions:

   Width 12 mm _____

   Depth 12 mm _____

   Root opening 2 mm _____

   Thickness 15 mm _____

   Weld length 3600 mm _____

INSTRUCTOR'S COMMENTS _____

_____

_____

■ PRACTICE 20-2

Name _____ Date _____

Class _____ Instructor _____ Grade _____

OBJECTIVE: After completing this practice, you will be able to calculate the weight of metal required for each of the welds described in Practice 20-1. Calculate the weight for both steel and aluminum base and filler metals.

EQUIPMENT AND MATERIALS NEEDED FOR THIS PRACTICE

Paper, pencil, calculator, and from the textbook, Table 20-7 Densities of Metals.

INSTRUCTIONS

Each example given is of a groove-type weld. Therefore, use Formula 20-4 and determine the GV (groove volume). Multiply the resulting groove volume by the appropriate metal density given in Table 20-7. Figure each weld groove for both aluminum and steel.

INSTRUCTOR'S COMMENTS _____

_____

_____

■ PRACTICE 20-3

Name _____ Date _____

Class _____ Instructor _____ Grade _____

OBJECTIVE: After completing this practice, you will be able to calculate the cost per pound of filler metal as it would be used to create a bill of materials.

EQUIPMENT AND MATERIALS NEEDED FOR THIS PRACTICE

Paper, pencil, and calculator.

INSTRUCTIONS

Use the data supplied below in A. and B. to fill out the information in the Total Cost per Pound of Weld Metal Deposited.

Total Cost per Pound of Weld Metal Deposited.
Welding Process:_____
Filler metal: AWS Number _____ Diameter _____
Welding Current: Amperage _____ Voltage _____

LABOR and OVERHEAD	(Labor & Overhead Cost/hr) _____ (Deposition Rate [ib/hr]) X (Operating Factor)	=	_____
ELECTRODE	(Electrode Cost/ib) _____ (Deposition Efficiency)	=	_____
GAS	(Gas Flow Rate [cfh]) _____ X (Gas Cost/cu ft) _____ (Deposition Rate [ib/hr])	=	_____
	Total Cost/ib of Weld Metal Deposited (Sum of the Above)		_____

A.

	FCAW
Electrode Type	.045-in. dia. E71T-2
Labor & Overhead	$25.00/hr
Welding Current	200 amperes
Deposition Rate	5.5 lb/hr
Operation Factor	45%
Electrode Cost	$1.55/lb.
Deposition Efficiency	85%
Gas Flow Rate	35 cfh
Gas Cost per Cu Ft	$.04 CO_2

B.

SMAW

Electrode Type	1/8-in. dia. E7018
Labor & Overhead	$23.00/hr
Welding Current	120 amperes
Deposition Rate	2.58 lb/hr
Operation Factor	30%
Electrode Cost	$0.97/lb
Deposition Efficiency	71.6%
Gas Flow Rate	Not Applicable
Gas Cost per Cu Ft	Not Applicable

INSTRUCTOR'S COMMENTS _____

_____

_____

CHAPTER 20: SHOP MATH AND WELD COST QUIZ

Name _____ Date _____

Class _____ Instructor _____ Grade _____

Instructions: Carefully read Chapter 20 in the text and answer the following questions.

A. SHOP MATH

Calculate the answers to the following math questions.

1. add 1/2 + 5/8 = _____, 7/16 + 1 1/8 = _____, 2 9/16 + 1 15/16 = _____

2. add 2.5 + 6.4 = _____, 0.3 + 8.5 = _____, 0.6 + 0.84 = _____

3. subtract 7/8 − 3/8 = _____, 7 5/16 − 3 7/16 = _____, 5 3/8 − 2 3/4 = _____

4. subtract 6.8 − 3.4 = _____, 5.5 − 3.7 = _____, 5.8 − 4.75 = _____

5. reduction 1 1/8 = _____, 1 15/4 = _____, 8 19/16 = _____

6. rounding (2 decimal places) 3.789 = _____, 4.3514 = _____, 7.958 = _____

7. fraction to decimals 3/4 = _____, 5/8 = _____, 11/16 = _____

8. decimal to fractions 0.25 = _____, 0.375 = _____, 0.76 = _____

9. in. to mm 1 in. = _____ mm, 1.5 in. = _____ mm, 1 3/4 in. = _____ mm

10. mm to in. 13 mm = _____ in., 8 mm = _____ in., 300 mm = _____ in.

B. MATCHING

In the blank space provided, write the letter from Column B that best answers the question or completes the statement in Column A.

Column A	Column B
11. A(n) _____ is a mathematical statement in which both sides are equal to each other.	a. English
12. In the equation from the textbook hrs × $ = T, the symbol T stands for _____.	b. SI
13. A _____ is a mathematical statement of the relationship of items _____.	c. total labor bill
14. Most welding shops use this system for dimensioning _____.	d. formula
15. The abbreviation for le Système international d'unités is _____.	e. equation

C. SOLVE THE FOLLOWING

Calculate the weights for the following pieces of steel assuming that the steel weighs 490 lb (222.215 kg) per cubic foot:

a. 10 ft long by 48 in. wide and 1 in. thick: _____

b. 8 ft long by 36 in. wide by 3/8 in. thick: _____

c. 120 in. long by 48 in. wide by 1/2 in. thick: _____

d. 96 in. long by 36 in. wide by 5/8 in. thick: _____

INSTRUCTOR'S COMMENTS _____

_____

_____

Chapter 21

Reading Technical Drawings

■ PRACTICE 21-1

Name _____ Date _____

Class _____ Instructor _____ Grade _____

OBJECTIVE: After completing this practice, you will be able to sketch a series of 6-in. long straight lines.

EQUIPMENT AND MATERIALS NEEDED FOR THIS PRACTICE

You will need a pencil and unlined paper.

INSTRUCTIONS

Using a pencil and unlined paper, you are going to sketch a series of 6-in. long horizontal straight lines spaced approximately 1/2 in. to 3/4 in. apart. Practice sketching from the right to the left and left to the right. The direction you sketch is not as important as your ability to make straight lines.

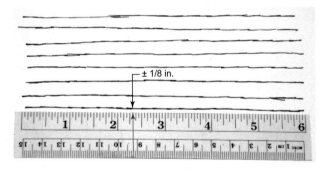

Once you have completed six or eight lines, lay a straightedge next to the lines and see how straight you were able to make them. Keep practicing sketching straight lines until you are able to make 6-in. long lines that are within ±1/8 in. of being straight.

INSTRUCTOR'S COMMENTS _____

_____

_____

■ PRACTICE 21-2

Name _____ Date _____

Class _____ Instructor _____ Grade _____

OBJECTIVE: After completing this practice, you will be able to sketch a series of circles.

EQUIPMENT AND MATERIALS NEEDED FOR THIS PRACTICE

You will need a pencil and unlined paper.

INSTRUCTIONS

1. Start by sketching two light construction lines that cross at right angles.

2. Make two marks on each of the construction lines about 1/2 in. from the center point. These points will serve as your aiming points as you sketch the circle. The circle you sketch will be tangent to these points. A tangent straight line is one that meets a circle at a point where the circular line and straight line are going in the same direction, much like placing a 12-in. ruler on a round pipe; where the ruler and pipe meet is the tangent point. If you were to make a short, straight line at the tangent point and keep doing this all the way around the pipe, you would wind up with a circle drawn from a series of short, straight lines.

3. Sketching a tangent line starting at the top mark, keep sketching and gradually turn the line toward the mark on the next construction line. Once you have completed the first quarter of the circle, you may find it easier to continue if you turn the paper.

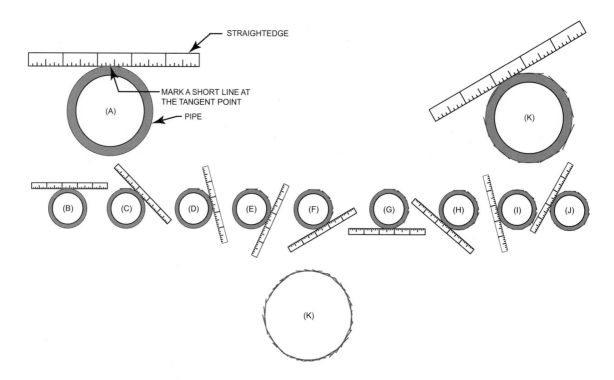

4. Repeat the sketching process until you have completed sketching the circle.

5. Repeat this process making several different size circles. Use a circle template to check your circles for accuracy.

INSTRUCTOR'S COMMENTS _____

_____

_____

■ PRACTICE 21-3

Name _____ Date _____

Class _____ Instructor _____ Grade _____

OBJECTIVE: After completing this practice, you will be able to sketch a mechanical drawing showing three views of a block, as shown in Figure 21-21 in the text.

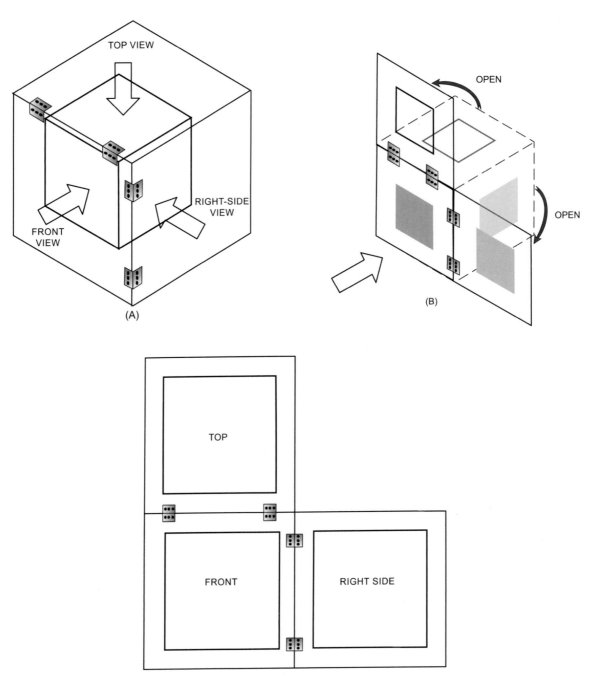

EQUIPMENT AND MATERIALS NEEDED FOR THIS PRACTICE

You will need a pencil and unlined paper.

INSTRUCTIONS

1. Start by sketching construction lines, as shown in Figure 21-22A in the text. These lines will form the boxes for the front, top, and right-side views.

   Darken the lines that make up the object's lines so it is easier to see.

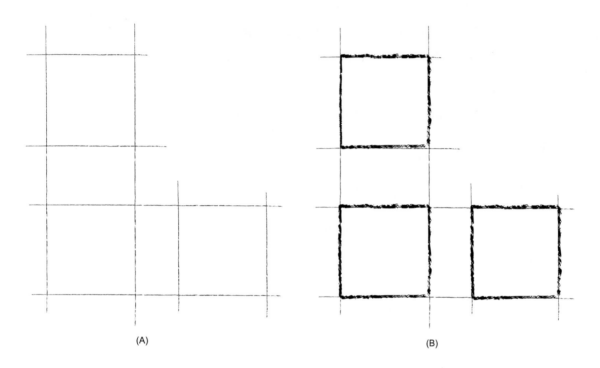

(A)   (B)

■ PRACTICE 21-4

Name _____ Date _____

Class _____ Instructor _____ Grade _____

OBJECTIVE: After completing this practice, the student will be able to sketch a three-view mechanical drawing of a candlestick holder.

EQUIPMENT AND MATERIALS NEEDED FOR THIS PRACTICE

You will need a pencil and unlined paper.

INSTRUCTIONS

1. Using Figure 21-24, in the text, you will sketch a three-view mechanical drawing of the candlestick holder.

2. To lay out the angle for the candlestick holder, draw a vertical centerline that is 8 in. long.

3. Measure down the correct distance from the top, and draw the 4-in. long bottom line centered on the centerline.

4. Connecting the endpoints of the top and bottom lines will automatically give the angle.

5. Repeat the process using the candlestick holder shown in Figure 21-25 in the text.

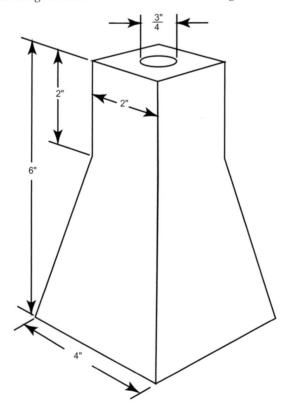

INSTRUCTOR'S COMMENTS _____

_____

_____

■ PRACTICE 21-5

Name _____ Date _____

Class _____ Instructor _____ Grade _____

OBJECTIVE: After completing this practice, the student will be able to sketch irregular shapes and curves using graphing paper.

EQUIPMENT AND TOOLS NEEDED FOR THIS PRACTICE

You will need a pencil and graphing paper.

INSTRUCTIONS

1. Locate a series of points on the graph paper that coincide with points on the curve you are copying. Start with the easy points where the lines on the paper cross at a point on the object.

   For example, one end of the curve starts at the intersection of lines E-3, so put a dot there. Next, the curve is tangent to lines F-2, so put a dot there. Follow the curve around, putting additional dots at the other intersecting points.

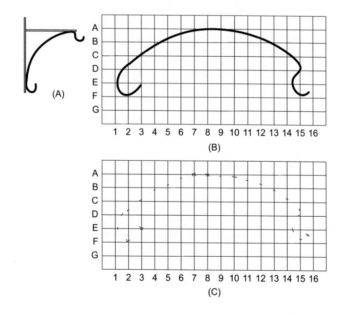

2. Once all of the easy dots are located, make some estimates for the next series of dots. For example, the curve almost touches the 1 line as it crosses the E line. Put a dot there and at similar points where the curve crosses other lines.

3. After you have located all of the points for the curve, sketch a line through all of the points. Refer to the figure below to see how the line should pass through the point. If you are not sure, you can always add additional points that are not on lines to help guide your sketching.

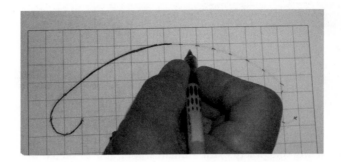

4. Repeat this practice and make a three-view drawing of the candlestick holder shown in Figure 21-31.

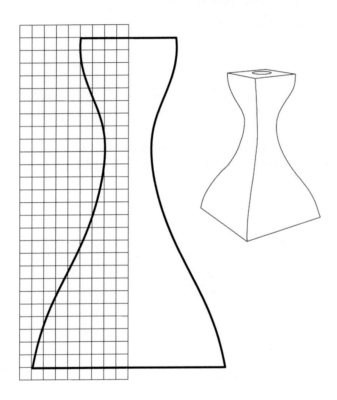

INSTRUCTOR'S COMMENTS _____

_____

_____

CHAPTER 21: READING TECHNICAL DRAWINGS QUIZ

Name _____ Date _____

Class _____ Instructor _____ Grade _____

Instructions: Carefully read Chapter 21 in the text and answer the following questions.

A. MATCHING

In the space provided to the left, write the letter from Column B that best answers the question or completes the statement in Column A.

Column A

Column B

1. A drawing that appears as though you were looking through the sides of a glass box at the object and tracing its shape on the glass is called a _____.

a. isometric drawing

2. A drawing that is drawn at a 30° angle so that it appears that you are looking at one corner is a _____.

b. front view

3. _____ drawings present the object in a more realistic or understandable form.

c. orthographic projection

4. The _____ of an object gives the best overall description of the object.

d. pictorial

B. ESSAY

List the names of 10 different alphabet of lines.

5. _____

6. _____

7. _____

8. _____

9. _____

10. _____

11. _____

12. _____

13. _____

14. _____

C. SHORT ANSWER

15. What is the purpose of a section view? _____

16. What is the purpose of a detail view?_____

17. Length dimensions can be found on what views? _____

18. Height dimensions can be found on what views? _____

19. To aid in making and reading scaled drawings, what type of drafting tool is used?

_____

20. Architectural scales are divided into _____.

INSTRUCTOR'S COMMENTS _____

_____

_____

Chapter 22

Welding Joint Design and Welding Symbols

■ PRACTICE 22-1

Name _____ Date _____

Class _____ Instructor _____ Grade _____

OBJECTIVE: After completing this practice, you should be able identify the elements that make up a welding symbol.

EQUIPMENT AND MATERIALS NEEDED FOR THIS PRACTICE

Pencil.

INSTRUCTIONS

In the space provided, identify the items shown in the illustration.

_____ 1. Field weld symbol

_____ 2. Groove angle; included angle of countersink for plug welds

_____ 3. Reference line

_____ 4. Basic weld symbol or detail reference

_____ 5. Pitch (center-to-center spacing) of welds

_____ 6. Specification, process, or other reference of filling

_____ 7. Finish symbol

_____ 8. Tail

_____ 9. Length of weld

_____ 10. Arrow connecting reference line to arrow side member of joint

_____ 11. Weld-all-around symbol

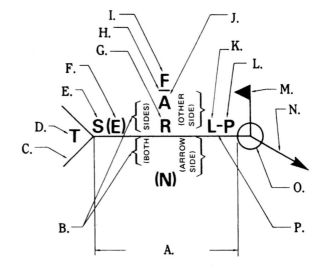

_____ 12. Depth of preparation; size or strength for certain welds

_____ 13. Elements in this area remain as shown when the tail and arrow are reversed

_____ 14. Root opening; depth for plug and slot welds

_____ 15. Contour symbol

_____ 16. Effective throat

INSTRUCTOR'S COMMENTS _____

_____

_____

CHAPTER 22: WELDING JOINT DESIGN AND WELDING SYMBOLS QUIZ

Name _____ Date _____

Class _____ Instructor _____ Grade _____

Instructions: Carefully read Chapter 22 in the text and answer the following questions.

A. IDENTIFICATION

In the space provided, identify the items shown in the illustration.

_____ 1. Tail

_____ 2. Basic weld symbol or detail reference

_____ 3. Elements in this area remain as shown when tail and arrow are reversed

_____ 4. Groove angle; included angle of countersink for plug welds

_____ 5. Effective throat

_____ 6. Root opening; depth for plug and slot welds

_____ 7. Depth of preparation; size or strength for certain welds

_____ 8. Field weld symbol

_____ 9. Weld all around symbol

_____ 10. Contour symbol

_____ 11. Length of weld

_____ 12. Finish symbol

_____ 13. Reference line

_____ 14. Specification, process, or other reference of filling

_____ 15. Arrow connecting reference line to arrow side member of joint

_____ 16. Pitch (center to center spacing) of welds

B. SHORT ANSWER

Write a brief answer in the space provided that will answer the question or complete the statement.

17. Identify the following abbreviations.

    a. MT _____

    b. PT _____

    c. VT _____

    d. NRT _____

    e. PRT _____

    f. UT _____

    g. ET _____

    h. RT _____

    i. DPT _____

    j. AET _____

    k. LT _____

    l. FPT _____

18. List eight classifications of welds.

    a. _____

    b. _____

    c. _____

    d. _____

    e. _____

    f. _____

    g. _____

    h. _____

19. List the 5 basic pipe welding positions?

    a. _____

    b. _____

    c. _____

    d. _____

    e. _____

C. DRAW

In the space provided, make a pencil drawing to illustrate the question.

20. From the following weld symbols, draw the welds in their appropriate locations by making a sketch of the joint.

A.
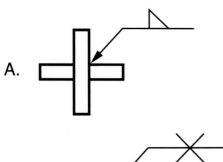

B.

C.
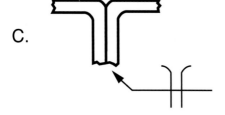

D.

E.
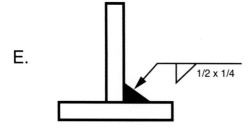

F.
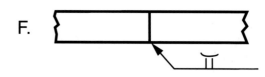

G.
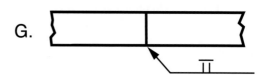

H.
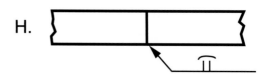

I.

J.
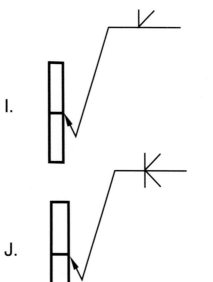

21. Sketch a V-grooved butt joint and label.

22. Sketch a weld on plates in the 1G and 1F positions.

23. Sketch a weld on plates in the 2G and 2F positions.

24. Sketch a weld on plates in the 3G and 3F positions.

25. Sketch a weld on plates in the 4G and 4F positions.

26. Sketch a weld on a pipe in the 1G position.

27. Sketch a weld on a pipe in the 5G position.

28. Sketch a weld on a pipe in the 2G position.

29. Sketch a weld on a pipe in the 6G position.

30. Sketch a weld on a pipe in the 6GR position

D. ESSAY

Provide complete answers for all of the following questions.

31. What are two effects that joint design has on the completed weld?

   a. _____

   b. _____

32. Why do joint designs require compromises?

   _____

   _____

33. How are welding symbols like a language for welders?

   _____

   _____

34. What are some of the specific problems metals can have that would require careful weld joint selection?

   _____

   _____

35. When is the tail omitted from a welding symbol?

   _____

   _____

36. What information could be provided in the tail of a welding symbol?

_____

_____

37. How is the size of a fillet weld shown on a welding symbol?

_____

_____

38. What is an intermittent fillet weld?

_____

_____

39. What is the main purpose of the root face?

_____

_____

40. What types of metal might use a flange weld?

_____

_____

INSTRUCTOR'S COMMENTS _____

_____

_____

■ PRACTICE 23-2

Name _____ Date _____

Class _____ Instructor _____ Grade _____

OBJECTIVE: After completing this practice, you should be able to lay out circles, arcs, and curves to within ±1/16 in. tolerance in both standard units and S.I. units. Refer to Figure 23-20 in the textbook.

EQUIPMENT AND MATERIALS NEEDED FOR THIS PRACTICE

Piece of metal or paper, soapstone or pencil, tape measure, compass or circle template, square.

INSTRUCTIONS

1. If you are using a piece of paper, draw two lines to form a 90° (right) angle near one corner of the paper. This will be your baseline.

2. Measure along the baseline or edge of the metal and lay out the lengths and centerline for each part.

3. Use the square or squared edge of the metal and lay out the widths and centerline for each part.

4. Connect the marks you made for each part by drawing a line between the laid out points.

5. Use a compass or circle template to lay out the circles and arcs. Be sure that the arcs meet the straight object lines so that they form a smooth tangent.

6. Convert the dimensions into S.I. units and repeat the layout process.

INSTRUCTOR'S COMMENTS _____

_____

_____

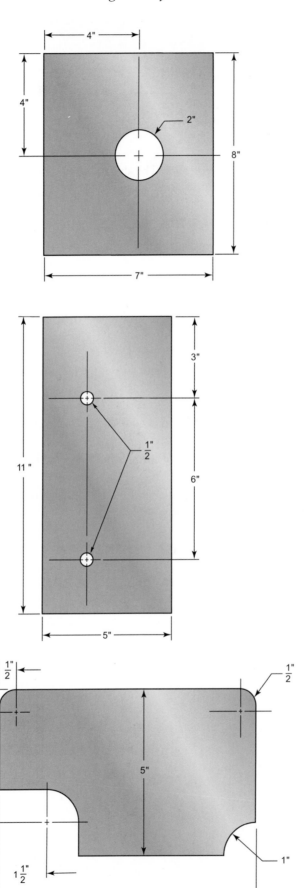

■ PRACTICE 23-3

Name _____ Date _____

Class _____ Instructor _____ Grade _____

OBJECTIVE: After completing this practice, you should be able to lay out parts, nesting them together so that the least amount of material will be wasted. Refer to Figure 23-22 in the textbook.

EQUIPMENT AND MATERIALS NEEDED FOR THIS PRACTICE

Metal or paper that is 8 1/2 in. × 11 in., soapstone or pencil, tape measure, square.

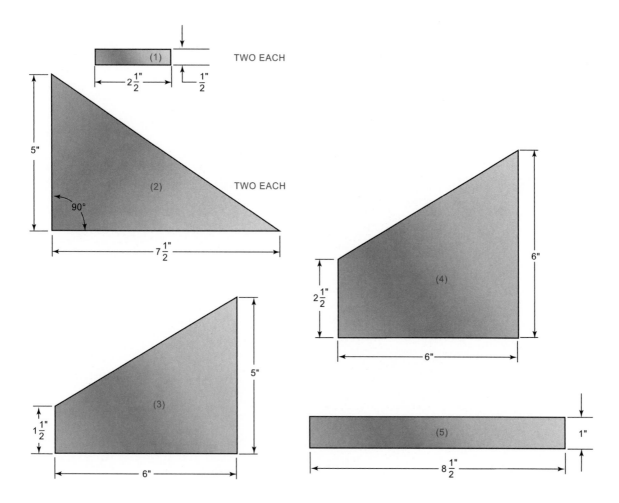

INSTRUCTIONS

1. Sketch the parts using different arrangements until you can arrive at the layout that requires the least space. Assume a 0 in. kerf width.

2. If you are using a piece of paper, draw two lines to form a 90° (right) angle near one corner of the paper. This will be your baseline.

3. Measure along the baseline or edge of the metal and lay out the lengths of each part.

4. Use the square or squared edge of the metal and lay out the widths of each part.

5. Connect the marks you made for each part by drawing a line between the laid out points.

6. Convert the dimensions into S.I. units and repeat the layout process.

INSTRUCTOR'S COMMENTS _____

_____

_____

■ PRACTICE 23-4

Name _____ Date _____

Class _____ Instructor _____ Grade _____

OBJECTIVE: After completing this practice, you should be able to fill out a Bill of Materials. Refer to Table 23-1 in the textbook.

EQUIPMENT AND MATERIALS NEEDED FOR THIS PRACTICE

Paper and pencil.

BILL OF MATERIALS				
Part	Number Required	Type of Material	Size Standard Units	SI Units

INSTRUCTIONS

1. Assume that all of the parts are made out of 1/2 in. (13 mm) low carbon steel plate and that one set is needed.

2. Write the part number in the first left-hand column.

3. Write the number of units required in the second column.

4. Write the type of material in the center column.

5. Write the overall dimension of the material in the next column.

6. Convert the standard units into S.I. units and write this in the right-hand column.

7. Total the amount of each type of material.

INSTRUCTOR'S COMMENTS _____

_____

_____

■ PRACTICE 23-5

Name _____ Date _____

Class _____ Instructor _____ Grade _____

OBJECTIVE: After completing this practice, you should be able to lay out a shape, allowing space for the material that will be removed in the cut's kerfs.

EQUIPMENT AND MATERIALS NEEDED FOR THIS PRACTICE

Pencil, 8 1/2 in. × 11 in. paper, measuring tape or rule, and square.

INSTRUCTIONS

1. If you are using a piece of paper, draw two lines to form a 90° (right) angle near one corner of the paper. This will be your baseline.

2. Measure along the baseline or edge of the metal and lay out the lengths of each part, adding 3/32 in. to each part for the kerf.

3. Use the square or squared edge of the metal and lay out the widths of each part, adding 3/32 in. to each part for the kerf.

4. Connect the marks you made for each part by drawing a line between the laid out points.

INSTRUCTOR'S COMMENTS _____

_____

_____

CHAPTER 23 FABRICATING TECHNIQUES AND PRACTICES QUIZ

Name _____ Date _____

Class _____ Instructor _____ Grade _____

Instructions: Carefully read Chapter 23 in the text and answer the following questions.

A. LAYOUT

Lay out the following parts on an 8 1/2-in. × 11-in. piece of paper. All layouts must be nested and assume a 0 in. kerf width unless otherwise noted.

1.

PART	NUMBER REQUIRED
1	4
2	1

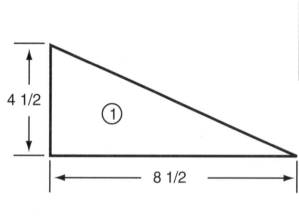

2.

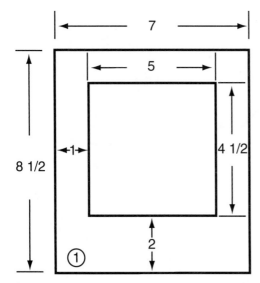

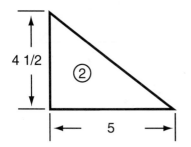

PART	NUMBER REQUIRED
1	1
2	2
3	1
4	1

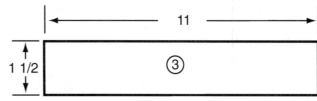

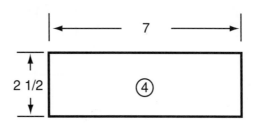

3.

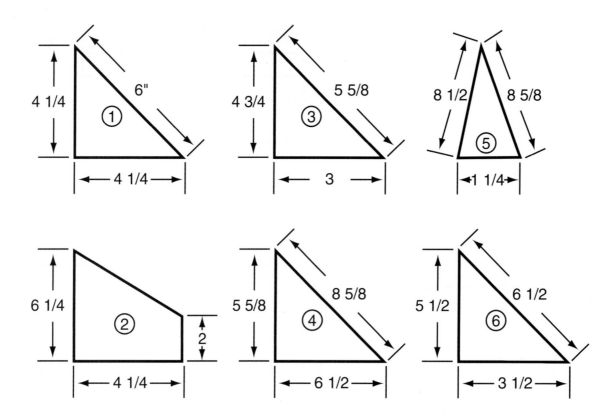

PART	NUMBER REQUIRED
1	2
2	2
3	1
4	1
5	1
6	1

4.

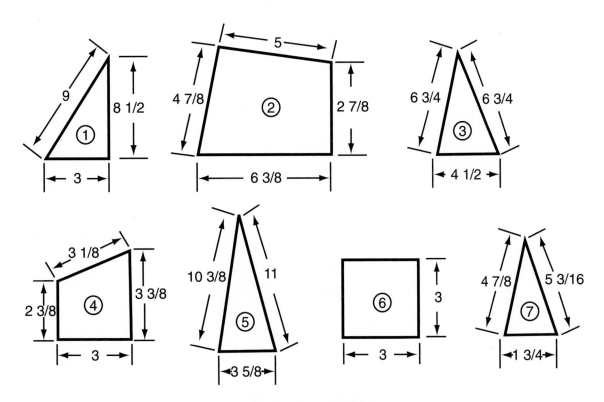

PART	NUMBER REQUIRED
1	1
2	1
3	1
4	1
5	1
6	1
7	2

Chapter 24

Welding Codes and Standards

WELDER AND WELDING OPERATOR QUALIFICATION TEST RECORD (WQR)

Welder or welding operator's name _____ Identification no. _____

Welding process _____ Manual _____ Semiautomatic _____ Machine _____
Position _____

(Flat, horizontal, overhead, or vertical — if vertical, state whether up or down)
In accordance with welding procedure specification no. _____

Material specification _____

Diameter and wall thickness (if pipe) — otherwise, joint thickness _____

Thickness range this qualifies _____

FILLER METAL

Specification No. _____ Classification _____ F-number _____

Describe filler metal (if not covered by AWS specification) _____
_____

Is backing strip used? _____

Filler metal diameter and trade name _____ Flux for submerged arc or gas
for gas metal arc or flux-cored arc welding _____

Guided-Bend Test Results

Appearance _____ Weld Size _____

    Type   Result   Type   Result

Test conducted by _____ Laboratory test no. _____

Fillet Test Results

Appearance _____ Fillet Size _____

Fracture test root penetration _____ Macroetch _____

(Describe the location, nature, and size of any crack or tearing of the specimen.)
Test conducted by _____ Laboratory test no. _____

Radiographic Test Results

Film Identification	Results	Remarks	Film Identification	Results	Remarks

Test conducted by _____ Laboratory test no. _____

We, the undersigned, certify that the statements in this record are correct and that the welds were prepared and tested in accordance with these requirements.

Manufacturer or contractor _____

Authorized by _____

Date _____

■ PRACTICE 24-1

Name _____ Date _____

Class _____ Instructor _____ Grade _____

OBJECTIVE: After completing this practice, you should be able to write a Welder and Welding Operator Qualification Test Record (WQR).

EQUIPMENT AND MATERIALS NEEDED FOR THIS PRACTICE

You will need the WQR form, which can be found on the previous pages.

INSTRUCTIONS

Using the form provided and following the example, you are going to write a Welder and Welding Operator Qualification Test Record. Most of the unique information is provided in this short outline. You will need to refer to some of the chapters on welding and to your notes to establish the actual limits of the welding variables (voltage, amperage, gas flow rates, nozzle size, etc.).

1. The joint tolerances will be almost the same as the ones in the practice welds in this text, or you can use a codebook to select a prequalified joint design. In either case, sketch the joint and include the dimension tolerances.

2. This information is included in each chapter where weld testing is required on practice welds.

3. Most of the preheat statement will be the same for all the WQRs you will be writing. You can copy the one on the sample or write your own.

4. This will require some of your knowledge and information found in chapters dealing with welding processes.

5. The same statement will fit almost all of the practices in the text. Copy the example or develop your own.

6. Follow the recommendations in the chapters covering the material you are welding and the process you are using.

7. Follow the procedures outlined in this chapter.

8. For most WQRs there are no repairs or defects allowed.

9. Sketch a finished weld as it should look.

INSTRUCTOR'S COMMENTS _____

_____

_____

■ PRACTICE 24-2

Name _____ Date _____

Class _____ Instructor _____ Grade _____

PROCEDURE QUALIFICATION RECORD (PQR)

Welding Qualification Record No: _____(1)_____ WPS No: _____(2)_____ Date: _____(3)_____
Material specification _____(4)_____ to _____
P-No. _____(5)_____ to P-No. _____ Thickness and O.D. _____(6)_____
Welding process: Manual _____(7)_____ Automatic _____(8)_____
Thickness Range _____(9)_____

Filler Metal

Specification No. _____(10)_____ Classification _____(11)_____ F-number _____(12)_____
A-number _____(13)_____ Filler Metal Size _____(14)_____ Trade Name _____(15)_____
Describe filler metal (if not covered by AWS specification) _____(16)_____
_____

Flux or Atmosphere

Shielding Gas _____(17)_____ Flow Rate _____(18)_____ Purge _____(19)_____
Flux Classification _____(20)_____ Trade Name _____(21)_____

Welding Variables

Joint Type _____(22)_____	Position _____(29)_____
Backing _____(23)_____	Preheat _____(30)_____
Passes and Size _____(24)_____	Bead Type _____(31)_____
No. of Arcs _____(25)_____	Current _____(32)_____
Ampere _____(26)_____	Volts _____(33)_____
Travel Speed _____(27)_____	Oscillation _____(34)_____

Interpass Temperature Range _____(28)_____

Weld Results

Appearance _____(35)_____ Weld Size _____(36)_____

Guided-Bend Test

Type	Result	Type	Result
(37)	(38)		

Tensile Test

Specimen No.	Dimensions Width|Thickness	Area	Ultimate Total Load, lb.	Ultimate Unit Stress, psi	Character of Failure and Location
(39)	(40)	(41)	(42)	(43)	(44)

Welder's Name _____(45)_____ Identification No. _____(46)_____ Laboratory Test No. _____
By virtue of these test welder meets performance requirements.
Test Conducted by _____(47)_____ Address _____
per _____(48)_____ Date _____(49)_____
We certify that the statements in this record are correct and that the test
welds performed and tested are in accordance with the WPS.

Manufacture _____(50)_____
Signed by _____
Date _____

OBJECTIVE: After completing this practice, you should be able to use a Procedure Qualification Record (PQR).

EQUIPMENT AND MATERIALS NEEDED FOR THIS PRACTICE

You will need the PQR form, which can be found on the previous page.

INSTRUCTIONS

Following the procedure you wrote in Practice 24-1, you are going to make the weld to see if your tentative welding procedure and specification can be certified.

INSTRUCTOR'S COMMENTS _____

_____

_____

CHAPTER 24: WELDING CODES AND STANDARDS QUIZ

Name _____ Date _____

Class _____ Instructor _____ Grade _____

Instructions: Carefully read Chapter 24 in the text and answer the following questions.

A. MATCHING

In the space provided to the left, write the letter from Column B that best answers the question or completes the statement in Column A.

Column A

Column B

_____ 1. The code most commonly used for buildings and other structural steel.

a. Local, state, or federal government regulations

_____ 2. The code most commonly used for pipe lines.

b. ASME Section IX

_____ 3. The code most commonly used for pressure vessels.

c. AWS D1.1

_____ 4. Many governing agencies require that a specific code or standard be followed.

d. API 1104

_____ 5. Lists all of the parameters required to produce a sound weld to the specific code, specification, or definition.

e. Welding Procedure Specification

B. ESSAY

Provide complete answers for all of the following questions.

6. What is a major area that the following codes are used for?

a. API 1104 _____

b. ASME Section IX_____

c. AWS D1.1 _____

7. Explain the differences between the following terms:

a. Welding Procedure Specification and Welding Schedule

_____

_____

b. Welding Schedule and Welding Procedure

_____

_____

8. List eight specific parameters found in WPSs.

   a. _____

   b. _____

   c. _____

   d. _____

   e. _____

   f. _____

   g. _____

   h. _____

9. How does a welder become "qualified?"

   _____

   _____

10. When can a welder be "certified?"

    _____

    _____

11. How long is a WPS in effect before it must be retested?

    _____

    _____

12. Why should supplementary information be included in the WPS?

    _____

    _____

13. What are some of the essential variables of WPSs?

    _____

    _____

INSTRUCTOR'S COMMENTS _____

_____

_____

Chapter 25

Testing and Inspection

Etching Solutions for Microscopic Examination of Metals	
Metal to Be Etched	**Etchant Mixture**
Aluminum	45 cc hydrochloric acid 15 cc nitric acid 15 cc hydrofluoric acid 25 cc water
Brass and copper alloys	25 cc ammonium hydroxide 25 cc hydrogen peroxide 25 cc water
Iron and steel	5 cc nitric acid 100 cc ethyl alcohol
Nickel and its alloys	50 cc of 70% nitric acid 50 cc of 50% acetic acid
Stainless steels	30 cc hydrochloric acid 10 cc nitric acid 80 cc glycerol

Mixing chemicals may be hazardous. Do not attempt to mix these solutions without proper supervision and instructions by someone trained to handle these materials, such as a chemist.

SAFETY PRECAUTIONS

When working with etchants, use extreme caution. Protective clothing is required (safety glasses, face shield, gloves, and an apron). Any mixing, handling, or use of an etchant is to be done in a well-ventilated area.

> **CAUTION** **As an added safety precaution always start the mixture by pouring the water into the mixing container first. THEN pour the acid slowly and carefully while stirring the mixture.**

CHAPTER 25: TESTING AND INSPECTION OF WELDS QUIZ

Name _____ Date _____

Class _____ Instructor _____ Grade _____

Instructions: Carefully read Chapter 25 in the text and answer the following questions.

A. MATCHING

In the space provided to the left, write the letter from Column B that best answers the question or completes the statement in Column A.

Column A

_____ 1. What is the resistance of metal to penetration and is an index to wear resistance and strength of the metals?

_____ 2. What is the name of a weld that has a flaw within the tolerance?

_____ 3. What is the interruption of the typical structure of a weldment?

_____ 4. A cylindrical porosity.

_____ 5. What is the lack of fusion between the molten filler metal and the base metal?

_____ 6. What is a discontinuity that will render a product unable to meet the minimum applicable acceptance standard or specification?

_____ 7. What is the difference between acceptable and perfection?

_____ 8. What is weld metal that flows over a surface without fusing to it?

_____ 9. What is the lack of coalescence between molten filler metal and the base metal?

_____ 10. What can be expected if prior welds were improperly cleaned?

Column B

a. wormhole

b. defect

c. incomplete fusion

d. hardness

e fit for service

f. discontinuity

g. overlap

h. inclusions

i. cold lap

j. tolerance

B. SHORT ANSWER

Write a brief answer in the space provided that will answer the question or complete the statement.

11. What are the two classifications of methods used in product quality control?

    a. _____

    b. _____

12. List three factors that you should consider when evaluating a weld with discontinuities.

    a. _____

    b. _____

    c. _____

13. List four factors that can affect the selection of radiographic equipment.

    a. _____

    b. _____

    c. _____

    d. _____

14. What are the two types of ultrasonic equipment used in the welding industry?

    a. _____

    b. _____

C. FILL IN THE BLANK

Fill in the blank with the correct word. Answers may be more than one word.

15. The basis for the types of inspection and the criteria for acceptance are all based on

    _____ accepted codes and standards.

16. _____ (_____) methods, except for hydrostatic testing, result in the product

    being destroyed.

17. _____ (_____) does not destroy the part being tested.

18. Ideally, a weld should not have any _____, but that is practically impossible.

19. Cylindrical porosity is called _____.

20. _____ can resemble porosity but, unlike porosity, they are generally not spherical.

21. The lack of fusion between the filler metal and previously deposited weld metal is called interpass

    _____.

22. Arc strikes, even when ground flush for a guided bend, will open up to form small

    _____.

23. _____, such as incorrect electrode angle or excessive weave, can also cause undercut.

24. The slag and oxides in the pipe are rolled out with the steel, producing the _____.

25. A solution to the problem of _____, is to redesign the joints in order to impose the lowest possible strain throughout the plate thickness.

26. In a tensile tests, if the weld metal is _____ than the plate, failure occurs in the plate; if the weld is _____, failure occurs in the weld.

27. In the fatigue test, the part is subjected to repeated changes in _____.

28. In a nick-break test, a force is then applied, and the specimen is _____ by one or more blows of a hammer.

29. In a guided-bend test, when the specimens are prepared, caution must be taken to ensure that all grinding marks run _____ to the specimen so that they do not cause stress cracking.

30. The most commonly used _____ are hydrochloric acid, ammonium persulphate, and nitric acid.

31. The major differences between the tests are that the Izod test specimen is gripped on one end and is held _____ and usually tested at room temperature and the Charpy test specimen is held _____, supported on both ends, and is usually tested at a specific temperature.

32. Visual inspection can be used to _____ welds that have excessive surface discontinuities that will not pass the code or standards being used.

33. If something _____ than the weld is present, such as a pore or lack of fusion defect, fewer X-rays are absorbed, darkening the film.

34. These _____ devices operate very much like depth sounders, or "fish finders."

D. ESSAY

Provide complete answers for all of the following questions.

35. What is the difference between destructive testing and nondestructive testing?

_____

_____

36. What is the maximum strength of a weld having a maximum load of 50,000 lb (22,679 kg) and a cross-sectional area of .0125 in. (0.3 mm)?

_____

_____

37. What is fatigue testing?

_____

_____

38. List two methods of testing shear strength.

a. _____

b. _____

39. What is the difference between a guided-bend and a free-bend test?

_____

_____

40. Why are the corners rounded on guided-bend test specimens?

_____

_____

41. What is the purpose of testing welds by the etching method?

_____

_____

42. Explain the difference between Charpy and Izod impact tests.

_____

_____

E. DEFINITIONS

43. Define the following terms:

fitness for service _____

discontinuity _____

defect _____

NDT _____

etching _____

INSTRUCTOR'S COMMENTS _____

_____

_____

Chapter 26

Welding Metallurgy

MELTING POINTS OF METALS AND ALLOYS		
	Melting Point	
Metal or alloy	°F	°C
Aluminum, cast (8% copper)	1175	635
Aluminum, pure	1220	660
Aluminum (5% silicon)	1118	603
Brass, naval	1625	885
Brass, yellow	1660	904
Bronze, aluminum	1905	1041
Bronze, manganese	1600	871
Bronze, phosphor	1830 to 1922	999 to 1050
Bronze, tobin	1625	885
Chromium	2740	1504
Copper	1981	1083
Iron, cast	2300	1260
Iron, malleable	2300	1260
Iron, pure	2786	1530
Iron, wrought	2750	1510
Lead	620	327
Magnesium	1200	649
Manganese	2246	1230
Molybdenum	4532	2500
Monel metal	2480	1360
Nickel	2686	1452
Nickel silver (18% nickel)	2030	1110
Silver, pure	1762	961
Silver solders (50% silver)	1160 to 1275	627 to 690
Solder (50–50)	420	216
Stainless steel (18–8)	2550	1399
Stainless steel, low carbon (18–8)	2640	1449
Steel, high carbon (0.55–0.83% carbon)	2500 to 2550	1371 to 1399
Steel, low carbon (maximum 0.30% carbon)	2600 to 2750	1427 to 1510
Steel, medium carbon (0.30–0.55% carbon)	2550 to 2600	1399 to 1427
Steel, cast	2600 to 2750	1427 to 1510
Steel, manganese	2450	1343
Steel, nickel (3.5% nickel)	2600	1427
Tantalum	5160	2849
Tin	420	232
Titanium	3270	1799
Tungsten	6152	3400
Vanadium	3182	1750
White metal	725	385
Zinc	786	419

HEAT COLORS AT GIVEN TEMPERATURES OF FERROUS MATERIALS		
Temper Color	**Temperature**	
	°F	**°C**
Faint straw	400	204
Straw	440	227
Dark straw	460	238
Very deep straw	480	249
Brown yellow	500	260
Bronze or brown purple	520	271
Peacock or full purple	540	282
Bluish purple	550	288
Blue	570	299
Full blue	590	310
Very dark blue	600	316
Light blue	640	338
Faint red (visible in dark)	750	399
Faint red	900	482
Blood red	1050	565
Dark cherry	1175	635
Medium cherry	1250	677
Cherry or full red	1375	746
Bright cherry	1450	788
Bright red	1550	843
Salmon	1650	899
Orange	1725	940
Lemon	1825	996
Light yellow	1975	1079
White	2200	1204

SURFACE COLOR OF SOME COMMON METALS			
Metals	**Color of Unfinished, Unbroken Surface**	**Color and Structure of Newly Fractured Surface**	**Color of Freshly Filed Surface**
White cast iron	Dull gray	Silvery white; crystalline	Silvery white
Gray cast iron	Dull gray	Dark silvery; crystalline	Light silvery gray
Malleable iron	Dull gray	Dark gray; finely crystalline	Light silvery gray
Wrought iron	Light gray	Bright gray	Light silvery gray
Low carbon and cast steel	Dark gray	Bright gray	Bright silvery gray
High carbon steel	Dark gray	Light gray	Bright silvery gray
Stainless steel	Dark gray	Medium gray	Bright silvery gray
Copper	Reddish-brown to green	Bright red	Bright copper color
Brass and bronze	Reddish-yellow, yellow-green, or brown	Red to yellow	Reddish-yellow to yellowish-white
Aluminum	Light gray	White; finely crystalline	White
Monel metal	Dark gray	Light gray	Light gray
Nickel	Dark gray	Off white	Bright silvery white
Lead	White to gray	Light gray; crystalline	White
Copper-nickel (70–30)	Gray	Light gray	Bright silvery white

IDENTIFICATION OF METALS BY CHIP TEST	
Metals	**Chip Characteristics**
White cast iron	Chips are small, brittle fragments. Chipped surfaces not smooth.
Gray cast iron	Chips are about 1/8 in. in length. Metal not easily chipped, so chips break off and prevent smooth cut.
Malleable iron	Chips are 1/4 to 3/8 in. in length (larger than chips from cast iron). Metal is tough and hard to chip.
Wrought iron	Chips have smooth edges. Metal is easily cut or chipped; chip can be made as a continuous strip.
Low carbon and cast steel	Chips have smooth edges. Metal is easily cut or chipped; chip can be made as a continuous strip.
High carbon steel	Chips show a fine-grain structure. Edges of chips are lighter in color than chips of low carbon steel. Metal is hard, but can be chipped in a continuous strip.
Copper	Chips are smooth, with sawtooth edges where cut. Metal is easily cut; chip can be cut as a continuous strip.
Brass and bronze	Chips are smooth, with sawtooth edges. These metals are easily cut, but chips are more brittle than chips of copper. Continuous strip is not easily cut.
Aluminum and aluminum alloys	Chips are smooth, with sawtooth edges. Chip can be cut as a continuous strip.
Monel	Chips have smooth edges. Continuous strip can be cut. Metal chips easily.
Nickel	Chips have smooth edges. Continuous strip can be cut. Metal chips easily.
Lead	Chips of any shape may be obtained because the metal is so soft that it can be cut with a knife.

IDENTIFICATION OF METALS BY OXYACETYLENE TORCH TEST	
Metals	**Reactions When Heated by Oxyacetylene Torch**
White cast iron	Metal becomes dull red before melting. Melts at moderate rate. A medium tough film of slag develops. Molten metal is watery, reddish-white in color, and does not show sparks. When flame is removed, depression in surface of metal under flame disappears.
Gray cast iron	Pool of molten metal is quiet, rather watery, but with heavy, tough film forming on surface. When torch flame is raised, depression in surface of metal disappears instantly. Molten pool takes time to solidify, and gives off no sparks.
Malleable iron	Metal becomes red before melting; melts at moderate rate. A medium tough film of slag develops, but can be broken up. Molten pool is straw-colored, watery, and leaves blow holes when it boils. Center of pool does not give off sparks, but the bright outside portion does.
Wrought iron	Metal becomes bright red before it melts. Melting occurs quietly and rapidly, without sparking. There is a characteristic slag coating, greasy or oily in appearance, with white lines. The straw-colored molten pool is not viscous, is usually quiet but may have a tendency to spark; is easily broken up.
Low carbon and cast steel	Melts quickly under the torch, becoming bright red before it melts. Molten pool is liquid, straw-colored, gives off sparks when melted, and solidifies almost instantly. Slag is similar to the molten metal and is quiet.
High carbon steel	Metal becomes bright red before melting, melts rapidly. Melting surface has cellular appearance, and is brighter than molten metal of low carbon steel; sparks more freely, and sparks are whiter. Slag is similar to the molten metal and is quiet.
Stainless steels	Reactions vary depending upon the composition.
Copper	Metal has high heat conductivity; therefore, larger flame is required to produce fusion than would be required for the same size piece of steel. Copper color may become intense before metal melts; metal metals slowly, and may turn black and then red. There is little slag. Molten pool shows mirror-like surface directly under flame, and tends to bubble. Copper that contains small amounts of other metals melts more easily, solidifies more slowly than pure copper.
Brass and bronze	These metals melt very rapidly, becoming noticeably red before melting. True brass gives off white fumes when melting. Bronze flows very freely when melting, and may fume slightly.
Aluminum and aluminum alloys	Melting is very rapid, with no apparent change in color of metal. Molten pool is same color as unheated metal and is fluid; stiff black scum forms on surface, tends to mix with the metal, and is difficult to remove.
Monel	Melts more slowly than steel, becoming red before melting. Slag is gray scum, quiet, and hard to break up. Under the scum, molten pool is fluid and quiet.
Nickel	Melts slowly (about like Monel), becoming red before melting. Slag is gray scum, quiet, and hard to break up. Under the scum, molten pool is fluid and quiet.
Lead	Melts at very low temperature, with no apparent change in color. Molten metal is white and fluid under a thin coat of dull gray slag. At higher temperature, pool boils and gives off poisonous fumes.

CHAPTER 26: WELDING METALLURGY QUIZ

Name _____ Date _____

Class _____ Instructor _____ Grade _____

Instructions: Carefully read Chapter 26 in the text and answer the following questions.

A. MATCHING

In the space provided to the left, write the letter from Column B that best answers the question or completes the statement in Column A.

Column A

_____ 1 The SI unit of heat.

_____ 2. The heat required to change matter from one state to another.

_____ 3. These have no orderly arrangement in the crystalline structure.

_____ 4. The fundamental building blocks of all metals.

_____ 5. Two or more metals dissolved into each other after cooling.

_____ 6. A metal that can exist in two or more crystalline structures.

_____ 7. A compound of iron and carbon, iron carbide.

Column B

a. Tempering

b. Allotropic

c. Solid solution

d. Joule

e. Cementite

f. Amorphic

g. Unit cells

h. Latent heat

B. SHORT ANSWER

Write a brief answer in the space provided that will answer the question or complete the statement.

8. _____ is the property of a material to withstand a twisting force.

9. _____ is the ability of a material to return to its original form after the removal of a load.

10. _____ is the measure of how well a part can withstand forces acting to cut or slice it apart.

11. _____ is the property that allows a metal to withstand forces, sudden shock, or bends without fracturing.

12. _____ is the resistance to penetration.

13. _____ problems are avoidable by keeping organic materials away from weld joints, keeping the welding consumables dry, and preheating the components to be welded.

14. _____ is used to reduce the rate at which welds cool.

15. When _____ is viewed under a microscope, it has an acicular grain structure.

16. _____ is a process of reheating a part after it has been heated and quenched for the purpose of increasing toughness and improving the tensile strength.

17. _____ metals have the ability to exist in two or more different crystalline structures in their solid state.

18. Many of the problems and defects associated with welding are due to hydrogen. Place an "X" in the blanks provided by the products in the following list that contain hydrogen.

_____ a. Aluminum oxide

_____ b. Moisture in electrode coatings

_____ c. Carbon

_____ d. Moisture in humid air or in weld joints

_____ e. Silicon

_____ f. Organic lubricants

_____ g. Rust in weld joints

_____ h. Organic particles such as paint

_____ i. Cloth fibers

19. What are the four most common methods of quenching metals?

a. _____

b. _____

c. _____

d. _____

20. What three conditions must be present to cause hydrogen-induced cracking?

a. _____

b. _____

c. _____

21. What are the two methods with which ductility can be measured when performing a tensile test?

a. _____

b. _____

22. List four ways that strength can be measured in a material.

    a. _____

    b. _____

    c. _____

    d. _____

C. FILL IN THE BLANK

Fill in the blank with the correct word. Answers may be more than one word.

23. Welding operations heat the metals, and that heating will certainly change not only the metal's

    _____ but its _____ as well.

24. _____ is the amount of thermal energy in matter.

25. _____ is a measurement of the vibrating speed or frequency of the atoms

    in matter.

26. _____, _____, and _____ each

    affect the grain structure of metals in their own unique way.

27. As the mixture melts in a furnace, the _____ combine with the ore to separate

    the metal from impurities in the mixture.

28. The ingots are then placed in a furnace where they are slowly reheated so that the

    _____ the entire ingot is the same.

29. _____ is the most efficient way that molten steel is initially formed into a

    shape that can be formed into a finished product.

30. The individual grains grow toward the center of the _____.

31. This grain structure is strongest in the direction it _____.

32. The forging process results in a strong hard layer of fine grain structure _____

    grain structure.

33. Since _____ is proportional to strength, it is a quick way to determine strength.

34. Some types of cast iron are _____ and once broken will fit back together like a

    puzzle's parts.

35. _____is measured most often with the Charpy test.

36. _____is the amount of strain needed to permanently deform a test specimen.

37. _____is the ability of a material to return to its original form after removal of the load.

38. Phase diagrams are also known as _____or _____ diagrams, and the terms are used interchangeably.

39. Notice on the chart that although 100% lead becomes a liquid at _____°F (_____ °C) and 100% tin becomes a liquid at _____°F (_____ °C), a mixture of 38.1% lead and 61.9% tin becomes a liquid at _____°F (_____ °C).

40. This soft iron when alloyed with as little as _____% carbon can become tool steel.

41. _____, _____, and _____are all crystalline forms of iron and iron-carbon alloys.

42. _____is practically pure iron (in plain carbon steels) existing below the lower transformation temperature.

43. _____is the nonmagnetic form of iron and has the power to dissolve carbon and alloying elements.

44. It is possible to replace some of the atoms in the crystal lattice with atoms of another metal in a process called _____.

45. This process, called precipitation hardening or age hardening, is the heat treatment used to strengthen many _____.

46. The _____the metal cools, the greater the quenching effect.

47. The reheating during the tempering process _____some of the brittle hardness caused by the quenching, replacing it with _____and increased _____.

48. _____is the hardest of the transformation products of austenite.

49. Sheets, bars, and tubes are intentionally _____ to increase their strength, since cold working will strengthen almost all metals and their alloys.

50. The lowest possible temperature should be selected because preheat increases the size of the _____ and can damage some grades of _____ and _____ steels.

51. Heating to just under the critical temperature does offer a stress reduction of about _____%.

52. During the slow cooling process of annealing, the austenite transforms to _____and _____.

53. The speed of a weld and the temperature of the surrounding metal will affect the transformation of the grain structure in the surrounding area called the _____ (_____).

54. If the austenite is cooled too quickly, it will transform into _____, which is a very large, hard, and brittle grain structure with almost no ductility.

55. _____is a layered combination of cementite (Fe_3C) and ferrite (Fe).

56. The _____that occurs during welding is significant and can have a profound effect on the welds fitness for service.

57. Since _____improves the strength of stainless steels, it sometimes is added intentionally to shield gases.

58. About 2% of oxygen is added intentionally to stabilize the _____ process when welding steels with argon shielding.

59. The carbon in carbon dioxide is a potential contaminant that can cause problems with _____in the low carbon grades of _____.

60. Even in amounts as low as _____ parts per million, hydrogen can cause cold cracking in high-strength steels.

61. As metals cools and begins forming solid metal, the hydrogen atoms are no longer soluble, so they are forced out into the _____.

62. Severely concave welds may also cause _____cracking because the welds are not as strong.

63. The formation of _____depletes stainless steel of the free chromium needed for protection.

D. ESSAY

Provide complete answers for all of the following questions.

64. Explain why welders should have some understanding of metallurgy.

_____

_____

65. Describe the changes that occur in an alloy of 30% tin and 70% lead as it cools slowly from 600°F to 200°F (316°C to 93°C).

_____

_____

66. Explain four methods that can be used for strengthening metals.

   a. _____

   b. _____

   c. _____

   d. _____

67. Explain the differences between preheat and stress relief anneal.

_____

_____

68. Describe some of the welding problems caused by air.

_____

_____

69. Describe how poor welding techniques can produce defective welds.

_____

_____

70. How can the defects discussed in Question 43 be avoided?

_____

_____

INSTRUCTOR'S COMMENTS _____

_____

_____

Chapter 27

Weldability of Metals

STANDARD STEEL AND STEEL ALLOY NUMBERING SYSTEM	
Class	**Number**
Plain carbon steels	10XX
Free-cutting carbon steels	11XX
Manganese steels	13XX
Nickel steels	20XX
Nickel-chromium steels	30XX
Molybdenum steels	40XX
Chrome-molybdenum steels	41XX
Nickel-chrome-molybdenum steels	43XX, 47XX
Nickel-molybdenum steels	46XX, 48XX
Chromium steels	50XX
Chromium-vanadium steels	60XX
Heat-resisting casting alloys	70XX
Nickel-chrome-molybdenum steels	80XX, 93XX, 98XX
Silicon-manganese steels	90XX

STANDARD ALUMINUM AND ALUMINUM ALLOY NUMBER DESIGNATIONS	
Major Alloying Element	**Number**
Aluminum (99% minimum)	1XXX
Copper	2XXX
Manganese	3XXX
Silicon	4XXX
Magnesium	5XXX
Magnesium-silicon	6XXX
Zinc	7XXX
Other element	8XXX
Unused class	9XXX

STANDARD WROUGHT ALUMINUM AND MAGNESIUM COMPOSITIONS TEMPER DESIGNATION

Designation	Temper
F	As fabricated
O	Annealed
H	Strain hardened
H1	Strain hardened only
H2	Strain hardened, then partially annealed
H3	Strain hardened, then stabilized
W	Solution heat treated but with unstable temper
T	Thermally treated to produce stable tempers other than F, O, or H
T2	Annealed (for castings only)
T3	Solution heat treated, then cold worked
T4	Solution heat treated, then naturally aged
T5	Artificially aged
T6	Solution heat treated, then artificially aged
T7	Solution heat treated, then stabilized
T8	Solution heat treated, then cold worked, then artificially aged
T9	Solution heat treated, then artificially aged, then cold worked
T10	Artificially aged, then cold worked

STANDARD COPPER AND COPPER ALLOY SERIES DESIGNATIONS

Series	Alloying Element
100	None or very slight amount
200, 300, 400, and 665 to 699	Zinc
500	Tin
600 to 640	Aluminum
700 to 735	Nickel
735 to 799	Nickel and zinc

LETTERS USED TO IDENTIFY ALLOYING ELEMENTS IN MAGNESIUM ALLOYS

Letter	Alloying Element
A	Aluminum
B	Bismuth
C	Copper
D	Cadmium
E	Rare earth
F	Iron
H	Thorium
K	Zirconium
L	Beryllium
M	Manganese
N	Nickel
P	Lead
Q	Silver
R	Chromium
S	Silicon
T	Tin
Z	Zinc

STANDARD STAINLESS STEEL NUMBERING SYSTEM	
Austenitic	**Common Application and Characteristics**
301	A general utility stainless steel, easily worked
302	Readily fabricated, for decorative or corrosion resistance
304, 304LC	A general utility stainless steel, easily worked
308	Used where corrosion resistance better than 1800 is needed
309	High-scaling resistance and good strength at high temperatures
310	More chromium and nickel for greater resistance to scaling in high heat
316, 316LC	Excellent resistance to chemical corrosion
317	Higher alloy than 316 for better corrosion resistance
321	Titanium stabilized to prevent carbide precipitation
347	Columbian stabilized to prevent carbide precipitation
Martensitic	
403	Used for forged turbine blades
410	General purpose, low priced, heat treatable
414	Nickel added; for knife blades, springs
416	Free machining
420	Higher carbon for cutlery and surgical instruments
431	High mechanical properties
440A	For instruments, cutlery, valves
440B	Higher carbon than 440A
440C	Higher carbon than 440A or B for high hardness
501	Less resistance to corrosion than chromium nickel types
502	Less resistance to corrosion than chromium nickel types
Ferritic	
405	Nonhardening when air cooled from high temperatures
406	For electrical resistances
430	Easily formed alloy, for automobile trim
430F	Free machining variety of 430 grade
446	High resistance to corrosion and scaling up to 215°F

CHAPTER 27: WELDABILITY OF METALS QUIZ

Name _____ Date _____

Class _____ Instructor _____ Grade _____

Instructions: Carefully read Chapter 27 in the text and answer the following questions.

A. MATCHING

In the space provided to the left, write the letter from Column B that best answers the question or completes the statement in Column A.

Using the SAW and AISI classification systems, match the following numbers with the proper type of steel.

Column A

Column B

_____ 1. 1XXX

a. tungsten

_____ 2. 2XXX

b. chromium

_____ 3. 3XXX

c. molybdenum

_____ 4. 4XXX

d. carbon

_____ 5. 5XXX

e. silicon-manganese

_____ 6. 6XXX

f. nickel-chromium-molybdenum

_____ 7. 7XXX

g. nickel

_____ 8. 8XXX

h. chromium-vanadium

_____ 9. 9XXX

i. nickel-chrome

Match the following terms with their definitions.

Column A

Column B

_____ 10. What is a silvery-gray metal weighing about half as much as steel, or about one and one-half times as much as aluminum?

a. tool steels

b. cast iron

_____ 11. What is an extremely light metal having a silvery-white color? Its weight is one-fourth that of steel and approximately two-thirds that of aluminum.

c. titanium

d. malleable cast iron

_____ 12. This type of steel is used for high-temperature service and for aircraft parts.

e. high-manganese steels

f. magnesium

_____ 13. What is white cast iron that has undergone a transformation as the result of a long heat-treating process to reduce the brittleness?

g. chromium-molybdenum

_____ 14. What has from 0.30% to 0.50% carbon content?

h. medium carbon steels

i. high carbon steels

_____ 15. What has a carbon content from 0.8% to 1.50%?

_____ 16. What has a carbon content from 0.50% to 0.90%?

_____ 17. These steels are used for wear resistance in applications involving impact.

_____ 18. All grades of what metal have a high carbon content, usually ranging from 1.7% to 4%?

B. FILL IN THE BLANK

Fill in the blank with the correct word. Answers may be more than one word.

19. Good _____ means that almost any process can be used to produce acceptable welds and that little effort is needed to control the procedures.

20. The heating and cooling cycles can set up _____ and _____ in the weld.

21. If the wrong filler metal is selected, the weld can have _____and not be

    _____.

22. _____the part before starting the weld will reduce the stress caused by the weld and will help the filler metal flow.

23. When large welds are needed, it is better to make _____welds than

    _____ welds.

24. The last two or three digits of the steel classification system refer to the approximate permissible range of _____ content.

25. Plain carbon steel is basically an alloy of _____and _____.

26. Low carbon (mild) steels can be welded readily by the _____method.

27. The gas tungsten arc process is slow and will cause severe porosity in the weld if the steel is not fully _____.

28. The use of an electrode with a _____may be necessary to reduce the tendency toward underbead cracking of medium carbon steels.

29. In arc welding high carbon steel, _____shielded metal arc electrodes are generally used.

30. Recommended practice to weld tool steel is to _____the metal, followed by a slow _____after the welding.

31. Martensite is very hard and brittle and must be checked for cracks _____.

32. The combining of chromium and carbon lowers the chromium that is available to provide _____ in the metal.

33. To identify the low carbon from the standard AISI number, the _____ is added as a suffix.

34. Since austenitic stainless steels are not _____ by quenching, the weld can be cooled using a chill plate to prevent chromium carbides from forming.

35. Martensitic stainless steels are used in applications requiring both _____ resistance and _____ resistance.

36. Cast iron is _____and _____, which makes it ideal for any size casing or frame that must hold its shape even under heavy loads.

37. Gray cast iron is easily welded, but because it is somewhat porous it can absorb _____ into the surface, which must be baked out before welding.

38. Malleable cast iron can _____ be welded.

39. Unless the heating and cooling cycles are _____and _____, stresses within brittle materials will cause them to crack.

40. Thick aluminum casting must be preheated to about _____°F (_____°C) before welding.

41. Two of the most important properties of titanium are its _____ (in alloy form) and its generally excellent _____.

42. Magnesium alloys may be classified as _____or _____types.

43. The spark test should be done using a _____grinding stone.

C. SHORT ANSWER

Write a brief answer in the space provided that will answer the question or complete the statement.

44. _____ occurs when alloys containing both chromium and carbon are heated.

45. The four groups of stainless steel are _____ , _____ , _____ , and _____ .

46. _____ stainless steels are not hardenable by quenching.

47. The most common grades of cast iron contain _____ to _____ total carbon.

48. Define the term *weldability*.

    _____

    _____

49. What effect does controlled heating and cooling have on metals with poor weldability?

    _____

    _____

50. What do the abbreviations AISI and SAE stand for?

    _____

    _____

51. Explain the significance of each of the four digits in the steel number 2140.

    _____

    _____

52. Why are tool steels difficult to weld?

    _____

    _____

53. Describe how to weld gray cast iron.

_____

_____

54. Why must titanium be kept clean during welding?

_____

D. DEFINITIONS

55. Define the following terms:

alloy steels _____

18-8 stainless _____

low alloy steels _____

INSTRUCTOR'S COMMENTS _____

_____

_____

Chapter 28

Filler Metal Selection

American Welding Society and American Society for Testing and Materials Specifications		
AWS Designation	**ASTM Designation**	**Title**
A5.1	A 233	Mild Steel Covered Arc Welding Electrodes
A5.2	A 251	Iron and Steel Gas Welding Rods
A5.3	B 184	Aluminum and Aluminum-Alloy Arc Welding Electrodes
A5.4	A 298	Corrosion-Resisting Chromium and Chromium-Nickel Steel Covered Welding Electrodes
A5.5	A 316	Low-Alloy Steel Covered Arc Welding Electrodes
A5.6	B 255	Copper and Copper-Alloy Arc Welding Electrodes
A5.7	B 259	Copper and Copper-Alloy Welding Rods
A5.8	B 260	Brazing Filler Metal
A5.9	A 371	Corrosion-Resisting Chromium and Chromium-Nickel Steel Welding Rods and Bare Electrodes
A5.10	B 285	Aluminum and Aluminum-Alloy Welding Rods and Bare Electrodes
A5.11	B 295	Nickel and Nickel-Alloy Covered Welding Electrodes
A5.12	B 297	Tungsten Arc Welding Electrodes
A5.13	A 399	Surfacing Welding Rods and Electrode
A5.14	B 304	Nickel and Nickel-Alloy Bare Welding Rods and Electrodes
A5.15	A 398	Welding Rods and Covered Electrodes for Welding Cast Iron
A5.16	B 232	Titanium and Titanium-Alloy Bare Welding Rods and Electrodes
A5.17	A 558	Bare Mild Steel Electrodes and Fluxes for Submerged Arc Welding
A5.18	A 559	Mild Steel Electrodes for Gas Metal Arc Welding

CHAPTER 28: FILLER METAL SELECTION QUIZ

Name _____ Date _____

Class _____ Instructor _____ Grade _____

Instructions: Carefully read Chapter 28 in the text and answer the following questions.

A. FILL IN THE BLANK

Fill in the blank with the correct word. Answers may be more than one word.

1. The AWS classification system is for _____ requirements within a grouping.

2. Yield point, psi (N/mm²)—the point in low and medium carbon steels at which the metal _____ when force (stress) is applied after which it will not return to its original length.

3. As the percentage of carbon increases, the tensile strength increases, the hardness _____, and ductility is _____.

4. As the percentage of chromium increases, tensile strength, hardness, and corrosion resistance _____with some _____in ductility.

5. As the percentage of molybdenum increases, tensile strength _____ at elevated temperatures; creep resistance and corrosion resistance all _____, too. It is also a ferrite and carbide former.

6. Welding fluxes can affect the _____ and _____ of the weld bead.

7. When making an electrode selection, _____ must be kept in mind, and the performance characteristics must be compared before making a final choice.

8. _____indicates a rod (_____) that is heated by some source other than electric current flowing directly through it.

9. The AWS specification for carbon steel–covered arc electrodes is _____, and for low alloy steel–covered arc electrodes it is _____.

10. E6010 electrodes have a forceful arc that results in _____ penetration and _____ metal transfer in the vertical and overhead positions.

11. E6012 electrodes have a(n) _____ arc that is not very forceful, resulting in a(n) _____ penetration characteristic.

12. E7018 electrodes have a(n) _____ flux with iron powder added.

13. The deoxidizers allow ER70S-2 wire to be used on metal that has _____ coverings of rust or oxides.

14. ER70S-3 wire produces high-quality welds on _____ and _____ steels.

15. E70T-1 and E71T-1 filler metal can be used for _____ or _____ pass welds.

16. E70T-4 and E71T-4 are _____, flux cored filler metal.

17. Metal cored electrodes classified as E70C-XX may be used for _____ and _____ pass welds.

18. Metal cored electrodes have a _____% or higher deposition rate due to their low nonmetallic and high metallic powered.

19. E70C-XX filler metals are frequently used in _____ and _____ welding applications.

20. The number _____ on a stainless steel covered electrode is used to indicate that there is a lime base coating, and the DCEP polarity welding current should be used.

21. The number _____ on a stainless steel covered electrode is used to indicate there is a titanium-type coating, and AC or DCEP polarity welding currents can be used.

22. 310 stainless steels are used for _____-temperature service where _____creep is desired.

23. The AWS specifications for aluminum and aluminum alloy filler metals are _____ for covered arc welding electrodes and _____ for bare welding rods and electrodes.

24. The 1100 aluminum filler wire is also relatively _____.

25. Aluminum bronze welding electrodes are used for _____bearing surfaces.

26. ECoCr-C electrodes are _____- and _____-resistant welding electrodes.

B. SHORT ANSWER

Write a brief answer in the space provided that will answer the question or complete the statement.

27. What welding currents can be used on the following electrodes?

a. EXXX0 _____

b. EXXX1 _____

c. EXXX2 _____

d. EXXX3 _____

e. EXXX4 _____

f. EXXX5 _____

g. EXXX6 _____

h. EXXX8 _____

28. The following electrodes are made up of letters and numbers. Explain what each bracket in the electrode means using the American Welding Society (AWS) numbering system.

A B C D   E F G H   I J K L M

E 60 1 2   E 70 S 3   E 7 1 T 2

A. _____

B. _____

C. _____

D. _____

E. _____

F. _____

G. _____

H. _____

I. _____

J. _____

K. _____

L. _____

M. _____

29. List two functions of the core wire in a shielded metal arc welding electrode.

a. _____

b. _____

30. Place an X beside each of the following flux cored electrodes that can be used in all positions.

_____ a. E70T-1

_____ b. E70T-2

_____ c. E70T-4

_____ d. E71T-2

_____ e. E71T-10

C. MATCHING

In the space provided to the left, write the letter from Column B that best answers the question or completes the statement in Column A.

Column A

_____ 31. As the percentage of this increases, tensile strength increases, and cracking may increase.

_____ 32. This is usually a contaminant and the percentage should be kept as low as possible. As the percentage of this increases, it can cause weld brittleness, reduced shock resistance, and increased cracking.

_____ 33. As the percentage of this increases, the tensile strength, toughness, and corrosion resistance increases.

_____ 34. As the percentage of this increases, tensile strength at elevated temperatures and corrosion resistance increase.

_____ 35. This is usually a contaminant and the percentage should be kept as low as possible, below .04%. As the percentage of carbon increases, this element can cause hot shortness and porosity.

_____ 36. As the percentage of this increases, the tensile strength hardness, resistance to abrasion, and porosity all increase; hot shortness is reduced.

_____ 37. As the percentage of this increases, tensile strength, hardness, and corrosion resistance increase, with some decrease in ductility.

_____ 38. As the percentage of this increases, the tensile strength increases, the hardness increases, and the ductility is reduced.

Column B

a. Ni

b. Mn

c. Cr

d. Mo

e. C

f. Si

g. S

h. P

D. ESSAY

Provide complete answers for all of the following questions.

39. What are the names of the organizations the following abbreviations stand for?

    a. AWS _____

    b. ASTM _____

    c. AISI _____

40. Why do some manufacturers choose to add one type of element to an electrode and another manufacturer makes a different choice?

    _____

    _____

41. Why do some manufacturers choose to make more than one electrode within a single classification?

    _____

    _____

42. List four characteristics of a filler metal that might be listed in a manufacturer's technical description.

    a. _____

    b. _____

    c. _____

    d. _____

43. List four properties of iron that change when the percentage of carbon changes.

    a. _____

    b. _____

    c. _____

    d. _____

44. List four properties of iron that change when the percentage of manganese changes.

    a. _____

    b. _____

    c. _____

    d. _____

45. List two properties of iron that change when the percentage of phosphorus changes.

    a. _____

    b _____

46. List two properties of iron that change when the percentage of silicon changes.

    a. _____

    b. _____

47. List two properties of iron that change when the percentage of nickel changes.

    a. _____

    b. _____

48. List two properties of iron that change when the percentage of molybdenum changes.

    a. _____

    b. _____

49. List two properties of iron that change when the percentage of copper changes.

    a. _____

    b. _____

50. What is meant by carbon equivalence (CE)?

    _____

    _____

51. List two functions of the core wire for stick electrodes.

    a. _____

    b. _____

52. List five functions of the flux covering for stick electrodes.

    a. _____

    b. _____

    c. _____

    d. _____

    e. _____

53. Why do some high-temperature slags solidify before the weld metal solidifies?

    _____

    _____

54. List six things that must be considered before selecting a SMAW electrode.

    a. _____

    b. _____

    c. _____

    d. _____

    e. _____

    f. _____

55. Explain the meaning of the following AWS classifications.

    1. E _____

    2. R _____

    3. ER _____

    4. EC _____

    5. B _____

    6. BR _____

    7. RG _____

    8. IN _____

    9. EW _____

    10. F _____

    11. S _____

    12. T _____

    13. L _____

56. Explain the "0" and "1" as they are used in the AWS numbering system for tubular wire.

    _____

    _____

57. Explain the AWS electrode identification system for stainless steel filler metals.

    _____

    _____

58. Explain this AWS electrode identification: "ECuAl."

    _____

    _____

59. Explain the AWS electrode identification system for aluminum and aluminum-alloy filler metals.

    _____

    _____

60. List three classifications or groups of hardfacing electrodes.

    a. _____

    b. _____

    c. _____

INSTRUCTOR'S COMMENTS _____

_____

_____

Chapter 29

Welding Automation and Robotics

Name _____ Date _____

Class _____ Instructor _____ Grade _____

Instructions: Carefully read Chapter 29 in the text and complete the following questions.

A. IDENTIFICATION

In the space provided, identify the items shown on the illustration.

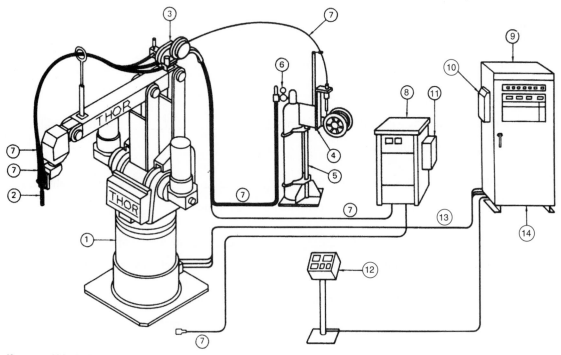

(Courtesy of Merrick Engineering, Inc.)

1. _____ arc welding robot

2. _____ robot control unit

3. _____ teaching box

4. _____ wire feeder

5. _____ welding power supply

6. _____ shielding gas flow regulator

7. _____ welding torch

B. FILL IN THE BLANK

Fill in the blank with the correct word. Answers may be more than one word.

8. The first industrial robots were mainly _____ robots used to move material, with little repetitive accuracy required.

9. _____ and microprocessors assist modern industry in producing high-quality products with a minimum waste of materials and time.

10. In welding applications, the increased use of robots, however, will never replace the tremendous need for _____.

11. The most commonly used manual arc (MA) welding process is _____.

12. The most commonly used semiautomatic arc (SA) welding processes are _____ and _____.

13. On some large machine welds, the operator may _____ with the welding head along the path of the weld.

14. Automatic welding or brazing is best suited to _____-volume production runs because of the expense involved in special jigs and fixtures.

15. A computer or microprocessor can _____ the robot's operation to positioners, conveyors, automatic fixtures, and other production machines.

16. When programming a robot using computer software, a computer _____ can be run to ensure that the program performs the operations correctly.

17. A system using a robot can be _____ % more productive than a system using manual welders.

18. The parts design must also take into consideration the higher _____ from the almost constant welding.

19. Weld spatter can stick to unprotected machine surfaces, causing the arm to jam or resulting in excessive wear.

C. SHORT ANSWER

Write a brief answer in the space provided that will answer the question or complete the statement.

20. List four factors to be considered for proper system planning.

    a. _____

    b. _____

    c. _____

    d. _____

21. List five safety precautions that are recommended for the use of automatic welding equipment and robots.

    i. _____

    ii. _____

    iii. _____

    iv. _____

    v. _____

D. ESSAY

Provide complete answers for all of the following questions.

22. What is meant by the abbreviation CAD/CAM?

    _____

    _____

23. What variables does the welder control during manual arc welding?

    _____

    _____

24. What variables does the welder control during semiautomatic arc welding?

    _____

    _____

25. What variables does the welder operator control during machine welding?

    _____

    _____

26. What is the direction of the X-axis?

    _____

    _____

27. What is the direction of the Y-axis?

_____

_____

28. What is the direction of the Z-axis?

_____

_____

29. How can a robot be programmed?

_____

_____

30. Why should the robotic arm be able to move at various speeds?

_____

_____

31. Explain why the robot should be interfaced with other equipment.

_____

_____

32. What is meant by the term work zone?

_____

_____

33. List five robot safety considerations.

a. _____

b. _____

c. _____

d. _____

e. _____

E. DEFINITIONS

34. Define the following terms:

duty cycle _____

pick and place _____

work cell _____

INSTRUCTOR'S COMMENTS _____

_____

_____

Chapter 30

Other Welding Processes

CHAPTER 30: OTHER WELDING PROCESSES QUIZ

Name _____ Date _____

Class _____ Instructor _____ Grade _____

Instructions: Carefully read Chapter 30 in the text and complete the following questions.

A. SHORT ANSWER

Write a brief answer in the space provided that will answer the question or complete the statement.

The following abbreviations refer to welding processes. Give the full name for each process.

1. RSEW _____

2. LBW _____

3. FW _____

4. SAW _____

5. PEW _____

6. PAW _____

7. RPW _____

8. THSP_____

9. UW _____

10. USW _____

11. SW _____

12. RW _____

13. EBW _____

14. RSW _____

15. ESW _____

16. PAC _____

17. List the seven materials that the AWS classifies as hardfacing metals.

    a. _____

    b. _____

    c. _____

    d. _____

    e. _____

    f. _____

    g. _____

18. What are the four functions that the timer controls on RSW equipment?

    a. _____

    b. _____

    c. _____

    d. _____

B. FILL IN THE BLANK

Fill in the blank with the correct word. Answers may be more than one word.

19. SAW, ESW, and EGW can be considered the workhorses of the fabrication industry. They can weld
    metal ranging from _____ in. (_____ mm) to more
    than _____ in. (_____ mm) thick in a single
    operation, depending on the process selected.

20. Mechanical travel can be provided by either moving the _____ along the
    joint or moving the work past a fixed _____.

21. SAW _____ are then cooled and ground into the desired granular size range.

22. SAW _____ are a mixture of fine particles of fluxing agents, deoxidizers,
    alloying elements, metal compounds, and a suitable binder that holds the mixture together in
    small, hard granules.

23. Even with a large number of SAW welders operating in confined spaces, _____
    is virtually eliminated.

24. Many codes permit SAW to be used on _____, _____,
    _____, and in many other critical applications.

25. Without some sort of _____, the SAW arc could easily move away from the joint being welded.

26. The ESW heat for welding is produced as a result of the _____ of the molten flux.

27. One drawback of the ESW process is that the metal produced during the starting process is _____ and must be removed and repaired manually.

28. During the resistance welding cycle, the mating surfaces of the parts are heated to a _____ state just before melting and are forced together.

29. The size and shape of the formed resistance spot welds are controlled somewhat by the size and contour of the _____.

30. Portable spot welders are used where work _____ to be moved.

31. Seam welding seam width should be about _____ of the sheet plus _____ in. (_____ mm).

32. RSEW-HF is used in the production of welded _____, _____, and _____ shapes.

33. The projections for resistance welding can be shape, such as _____.

34. Flash welding may be used to join dissimilar _____ alloys and to join _____ to other metals.

35. With percussion welding, immediately after or during the electrical discharge, _____ pressure is applied by a hammer blow or by the snap releasing of a spring.

36. Welds produced by the electron beam welding process are _____, which eliminates the usual fusion weld contaminants caused by water vapor, oxygen, nitrogen, hydrogen, and slag.

37. The ultrasonic welding tip oscillates in a plane _____ to the joint interface.

38. Ultrasonic welding is the most commonly used process to join the two halves of _____ packaging.

39. The inertia welding process produces a superior-quality, complete _____ weld.

40. Laser welding of high-thermal conductivity materials, such as _____, is not difficult to do.

41. Most laser beam hardening is performed on _____ parts that cannot be _____ in a normal manner or when only limited areas of a larger part need to be hardened.

42. In plasma arc welding (PAW), a _____ is produced by forcing gas to flow along an arc restricted electromagnetically as it passes through a nozzle.

43. The stud welding _____ and an area on the workpiece.

44. Thermal spraying techniques are used in _____ that involve wear and high-temperature problems.

45. Thermite compound consists of finely divided _____ and _____ mixed in a ratio of 1 to 3 by weight.

46. Hardfacing may involve _____ surfaces that have become worn.

47. Most hardfacing metals have a base of _____, _____, _____, or _____.

48. In all types of surfacing operations, the metal should _____ of all loose scale, rust, dirt, and other foreign substances before the alloy is applied.

49. The factor of _____ must be carefully considered because the composition of the added metal will differ from the base metal.

50. Tungsten _____ and _____ electrodes are included in the first group.

51. Care must be exercised when using the GMA, FCA, and GTA welding processes for hardfacing to avoid _____ of the weld.

52. To make an FS weld, the base metal must be _____ together with a backing plate held tightly behind the joint.

53. As the magnetic field of magnetic pulse welding flows across the fly plate it is forced downward at an angle called the _____ angle.

C. ESSAY

Provide complete answers for all of the following questions.

54. What is resistance welding?

_____

_____

55. How is a seam weld produced?

_____

_____

56. Describe the flash welding process.

_____

_____

57. What type of process is flash welding?

_____

_____

58. What is electron beam welding?

_____

_____

59. How does ultrasonic welding differ from resistance welding?

_____

_____

60. Explain how inertia welding is performed.

_____

_____

61. How is the concentrated coherent light beam produced in a laser?

_____

_____

62. How is plasma generated for plasma welding?

_____

_____

63. What is thermite welding?

_____

_____

64. What is hardfacing?

_____

_____

65. List four advantages of shielded metal arc hardfacing.

a. _____

b. _____

c. _____

d. _____

66. List four advantages of thermal spraying.

a. _____

b. _____

c. _____

d. _____

67. In the electroslag welding process, how is the welding heat produced?

_____

_____

D. DEFINITIONS

68. Define the following terms:

plasma _____

strip electrodes _____

stud welding _____

optical viewing system _____

upset welding _____

bonded fluxes _____

INSTRUCTOR'S COMMENTS _____

_____

_____

Chapter 31

Oxyfuel Welding and Cutting Equipment, Setup, and Operation

■ PRACTICE 31-1

Name _____ Date _____

Class _____ Instructor _____ Grade _____

OBJECTIVE: After completing this practice, you will be able to safely set up an oxyfuel torch set.

EQUIPMENT AND MATERIALS NEEDED

Disassembled oxyfuel torch set, two regulators, two reverse flow valves, one set of hoses, a torch body, a welding tip, two cylinders, a portable cart or supporting wall with safety chain, and a wrench.

INSTRUCTIONS

1. Safety chain the cylinders in the cart or to a wall. Then remove the valve protection caps.

2. Crack the cylinder valve on each cylinder for a second to blow away dirt that may be in the valve.

    CAUTION Always stand to one side. Point the valve away from anyone in the area and be sure there are no sources of ignition when cracking the valve.

3. Attach the regulators to the cylinder valves. The regulator nuts should be started by hand and then tightened with a wrench.

4. Attach a reverse flow valve or flashback arrestor, if the torch does not have them built in, to the hose connection on the regulator or to the hose connection on the torch body, depending on the type of reverse flow valve in the set. Occasionally, test each reverse flow valve by blowing through it to make sure it works properly.

5. Connect the hoses. The red hose has a left-hand nut and attaches to the fuel gas regulator. The green hose has a right-hand nut and attaches to the oxygen regulator.

6. Attach the torch to the hoses. Connect both hose nuts finger tight before using a wrench to tighten either one.

7. Check the tip seals for nicks or O rings, if used, for damage. Check the owner's manual, or a supplier, to determine if the torch tip should be tightened by hand only or should be tightened with a wrench.

CAUTION **Tightening a tip the incorrect way may be dangerous and might damage the equipment. Check all connections to be sure they are tight. The oxyfuel equipment is now assembled and ready for use.**

CAUTION **Connections should not be overtightened. If they do not seal properly, repair or replace them.**

CAUTION **Leaking cylinder valve stems should not be repaired. Turn off the valve, disconnect the cylinder, mark the cylinder, and notify the supplier to come and pick up the bad cylinder.**

The assembled oxyfuel welding equipment is now tested and ready to be ignited and adjusted.

INSTRUCTOR'S COMMENTS _____

_____

_____

■ PRACTICE 31-2

Name _____ Date _____

Class _____ Instructor _____ Grade _____

OBJECTIVE: After completing this practice, you should be able to safely turn on and test oxyfuel equipment for gas leaks.

EQUIPMENT AND MATERIALS NEEDED FOR THIS PRACTICE

1. Oxyfuel equipment that is properly assembled.

2. A nonadjustable tank wrench.

3. A leak-detecting solution.

INSTRUCTIONS

1. Back out the regulator pressure adjusting screws until they are loose.

2. Standing to one side of the regulator, open the cylinder valve SLOWLY so that the pressure rises on the gauge slowly.

   CAUTION **If the valve is opened quickly, the regulator or gauge may be damaged, or the gauge may explode.**

3. Open the oxygen valve all the way until it stops turning.

4. Open the acetylene or other fuel gas valve 1/4 turn or just enough to get gas pressure. If the cylinder valve does not have a handwheel, use a nonadjustable wrench and leave it in place on the valve stem while the gas is on.

   CAUTION **The acetylene valve should never be opened more than 1 1/2 turns, so that in an emergency it can be turned off quickly.**

5. Open one torch valve and point the tip away from any source of ignition. Slowly turn in the pressure adjusting screw until gas can be heard escaping from the torch. The gas should flow long enough to allow the hose to be completely purged (emptied) of air and replaced by the gas before the torch valve is closed. Repeat this process with the other gas.

6. After purging is completed, and with both torch valves off, adjust both regulators to read 5 psig (35 kPag).

7. Spray a leak-detecting solution on each hose and regulator connection and on each valve stem on the torch and cylinders. Watch for bubbles which indicate a leak. Turn off the cylinder valve before tightening any leaking connections.

CAUTION **Connections should not be overtightened. If they do not seal properly, repair or replace them.**

CAUTION **Leaking cylinder valve stems should not be repaired. Turn off the valve, disconnect the cylinder, mark the cylinder, and notify the supplier to come and pick up the bad cylinder.**

The assembled oxyfuel welding equipment is now tested and ready to be ignited and adjusted.

INSTRUCTOR'S COMMENTS _____

_____

_____

■ PRACTICE 31-3

Name _____ Date _____

Class _____ Instructor _____ Grade _____

OBJECTIVE: After completing this practice, you should be able to safely light and adjust an oxyacetylene flame.

EQUIPMENT AND MATERIALS NEEDED FOR THIS PRACTICE

1. Oxyfuel welding equipment that is properly assembled and tested.

2. A spark lighter.

3. Gas welding goggles.

4. Gloves and proper protective clothing.

INSTRUCTIONS

1. Wearing proper clothing, gloves, and gas welding goggles, turn both regulator adjusting screws in until the working pressure gauges read 5 psig (35 kPag). If you mistakenly turn on more than 5 psig (35 kPag), open the torch valve to allow the pressure to drop as the adjusting screw is turned outward.

2. Turn on the torch fuel-gas valve just enough so that some gas escapes.

   CAUTION Be sure the torch is pointed away from any sources of ignition or any object or person that might be damaged or harmed by the flame when it is lit.

3. Using a spark lighter, light the torch. Hold the lighter near the end of the tip but not covering the end.

   CAUTION A spark lighter is the only safe device to use when lighting any torch.

4. With the torch lit, increase the flow of acetylene until the flame almost stops smoking.

5. Slowly turn on the oxygen and adjust the torch to a neutral flame. Look for white acetylene cones, NOT BLUE.

   This flame setting uses the minimum gas flow rate for this specific tip. The fuel flow should never be adjusted to a rate below the point where the smoke stops. This is the minimum flow rate at which the cool gases will pull the flame heat out of the tip. If excessive heat is allowed to build up in a tip, it can cause a backfire or flashback.

The maximum gas flow rate gives a flame which, when adjusted to the neutral setting, does not settle back on the tip.

INSTRUCTOR'S COMMENTS _____

_____

_____

■ PRACTICE 31-4

Name _____ Date _____

Class _____ Instructor _____ Grade _____

OBJECTIVE: After completing this practice, you should be able to turn off and disassemble oxyfuel equipment.

EQUIPMENT AND MATERIALS NEEDED FOR THIS PRACTICE

1. Oxyfuel welding equipment that is properly assembled and tested.

2. A spark lighter.

3. Gas welding goggles.

4. Gloves and proper protective clothing.

INSTRUCTIONS

1. First, quickly turn off the torch fuel-gas valve. This action blows the flame out and away from the tip, ensuring that the fire is out. In addition, it prevents the flame from burning back inside the torch. On large tips or hot tips, turning the fuel off first may cause the tip to pop. The pop is caused by a lean fuel mixture in the tip.

   CAUTION **If you find that the tip pops each time you turn the fuel off first, turn the oxygen off first to prevent the pop. Be sure that the flame is out before putting the torch down.**

2. After the flame is out, turn off the oxygen valve.

3. Turn off the cylinder valves.

4. Open one torch valve at a time to bleed off the pressure.

5. When all of the pressure is released from the system, back both regulator adjusting screws out until they are loose.

6. Loosen both ends of both hoses and unscrew them.

7. Loosen both regulators and unscrew them from the cylinder valves.

8. Replace the valve protection caps.

INSTRUCTOR'S COMMENTS _____

_____

_____

CHAPTER 31: OXYFUEL WELDING AND CUTTING EQUIPMENT, SETUP, AND OPERATION QUIZ

Name _____ Date _____

Class _____ Instructor _____ Grade _____

Instructions: Carefully read Chapter 31 in the text and answer the following questions.

A. IDENTIFICATION

In the space provided, identify the items shown on the illustrations.

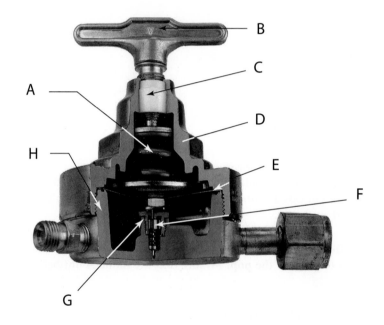

1. Name the parts of the regulator shown above.

   A. _____

   B. _____

   C. _____

   D. _____

   E. _____

   F. _____

   G. _____

   H. _____

2. From the picture below, are the fittings designed for fuel gas or oxygen? Circle the correct answer.

    a. fuel gas

    b. oxygen

3. From the picture used for Question 2, are the threads left- or right-hand threads? Circle the correct answer.

    a. left

    b. right

4. Identify each of the parts of a torch by inserting the correct letter from the illustration in the space provided.

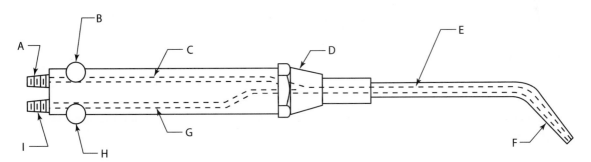

    a. _____ tip

    b. _____ mixer

    c. _____ torch body

    d. _____ control valves

    e. _____ hose connections

5. Identify each of the parts of a torch by inserting the correct letter from the illustration in the space provided.

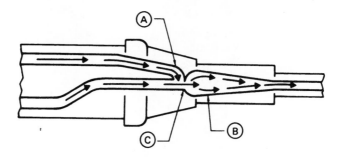

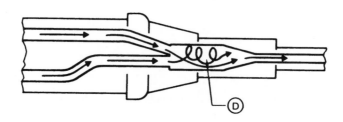

a. _____ injector mixing chamber

b. _____ equal-pressure mixing chamber

c. _____ injector

d. _____ venturi

B. SHORT ANSWER

Write a brief answer in the space provided that will answer the question or complete the statement.

6. Using the same illustration as in Question 2 and Question 3, how can you tell if the fitting is left- or right-hand threads?

_____

_____

7. All _____ processes use a high-heat and high-temperature flame produced by burning a fuel gas mixed with pure oxygen.

8. A _____ threaded fitting has a notched nut.

9. _____ a regulator may cause a fire or an explosion.

10. Oxygen hoses must be _____ in color and have right-hand threaded fittings.

C. FILL IN THE BLANK

Fill in the blank with the correct word. Answers may be more than one word.

11. The oxyfuel flame was used for fusion welding as early as the first half of the _____s when scientists developed the oxyhydrogen torch.

12. Today because of improvements in other processes, the oxyacetylene flame is seldom used on metal thicker than _____ in. (_____ mm).

13. Two-stage regulators have two sets of _____, _____ and _____.

14. The working pressure gauge shows the pressure at the _____, and not at the torch.

15. A safety release valve is made up of a small ball held tightly against a seat by a _____.

16. The advantage of using cryogenic gasses is that _____ can replace a lot of standard compressed gas cylinders.

17. Fittings should screw together _____ and require only light wrench pressure to be leak tight.

18. A gauge that gives a faulty reading or that is damaged can result in dangerous _____ settings.

19. If the regulator adjusting screw becomes tight and difficult to turn, it can be removed and cleaned with a _____.

20. The combination torch sets usually are more practical for portable welding since the one unit can be used for _____.

21. Fuel-air torches are often used by _____ and _____ technicians for brazing and soldering copper pipe and tubing.

22. The injector allows oxygen at the higher pressure to draw the fuel gas into the chamber, even when the fuel-gas pressure is as low as _____ oz/in.^{2} (_____ g/cm^2).

23. Torch tips may have _____ seals, or they may have _____ or a _____ between the tip and the torch seat.

24. When a flashback occurs, there is usually a serious problem with the equipment and a

_____.

25. The reverse flow valve is a spring-loaded _____ that closes when gas tries to flow backward through the torch valves.

26. The flashback arrestor is designed to quickly _____ the flow of gas during a flashback.

27. Fuel-gas hoses must be _____ and have left-hand threaded fittings. Oxygen hoses must be _____ and have right-hand threaded fittings.

28. When hoses are not in use, the gas _____ be turned off and the pressure bled off.

29. The crimping tool should be squeezed _____, the second time at right angles to the first.

30. Manifolds must be located _____ ft (6 m) or more from the actual work, or they must be located so that sparks cannot reach them.

31. Manifold systems should be tested for leaks at one and a half (1 1/2) times the _____ pressure.

32. If a station regulator is removed, a _____ must be put on the line so that air cannot enter the system.

33. A molecule of acetylene is made up of _____ hydrogen (H) and _____ carbon (C) atoms (C_2H_2).

34. In the secondary reaction, oxygen (O_2) from the surrounding air unites with free hydrogen (H) to form _____ (H_2O) and liberates more heat.

35. A homogeneous mixture of 50% acetylene and 50% oxygen has a burn rate of _____ ft per second (_____ m per second).

36. Acetylene is colorless, is lighter than air, and has a strong _____ smell.

37. The cylinder is filled with a porous material, and then acetone is added to the cylinder where it absorbs about _____ times its own weight in acetylene.

38. The high temperature produced by the oxyacetylene flame is concentrated around the
_____ cone.

39. High withdrawal rates of gas from liquefied gas cylinders will cause a drop in pressure, a
_____ of the cylinder temperature, and the possibility of freezing the regulator.

40. Because of propane and natural gas relatively low-temperature, low-heat flames, they are seldom used
for purposes other than _____ and as preheat fuels for _____.

41. Hydrogen has no smell, which makes it difficult to _____.

42. The short ends of both welding and brazing rods can be fused together so that the amount of
_____ metal can be minimized.

43. Mild steel and low alloy gas welding rods are classified by the AWS as _____,
_____, and _____.

D. ESSAY

Provide complete answers for all of the following questions.

44. Why should long hoses be larger in diameter than short hoses?

_____

_____

45. What is the difference between psi and psig (kg/cm^2 and kg/cm^2G)?

_____

_____

46. What does it mean if a regulator creeps?

_____

_____

47. How is a leaking regulator diaphragm tested?

_____

_____

48. What is a backfire?

_____

_____

49. What is a flashback?

_____

_____

50. Why can an injector torch system use low-pressure fuel gases?

_____

_____

51. How can a leak around a torch valve stem be stopped?

_____

_____

52. Why are hoses bled?

_____

_____

53. How far should acetylene valves be turned on? Why?

_____

_____

54. How does a regulator control gas pressure?

_____

_____

E. DEFINITIONS

55. Define the following terms: _____

    Bourdon tube _____

    safety-release valve _____

    safety disc _____

    mixing chamber _____

    injector _____

    manifold _____

    cylinder pressure _____

    diaphragm _____

    creep _____

    Siamese hose _____

    leak-detecting solution _____

INSTRUCTOR'S COMMENTS _____

_____

_____

Chapter 32

Oxyacetylene Welding

■ PRACTICE 32-1

Name _____ Date _____

Base Metal Thickness _____ Filler Metal Diameter_____

Class _____ Instructor _____ Grade _____

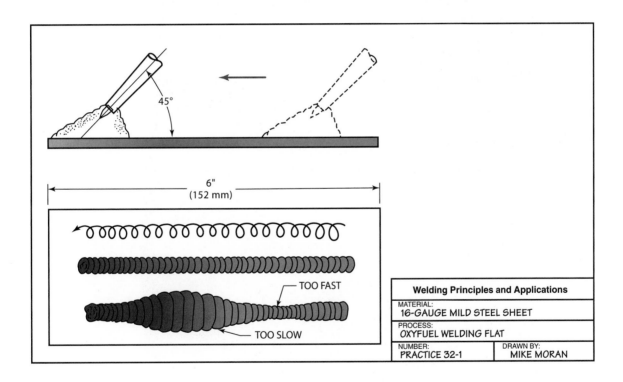

6"
(152 mm)

TOO FAST

TOO SLOW

Welding Principles and Applications

MATERIAL:	
16-GAUGE MILD STEEL SHEET	
PROCESS:	
OXYFUEL WELDING FLAT	
NUMBER:	DRAWN BY:
PRACTICE 32-1	MIKE MORAN

OBJECTIVE: After completing this practice, you should be able to push a weld pool on 16-gauge carbon steel sheet.

EQUIPMENT AND MATERIALS NEEDED FOR THIS PRACTICE

1. Properly set-up oxyacetylene welding equipment.

2. Proper safety protection (welding goggles, safety glasses, spark lighter, and pliers). Refer to Chapter 2 in the text for more specific safety information.

3. Two or more pieces of 16-gauge carbon steel sheet, approximately 3 in. (76 mm) wide by 6 in. (152 mm) long.

INSTRUCTIONS

1. Start at one end and hold the torch at a 45° angle in the direction of the weld.

2. When the metal starts to melt, move the torch in a circular pattern down the sheet toward the other end. If the size of the molten weld pool changes, speed up or slow down to keep it the same size all the way down the sheet.

3. Repeat this practice until you can keep the width of the molten weld pool uniform and the direction of travel in a straight line.

4. Turn off the gas cylinders and regulators, and clean up your work area when you are finished welding.

INSTRUCTOR'S COMMENTS _____

_____

_____

■ PRACTICE 32-2

Name _____ Date _____

Base Metal Thickness _____ Filler Metal Diameter_____

Class _____ Instructor _____ Grade _____

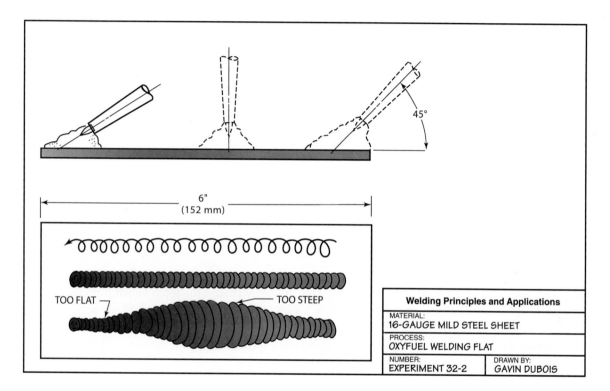

6" (152 mm)		
TOO FLAT		TOO STEEP

Welding Principles and Applications

MATERIAL: 16-GAUGE MILD STEEL SHEET	
PROCESS: OXYFUEL WELDING FLAT	
NUMBER: EXPERIMENT 32-2	DRAWN BY: GAVIN DUBOIS

OBJECTIVE: After completing this experiment, you should be able to demonstrate the controlling of the weld pool by changing the torch angle and torch height while pushing a pool on 16-gauge carbon steel sheet.

EQUIPMENT AND MATERIALS NEEDED FOR THIS EXPERIMENT

1. Properly set-up oxyacetylene welding equipment.

2. Proper safety protection (welding goggles, safety glasses, spark lighter, and pliers). Refer to Chapter 2 in the text for more specific safety information.

3. One or more pieces of 16-gauge carbon steel sheet, approximately 6 in. (152 mm) long.

INSTRUCTIONS

CHANGING TORCH ANGLE

1. Start at one end and hold the torch at a 45° angle in the direction of the weld.

2. When the metal starts to melt, move the torch in a circular pattern down the sheet toward the other end. As the weld is moved toward the other end, gradually change the torch angle from the starting 45° to an angle of 90° then to an ending angle of 30°.

3. Repeat this part of the experiment until you can control the width of the molten weld pool by changing the torch angle.

INSTRUCTOR'S COMMENTS _____

_____

_____

CHANGING TORCH HEIGHT

1. Start at one end and hold the torch with the inner cone about 1/8 in. (3 mm) above the metal surface at a 45° angle in the direction of the weld.

2. When the metal starts to melt, move the torch in a circular pattern down the sheet toward the other end. As the weld is moved toward the other end, gradually change the torch height from the starting 1/8 in. (3 mm) to an ending height of 1/2 in. (13 mm).

3. Repeat this part of the experiment until you can control the width of the molten weld pool by changing the torch height.

4. Turn off the gas cylinders and regulators, and clean up your work area when you are finished welding.

INSTRUCTOR'S COMMENTS _____

_____

_____

■ PRACTICE 32-3

Name _____ Date _____

Base Metal Thickness _____ Filler Metal Diameter_____

Class _____ Instructor _____ Grade _____

OBJECTIVE: After completing this practice, you should be able to make stringer beads on sheet steel in the flat position.

EQUIPMENT AND MATERIALS NEEDED FOR THIS PRACTICE

1. Properly set-up oxyacetylene welding equipment.

2. Proper safety protection (welding goggles, safety glasses, spark lighter, and pliers). Refer to Chapter 2 in the text for more specific safety information.

3. One or more pieces of 16-gauge carbon steel sheet, approximately 3 in. (76 mm) wide by 6 in. (152 mm) long.

4. 1/16 in. (2 mm), 3/32 in. (2.4 mm), and 1/8 in. (3 mm) diameter RG45 filler wire. Determine by trial and error the size of filler rod with which you are most comfortable.

INSTRUCTIONS

1. Use a clean piece of 16-gauge mild steel and a torch adjusted to a neutral flame.

2. Hold the torch at a 45° angle to the metal with the inner cone about 1/8 in. (3 mm) above the metal surface.

3. The end of the filler rod should always be kept inside the protective envelope of the flame. Experiment with the three different filler rods for your preference of filler wire.

4. The end of the filler rod should be dipped into the leading edge of the molten weld pool. If the filler rod touches the hot metal around the molten weld pool, it will stick. When this happens, move the flame directly to the end of the filler rod to melt and free it.

5. Move the torch in a circular pattern down the sheet toward the other end while adding filler rod.

6. The width and buildup of the welding bead should be uniform.

7. Repeat this practice until you can keep the width of the bead uniform and the direction of travel in a straight line.

8. Turn off the gas cylinders and regulators, and clean up your work area when you are finished welding.

INSTRUCTOR'S COMMENTS _____

_____

_____

■ PRACTICE 32-4

Name _____ Date _____

Base Metal Thickness _____ Filler Metal Diameter_____

Class _____ Instructor _____ Grade _____

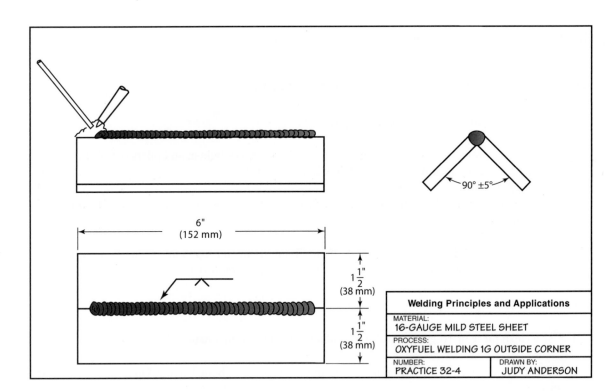

Welding Principles and Applications

| MATERIAL: |
| 16-GAUGE MILD STEEL SHEET |

| PROCESS: |
| OXYFUEL WELDING 1G OUTSIDE CORNER |

| NUMBER: | DRAWN BY: |
| PRACTICE 32-4 | JUDY ANDERSON |

OBJECTIVE: After completing this practice, you should be able to weld outside corner joints in the flat position.

EQUIPMENT AND MATERIALS NEEDED FOR THIS PRACTICE

1. Properly set-up oxyacetylene welding equipment.

2. Proper safety protection (welding goggles, safety glasses, spark lighter, and pliers). Refer to Chapter 2 in the text for more specific safety information.

3. Two or more pieces of 16-gauge carbon steel sheet, approximately 1 1/2 in. (38 mm) wide by 6 in. (152 mm) long.

4. 1/16 in. (2 mm), 3/32 in. (2.4 mm), and 1/8 in. (3 mm) diameter RG45 filler wire. Determine by trial and error the size of filler rod with which you are most comfortable.

INSTRUCTIONS

1. Place one of the pieces of metal in a jig or on a firebrick and hold or brace the other piece of metal horizontally on it.

2. Tack the ends of the two sheets together.

3. Then set it upright and put two or three more tacks on the joint.

4. Hold the torch at a 45° angle to the metal with the inner cone about 1/8 in. (3 mm) above the metal surface and make a uniform weld along the joint.

5. Repeat this weld until the weld can be made without defects.

6. Turn off the gas cylinders and regulators, and clean up your work area when you are finished welding.

INSTRUCTOR'S COMMENTS _____

_____

_____

■ PRACTICE 32-5

Name _____ Date _____

Base Metal Thickness _____ Filler Metal Diameter _____

Class _____ Instructor _____ Grade _____

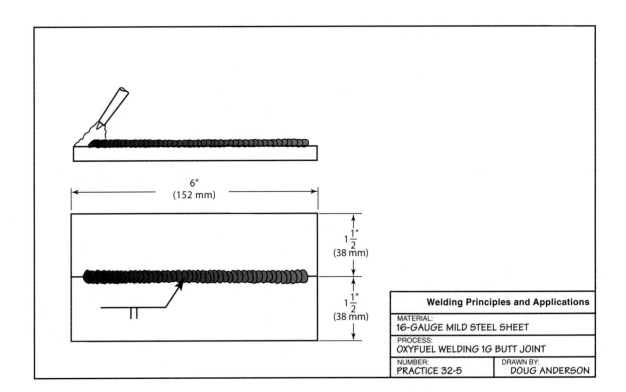

6"
(152 mm)

$1\frac{1}{2}"$
(38 mm)

$1\frac{1}{2}"$
(38 mm)

Welding Principles and Applications

MATERIAL:
16-GAUGE MILD STEEL SHEET

PROCESS:
OXYFUEL WELDING 1G BUTT JOINT

NUMBER:	DRAWN BY:
PRACTICE 32-5	DOUG ANDERSON

OBJECTIVE: After completing this practice, you should be able to weld butt joints in the flat position.

EQUIPMENT AND MATERIALS NEEDED FOR THIS PRACTICE

1. Properly set-up oxyacetylene welding equipment.

2. Proper safety protection (welding goggles, safety glasses, spark lighter, and pliers). Refer to Chapter 2 in the text for more specific safety information.

3. Two or more pieces of 16-gauge carbon steel sheet, approximately 1 1/2 in. (38 mm) wide by 6 in. (152 mm) long.

4. 1/16 in. (2 mm), 3/32 in. (2.4 mm), and 1/8 in. (3 mm) diameter RG45 filler wire. Determine by trial and error the size of filler rod with which you are most comfortable.

INSTRUCTIONS

1. Place the two pieces of metal in a jig or on a firebrick and tack weld both ends together.

2. The tack on the ends can be made by simply heating the ends and allowing them to fuse together or by placing a small drop of filler metal on the sheet and heating the filler metal until it fuses to the sheet. The latter way is especially convenient if you have to use one hand to hold the sheets together and the other to hold the torch.

3. After both ends are tacked together, place one or two small tacks along the joint to prevent warping during welding.

4. With the sheets tacked together, start welding from one end to the other.

5. Repeat this weld until you can make a welded butt joint that is uniform in width and reinforcement and has no visual defects. The penetration of this practice weld may vary.

6. Turn off the gas cylinders and regulators, and clean up your work area when you are finished welding.

INSTRUCTOR'S COMMENTS _____

_____

_____

■ PRACTICE 32-6

Name _____ Date _____

Base Metal Thickness _____ Filler Metal Diameter_____

Class _____ Instructor _____ Grade _____

OBJECTIVE: After completing this practice you should be able to make a butt joint in the flat position with 100% penetration.

EQUIPMENT AND MATERIALS NEEDED FOR THIS PRACTICE

1. Properly set-up oxyacetylene welding equipment.

2. Proper safety protection (welding goggles, safety glasses, spark lighter, and pliers). Refer to Chapter 2 in the text for more specific safety information.

3. Two or more pieces of 16-gauge carbon steel sheet, approximately 1 1/2 in. (38 mm) wide by 6 in. (152 mm) long.

4. 1/16 in. (2 mm), 3/32 in. (2.4 mm), and 1/8 in. (3 mm) diameter RG45 filler rods. Determine by trial and error the size of filler rod with which you are most comfortable.

INSTRUCTIONS

1. Place the two pieces of metal in a jig or on a firebrick and tack weld both ends together in a butt joint configuration.

2. After both ends are tacked together, place one or two additional tacks along the joint to prevent warping.

3. Apply heat for a long enough period of time so that a keyhole appears before adding filler metal. This will ensure 100% penetration of the joint.

4. Continue making keyholes all the way along the joint and filling them with filler metal.

5. Inspect the bottom side of the weld for 100% penetration and visual defects. Repeat this practice until you can consistently make welds with 100% penetration and that are visually defect free.

6. Turn off the gas cylinders and regulators and clean up your work area when you are finished.

INSTRUCTOR'S COMMENTS _____

_____

_____

■ PRACTICE 32-7

Name _____ Date _____

Base Metal Thickness _____ Filler Metal Diameter_____

Class _____ Instructor _____ Grade _____

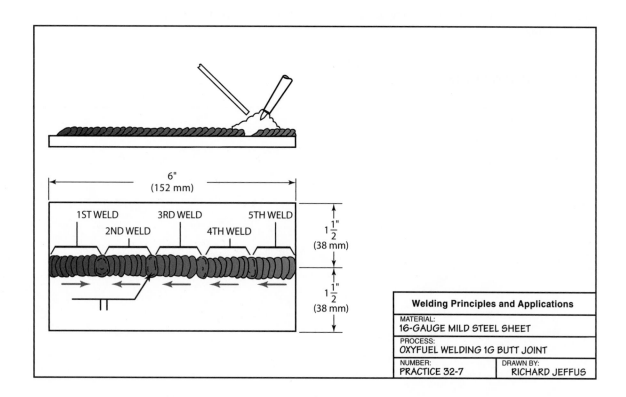

6"
(152 mm)

1ST WELD   3RD WELD   5TH WELD

2ND WELD   4TH WELD

$1\frac{1}{2}"$
(38 mm)

$1\frac{1}{2}"$
(38 mm)

Welding Principles and Applications

MATERIAL:
16-GAUGE MILD STEEL SHEET

PROCESS:
OXYFUEL WELDING 1G BUTT JOINT

NUMBER:
PRACTICE 32-7

DRAWN BY:
RICHARD JEFFUS

OBJECTIVE: After completing this practice, you should be able to weld butt joints on gauge material with minimum distortion.

EQUIPMENT AND MATERIALS NEEDED FOR THIS PRACTICE

1. Properly set-up oxyacetylene welding equipment.

2. Proper safety protection (welding goggles, safety glasses, spark lighter, and pliers). Refer to Chapter 2 in the text for more specific safety information.

3. Two or more pieces of 16-gauge carbon steel sheet, approximately 1 1/2 in. (38 mm) wide by 6 in. (152 mm) long.

4. 1/16 in. (2 mm), 3/32 in. (2.4 mm), and 1/8 in. (3 mm) diameter RG45 filler wire. Determine by trial and error the size of filler rod with which you are most comfortable.

INSTRUCTIONS

1. Tack weld the two plates together.

2. Use a back-stepping technique. Distortion can be controlled by back stepping, proper tacking, and clamping.

3. Practice this weld until you can pass a visual inspection for distortion.

4. Turn off the gas cylinders and regulators, and clean up your work area when you are finished welding.

INSTRUCTOR'S COMMENTS _____

_____

_____

■ PRACTICE 32-8

Name _____ Date _____

Base Metal Thickness _____ Filler Metal Diameter_____

Class _____ Instructor _____ Grade _____

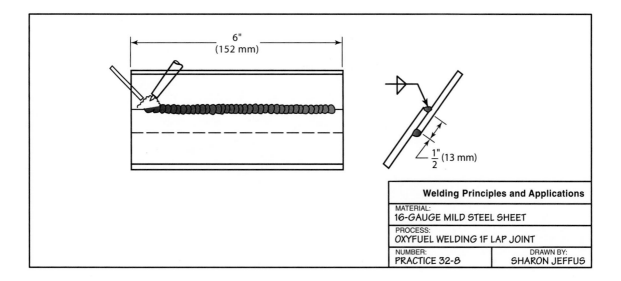

Welding Principles and Applications		
MATERIAL: 16-GAUGE MILD STEEL SHEET		
PROCESS: OXYFUEL WELDING 1F LAP JOINT		
NUMBER: PRACTICE 32-8	DRAWN BY: SHARON JEFFUS	

OBJECTIVE: After completing this practice, you should be able to weld lap joints on gauge material in the flat position.

EQUIPMENT AND MATERIALS NEEDED FOR THIS PRACTICE

1. Properly set-up oxyacetylene welding equipment.

2. Proper safety protection (welding goggles, safety glasses, spark lighter, and pliers). Refer to Chapter 2 in the text for more specific safety information.

3. Two or more pieces of 16-gauge carbon steel sheet, approximately 1 1/2 in. (38 mm) wide by 6 in. (152 mm) long.

4. 1/16 in. (2 mm), 3/32 in. (2.4 mm), and 1/8 in. (3 mm) diameter RG45 filler wire. Determine by trial and error the size of filler rod with which you are most comfortable.

INSTRUCTIONS

1. Place the two pieces of metal on a firebrick and tack both ends.

2. Starting at one end, make a uniform weld along the joint. Both sides of the joint can be welded.

3. Repeat this weld until the weld can be made without defects.

4. Turn off the gas cylinders and regulators, and clean up your work area when you are finished welding.

INSTRUCTOR'S COMMENTS _____

_____

_____

■ PRACTICE 32-9

Name _____ Date _____

Base Metal Thickness _____ Filler Metal Diameter_____

Class _____ Instructor _____ Grade _____

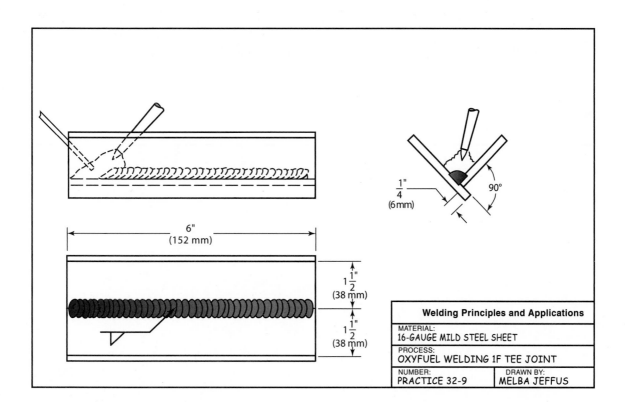

Welding Principles and Applications	
MATERIAL: 16-GAUGE MILD STEEL SHEET	
PROCESS: OXYFUEL WELDING 1F TEE JOINT	
NUMBER: PRACTICE 32-9	DRAWN BY: MELBA JEFFUS

OBJECTIVE: After completing this practice, you should be able to weld tee joints on gauge materials in the flat position.

EQUIPMENT AND MATERIALS NEEDED FOR THIS PRACTICE

1. Properly set-up oxyacetylene welding equipment.

2. Proper safety protection (welding goggles, safety glasses, spark lighter, and pliers). Refer to Chapter 2 in the text for more specific safety information.

3. Two or more pieces of 16-gauge carbon steel sheet, approximately 1 1/2 in. (38 mm) wide by 6 in. (152 mm) long.

4. 1/16 in. (2 mm), 3/32 in. (2.4 mm), and 1/8 in. (3 mm) diameter RG45 filler wire. Determine by trial and error the size of filler rod with which you are most comfortable.

INSTRUCTIONS

1. Place the first piece of metal flat on a firebrick and hold or brace the second piece vertically on the first piece. The vertical piece should be within 5° of square to the bottom sheet.

2. Tack the two sheets at the ends.

3. Starting at one end, make a uniform weld along the joint.

4. Repeat this weld until the weld can be made without defects.

5. Turn off the gas cylinders and regulators, and clean up your work area when you are finished welding.

INSTRUCTOR'S COMMENTS _____

_____

_____

■ PRACTICE 32-10

Name _____ Date _____

Base Metal Thickness _____ Filler Metal Diameter_____

Class _____ Instructor _____ Grade _____

OBJECTIVE: After completing this practice you should be able to make a stringer bead at a 45° angle.

EQUIPMENT AND MATERIALS NEEDED FOR THIS PRACTICE

1. Properly set-up oxyacetylene welding equipment.

2. Proper safety protection (welding goggles, safety glasses, spark lighter, and pliers). Refer to Chapter 2 in the text for more specific safety information.

3. One or more pieces of 16-gauge carbon steel sheet, approximately 3 in. (76 mm) wide by 6 in. (152 mm) long.

4. 1/16 in. (2 mm), 3/32 in. (2.4 mm), and 1/8 in. (3 mm) diameter RG45 filler rods. Determine by trial and error the size of filler rod with which you are most comfortable.

INSTRUCTIONS

1. Position the workpiece at a 45° angle to the surface of the table.

2. Use the torch and add filler metal as you did in Practice 32-3.

3. It may be necessary to flash the torch away from the metal to avoid overheating. Move the hotter primary flame away from the puddle while still keeping the puddle within the secondary flame.

4. Establish a rhythm of moving the torch and adding filler metal so as to keep the bead uniform.

5. Inspect the completed bead for uniformity and freedom from defects.

6. Repeat this practice until you can consistently make welds that are visually defect free.

7. Turn off the gas cylinders and regulators, and clean up your work area when you are finished.

INSTRUCTOR'S COMMENTS _____

_____

_____

■ PRACTICE 32-11

Name _____ Date _____

Base Metal Thickness _____ Filler Metal Diameter_____

Class _____ Instructor _____ Grade _____

OBJECTIVE: After completing this practice, you should be able to make a stringer bead in the vertical position.

EQUIPMENT AND MATERIALS NEEDED FOR THIS PRACTICE

1. Properly set-up oxyacetylene welding equipment.

2. Proper safety protection (welding goggles, safety glasses, spark lighter, and pliers). Refer to Chapter 2 in the text for more specific safety information.

3. One or more pieces of 16-gauge carbon steel sheet, approximately 3 in. (76 mm) wide by 6 in. (152 mm) long.

4. 1/16 in. (2 mm), 3/32 in. (2.4 mm), and 1/8 in. (3 mm) diameter RG45 filler rods. Determine by trial and error the size of filler rod with which you are most comfortable.

INSTRUCTIONS

1. Position the workpiece in a vertical position.

2. Use the torch and add filler metal as you did in Practice 32-3. Try making beads from the bottom up and from the top down.

3. It may be necessary to flash the torch away from the metal to avoid overheating. Move the hotter primary flame away from the puddle while still keeping the puddle within the secondary flame.

4. Establish a rhythm of moving the torch and adding filler metal so as to keep the bead uniform.

5. Inspect the completed bead for uniformity and freedom from defects.

6. Repeat this practice until you can consistently make welds that are visually defect free.

7. Turn off the gas cylinders and regulators, and clean up your work area when you are finished.

INSTRUCTOR'S COMMENTS _____

_____

_____

■ PRACTICE 32-12

Name _____ Date _____

Base Metal Thickness _____ Filler Metal Diameter_____

Class _____ Instructor _____ Grade _____

OBJECTIVE: After completing this practice you should be able to make a butt joint at a 45° angle.

EQUIPMENT AND MATERIALS NEEDED FOR THIS PRACTICE

1. Properly set-up oxyacetylene welding equipment.

2. Proper safety protection (welding goggles, safety glasses, spark lighter, and pliers). Refer to Chapter 2 in the text for more specific safety information.

3. Two or more pieces of 16-gauge carbon steel sheet, approximately 1 1/2 in. (38 mm) wide by 6 in. (152 mm) long.

4. 1/16 in. (2 mm), 3/32 in. (2.4 mm), and 1/8 in. (3 mm) diameter RG45 filler rods. Determine by trial and error the size of filler rod with which you are most comfortable.

INSTRUCTIONS

1. Place the two pieces of metal in a jig or on a firebrick and tack weld both ends together in a butt joint configuration.

2. After both ends are tacked together, place one or two additional tacks along the joint to prevent warping.

3. Position the tacked workpiece at a 45° angle to the surface of the table.

4. Use the torch and add filler metal as you did in Practice 32-5.

5. Establish a rhythm of moving the torch and adding filler metal so as to keep the bead uniform.

6. Inspect the completed bead for uniformity and freedom from defects.

7. Repeat this practice until you can consistently make welds that are visually defect free.

8. Turn off the gas cylinders and regulators, and clean up your work area when you are finished.

INSTRUCTOR'S COMMENTS _____

_____

_____

■ PRACTICE 32-13

Name _____ Date _____

Base Metal Thickness _____ Filler Metal Diameter_____

Class _____ Instructor _____ Grade _____

OBJECTIVE: After completing this practice you should be able to make a butt joint in a vertical position.

EQUIPMENT AND MATERIALS NEEDED FOR THIS PRACTICE

1. Properly set-up oxyacetylene welding equipment.

2. Proper safety protection (welding goggles, safety glasses, spark lighter, and pliers). Refer to Chapter 2 in the text for more specific safety information.

3. Two or more pieces of 16-gauge carbon steel sheet, approximately 1 1/2 in. (38 mm) wide by 6 in. (152 mm) long.

4. 1/16 in. (2 mm), 3/32 in. (2.4 mm), and 1/8 in. (3 mm) diameter RG45 filler rods. Determine by trial and error the size of filler rod with which you are most comfortable.

INSTRUCTIONS

1. Place the two pieces of metal in a jig or on a firebrick and tack weld both ends together in a butt joint configuration. After both ends are tacked together, place one or two additional tacks along the joint to prevent warping.

2. Position the tacked workpiece in a vertical position.

3. Use the torch and add filler metal as you did in Practice 32-5. Try making the bead from the bottom up and from the top down.

4. Establish a rhythm of moving the torch and adding filler rod so as to keep the bead uniform.

5. Inspect the completed bead for uniformity and freedom from visual defects.

6. Repeat this practice until you can consistently make welds with 100% penetration and that are visually defect free.

7. Turn off the gas cylinders and regulators, and clean up your work area when you are finished.

INSTRUCTOR'S COMMENTS _____

_____

_____

■ PRACTICE 32-14

Name _____ Date _____

Base Metal Thickness _____ Filler Metal Diameter_____

Class _____ Instructor _____ Grade _____

OBJECTIVE: After completing this practice you should be able to make a butt joint in a vertical position with 100% penetration.

EQUIPMENT AND MATERIALS NEEDED FOR THIS PRACTICE

1. Properly set-up oxyacetylene welding equipment.

2. Proper safety protection (welding goggles, safety glasses, spark lighter, and pliers). Refer to Chapter 2 in the text for more specific safety information.

3. Two or more pieces of 16-gauge carbon steel sheet, approximately 1 1/2 in. (38 mm) wide by 6 in. (152 mm) long.

4. 1/16 in. (2 mm), 3/32 in. (2.4 mm), and 1/8 in. (3 mm) diameter RG45 filler rods. Determine by trial and error the size of filler rod with which you are most comfortable.

INSTRUCTIONS

1. Place the two pieces of metal in a jig or on a firebrick and tack weld both ends together in a butt joint configuration. After both ends are tacked together, place one or two additional tacks along the joint to prevent warping.

2. Position the tacked workpiece in a vertical position.

3. Use the torch and add filler metal as you did in Practice 32-6. Try making the bead from the bottom up and from the top down.

4. Apply heat for a long enough period of time so that a keyhole appears before adding filler metal. This will ensure 100% penetration of the joint.

5. Inspect the bottom side of the weld for 100% penetration and visual defects.

6. Repeat this practice until you can consistently make welds with 100% penetration and that are visually defect free.

7. Turn off the gas cylinders and regulators, and clean up your work area when you are finished.

INSTRUCTOR'S COMMENTS _____

_____

_____

■ PRACTICE 32-15

Name _____ Date _____

Base Metal Thickness _____ Filler Metal Diameter_____

Class _____ Instructor _____ Grade _____

OBJECTIVE: After completing this practice you should be able to make a lap joint at a 45° angle.

EQUIPMENT AND MATERIALS NEEDED FOR THIS PRACTICE

1. Properly set-up oxyacetylene welding equipment.

2. Proper safety protection (welding goggles, safety glasses, spark lighter, and pliers). Refer to Chapter 2 in the text for more specific safety information.

3. Two or more pieces of 16-gauge carbon steel sheet, approximately 1 1/2 in. (38 mm) wide by 6 in. (152 mm) long.

4. 1/16 in. (2 mm), 3/32 in. (2.4 mm), and 1/8 in. (3 mm) diameter RG45 filler rods. Determine by trial and error the size of filler rod with which you are most comfortable.

INSTRUCTIONS

1. Place the two pieces of metal in a jig or on a firebrick and tack weld both ends together in a lap joint configuration. After both ends are tacked together, place one or two additional tacks along the joint to prevent warping.

2. Position the tacked workpiece at a 45° to the surface of the table.

3. Use the torch and add filler metal as you did in Practice 32-8.

4. It may be necessary to flash off the torch to control the heat of the molten weld pool.

5. Inspect the bead for uniformity and for visual defects.

6. Repeat this practice until you can consistently make welds that are visually defect free.

7. Turn off the gas cylinders and regulators, and clean up your work area when you are finished.

INSTRUCTOR'S COMMENTS _____

_____

_____

■ PRACTICE 32-16

Name _____ Date _____

Base Metal Thickness _____ Filler Metal Diameter_____

Class _____ Instructor _____ Grade _____

OBJECTIVE: After completing this practice you should be able to make a lap joint in the vertical position.

EQUIPMENT AND MATERIALS NEEDED FOR THIS PRACTICE

1. Properly set-up oxyacetylene welding equipment.

2. Proper safety protection (welding goggles, safety glasses, spark lighter, and pliers). Refer to Chapter 2 in the text for more specific safety information.

3. Two or more pieces of 16-gauge carbon steel sheet, approximately 1 1/2 in. (38 mm) wide by 6 in. (152 mm) long.

4. 1/16 in. (2 mm), 3/32 in. (2.4 mm), and 1/8 in. (3 mm) diameter RG45 filler rods. Determine by trial and error the size of filler rod with which you are most comfortable.

INSTRUCTIONS

1. Place the two pieces of metal in a jig or on a firebrick and tack weld both ends together in a lap joint configuration. After both ends are tacked together, place one or two additional tacks along the joint to prevent warping.

2. Position the tacked workpiece in a vertical position. Try making the bead from the bottom up and from the top down.

3. Use the torch and add filler metal as you did in Practice 32-8.

4. It may be necessary to flash off the torch to control the heat of the molten weld pool.

5. Inspect the bead for uniformity and for visual defects.

6. Repeat this practice until you can consistently make welds that are visually defect free.

7. Turn off the gas cylinders and regulators, and clean up your work area when you are finished.

INSTRUCTOR'S COMMENTS _____

_____

_____

■ PRACTICE 32-17

Name _____ Date _____

Base Metal Thickness _____ Filler Metal Diameter_____

Class _____ Instructor _____ Grade _____

OBJECTIVE: After completing this practice you should be able to make a tee joint at a 45° angle.

EQUIPMENT AND MATERIALS NEEDED FOR THIS PRACTICE

1. Properly set-up oxyacetylene welding equipment.

2. Proper safety protection (welding goggles, safety glasses, spark lighter, and pliers). Refer to Chapter 2 in the text for more specific safety information.

3. Two or more pieces of 16-gauge carbon steel sheet, approximately 1 1/2 in. (38 mm) wide by 6 in. (152 mm) long.

4. 1/16 in. (2 mm), 3/32 in. (2.4 mm), and 1/8 in. (3 mm) diameter RG45 filler rods. Determine by trial and error the size of filler rod with which you are most comfortable.

INSTRUCTIONS

1. Place the two pieces of metal in a jig or on a firebrick and tack weld them together in a tee joint configuration.

2. Position the tacked workpiece at a 45° angle to the surface of the table.

3. Use the torch and add filler metal as you did in Practice 32-9.

4. As you make the bead, be sure that it has uniform width and reinforcement.

5. Inspect the bead for uniformity and for visual defects.

6. Repeat this practice until you can consistently make welds that are visually defect free.

7. Turn off the gas cylinders and regulators, and clean up your work area when you are finished.

INSTRUCTOR'S COMMENTS _____

_____

_____

■ PRACTICE 32-18

Name _____ Date _____

Base Metal Thickness _____ Filler Metal Diameter_____

Class _____ Instructor _____ Grade _____

OBJECTIVE: After completing this practice you should be able to make a tee joint in a vertical position.

EQUIPMENT AND MATERIALS NEEDED FOR THIS PRACTICE

1. Properly set-up oxyacetylene welding equipment.

2. Proper safety protection (welding goggles, safety glasses, spark lighter, and pliers). Refer to Chapter 2 in the text for more specific safety information.

3. Two or more pieces of 16-gauge carbon steel sheet, approximately 1 1/2 in. (38 mm) wide by 6 in. (152 mm) long.

4. 1/16 in. (2 mm), 3/32 in. (2.4 mm), and 1/8 in. (3 mm) diameter RG45 filler rods. Determine by trial and error the size of filler rod with which you are most comfortable.

INSTRUCTIONS

1. Place the two pieces of metal in a jig or on a firebrick and tack weld them together in a tee joint configuration.

2. Position the tacked workpiece in a vertical position. Try making the bead from the bottom up and from the top down.

3. Use the torch and add filler metal as you did in Practice 32-9.

4. As you make the bead, be sure that it has uniform width and reinforcement.

5. Inspect the bead for uniformity and for visual defects.

6. Repeat this practice until you can consistently make welds that are visually defect free.

7. Turn off the gas cylinders and regulators, and clean up your work area when you are finished.

INSTRUCTOR'S COMMENTS _____

_____

_____

■ PRACTICE 32-19

Name _____ Date _____

Base Metal Thickness _____ Filler Metal Diameter_____

Class _____ Instructor _____ Grade _____

OBJECTIVE: After completing this practice you should be able to make a horizontal stringer bead at a 45° reclining angle.

EQUIPMENT AND MATERIALS NEEDED FOR THIS PRACTICE

1. Properly set-up oxyacetylene welding equipment.

2. Proper safety protection (welding goggles, safety glasses, spark lighter, and pliers). Refer to Chapter 2 in the text for more specific safety information.

3. One or more pieces of 16-gauge carbon steel sheet, approximately 3 in. (76 mm) wide by 6 in. (152 mm) long.

4. 1/16 in. (2 mm), 3/32 in. (2.4 mm), and 1/8 in. (3 mm) diameter RG45 filler rods. Determine by trial and error the size of filler rod with which you are most comfortable.

INSTRUCTIONS

1. Position the workpiece in a 45° horizontal reclining position.

2. Use the torch and add filler metal as you did in Practice 32-3.

3. Add the filler metal along the top leading edge of the molten weld pool.

4. You may have to flash the torch off the molten weld pool occasionally to control the heat and avoid sagging.

5. Inspect the bead for uniformity and for visual defects.

6. Repeat this practice until you can consistently make welds that are visually defect free.

7. Turn off the gas cylinders and regulators, and clean up your work area when you are finished.

INSTRUCTOR'S COMMENTS _____

_____

_____

■ PRACTICE 32-20

Name _____ Date _____

Base Metal Thickness _____ Filler Metal Diameter_____

Class _____ Instructor _____ Grade _____

OBJECTIVE: After completing this practice you should be able to make a horizontal stringer bead in the horizontal position.

EQUIPMENT AND MATERIALS NEEDED FOR THIS PRACTICE

1. Properly set-up oxyacetylene welding equipment.

2. Proper safety protection (welding goggles, safety glasses, spark lighter, and pliers). Refer to Chapter 2 in the text for more specific safety information.

3. One or more pieces of 16-gauge carbon steel sheet, approximately 3 in. (76 mm) wide by 6 in. (152 mm) long.

4. 1/16 in. (2 mm), 3/32 in. (2.4 mm), and 1/8 in. (3 mm) diameter RG45 filler rods. Determine by trial and error the size of filler rod with which you are most comfortable.

INSTRUCTIONS

1. Position the workpiece in a horizontal position.

2. Use the torch and add filler metal as you did in Practice 32-3.

3. Add the filler metal along the top leading edge of the molten weld pool.

4. You may have to flash the torch off the molten weld pool occasionally to control the heat and avoid sagging.

5. Inspect the bead for uniformity and for visual defects.

6. Repeat this practice until you can consistently make welds that are visually defect free.

7. Turn off the gas cylinders and regulators, and clean up your work area when you are finished.

INSTRUCTOR'S COMMENTS _____

_____

_____

■ PRACTICE 32-21

Name _____ Date _____

Base Metal Thickness _____ Filler Metal Diameter_____

Class _____ Instructor _____ Grade _____

OBJECTIVE: After completing this practice you should be able to make a butt joint in the horizontal position.

EQUIPMENT AND MATERIALS NEEDED FOR THIS PRACTICE

1. Properly set-up oxyacetylene welding equipment.

2. Proper safety protection (welding goggles, safety glasses, spark lighter, and pliers). Refer to Chapter 2 in the text for more specific safety information.

3. Two or more pieces of 16-gauge carbon steel sheet, approximately 1 1/2 in. (38 mm) wide by 6 in. (152 mm) long.

4. 1/16 in. (2 mm), 3/32 in. (2.4 mm), and 1/8 in. (3 mm) diameter RG45 filler rods. Determine by trial and error the size of filler rod with which you are most comfortable.

INSTRUCTIONS

1. Tack the two pieces together in a butt joint configuration and position the workpiece in a horizontal position.

2. Use the torch and add filler metal as you did in Practice 32-5.

3. Add the filler metal along the top leading edge of the molten weld pool.

4. You may have to flash the torch off the molten weld pool occasionally to control the heat and avoid sagging.

5. Inspect the bead for uniformity and for visual defects.

6. Repeat this practice until you can consistently make welds that are visually defect free.

7. Turn off the gas cylinders and regulators, and clean up your work area when you are finished.

INSTRUCTOR'S COMMENTS _____

_____

_____

■ PRACTICE 32-22

Name _____ Date _____

Base Metal Thickness _____ Filler Metal Diameter_____

Class _____ Instructor _____ Grade _____

OBJECTIVE: After completing this practice you should be able to make a lap joint in the horizontal position.

EQUIPMENT AND MATERIALS NEEDED FOR THIS PRACTICE

1. Properly set-up oxyacetylene welding equipment.

2. Proper safety protection (welding goggles, safety glasses, spark lighter, and pliers). Refer to Chapter 2 in the text for more specific safety information.

3. Two or more pieces of 16-gauge carbon steel sheet, approximately 1 1/2 in. (38 mm) wide by 6 in. (152 mm) long.

4. 1/16 in. (2 mm), 3/32 in. (2.4 mm), and 1/8 in. (3 mm) diameter RG45 filler rods. Determine by trial and error the size of filler rod with which you are most comfortable.

INSTRUCTIONS

1. Tack the two pieces together in a lap joint configuration and position the workpiece in a horizontal position.

2. Use the torch and add filler metal as you did in Practice 32-8.

3. Add the filler metal along the top leading edge of the molten weld pool.

4. You may have to flash the torch off the molten weld pool occasionally to control the heat and avoid sagging.

5. Inspect the bead for uniformity and for visual defects.

6. Repeat this practice until you can consistently make welds that are visually defect free.

7. Turn off the gas cylinders and regulators, and clean up your work area when you are finished.

INSTRUCTOR'S COMMENTS _____

_____

_____

■ PRACTICE 32-23

Name _____ Date _____

Base Metal Thickness _____ Filler Metal Diameter_____

Class _____ Instructor _____ Grade _____

OBJECTIVE: After completing this practice you should be able to make a tee joint in the horizontal position.

EQUIPMENT AND MATERIALS NEEDED FOR THIS PRACTICE

1. Properly set-up oxyacetylene welding equipment.

2. Proper safety protection (welding goggles, safety glasses, spark lighter, and pliers). Refer to Chapter 2 in the text for more specific safety information.

3. Two or more pieces of 16-gauge carbon steel sheet, approximately 1 1/2 in. (38 mm) wide by 6 in. (152 mm) long.

4. 1/16 in. (2 mm), 3/32 in. (2.4 mm), and 1/8 in. (3 mm) diameter RG45 filler rods. Determine by trial and error the size of filler rod with which you are most comfortable.

INSTRUCTIONS

1. Tack the two pieces together in a tee joint configuration and position the workpiece in a horizontal position.

2. Use the torch and add filler metal as you did in Practice 32-9.

3. Add the filler metal along the top leading edge of the molten weld pool.

4. You may have to flash the torch off the molten weld pool occasionally to control the heat and avoid sagging.

5. Inspect the bead for uniformity and for visual defects.

6. Repeat this practice until you can consistently make welds that are visually defect free.

7. Turn off the gas cylinders and regulators, and clean up your work area when you are finished.

INSTRUCTOR'S COMMENTS _____

_____

_____

■ PRACTICE 32-24

Name _____ Date _____

Base Metal Thickness _____ Filler Metal Diameter_____

Class _____ Instructor _____ Grade _____

OBJECTIVE: After completing this practice you should be able to make a stringer bead in the overhead position.

EQUIPMENT AND MATERIALS NEEDED FOR THIS PRACTICE

1. Properly set-up oxyacetylene welding equipment.

2. Proper safety protection (welding goggles, safety glasses, spark lighter, and pliers). Refer to Chapter 2 in the text for more specific safety information.

3. One or more pieces of 16-gauge carbon steel sheet, approximately 3 in. (76 mm) wide by 6 in. (152 mm) long.

4. 1/16 in. (2 mm), 3/32 in. (2.4 mm), and 1/8 in. (3 mm) diameter RG45 filler rods. Determine by trial and error the size of filler rod with which you are most comfortable.

INSTRUCTIONS

1. Position the workpiece in the overhead position.

2. Use the torch and add filler metal as you did in Practice 32-3.

3. It may be necessary to flash the torch away from the metal to avoid overheating and dripping. Move the hotter primary flame away from the puddle while still keeping the puddle within the secondary flame.

4. Establish a rhythm of moving the torch and adding filler metal so as to keep the bead uniform.

5. Inspect the completed bead for uniformity and freedom from defects.

6. Repeat this practice until you can consistently make welds that are visually defect free.

7. Turn off the gas cylinders and regulators, and clean up your work area when you are finished.

INSTRUCTOR'S COMMENTS _____

_____

_____

■ PRACTICE 32-25

Name _____ Date _____

Base Metal Thickness _____ Filler Metal Diameter_____

Class _____ Instructor _____ Grade _____

OBJECTIVE: After completing this practice you should be able to make a butt joint in the overhead position.

EQUIPMENT AND MATERIALS NEEDED FOR THIS PRACTICE

1. Properly set-up oxyacetylene welding equipment.

2. Proper safety protection (welding goggles, safety glasses, spark lighter, and pliers). Refer to Chapter 2 in the text for more specific safety information.

3. Two or more pieces of 16-gauge carbon steel sheet, approximately 1 1/2 in. (38 mm) wide by 6 in. (152 mm) long.

4. 1/16 in. (2 mm), 3/32 in. (2.4 mm), and 1/8 in. (3 mm) diameter RG45 filler rods. Determine by trial and error the size of filler rod with which you are most comfortable.

INSTRUCTIONS

1. Tack the two pieces together in a butt joint configuration and position the workpiece in the overhead position.

2. Use the torch and add filler metal as you did in Practice 32-5.

3. Add the filler metal along the leading edge of the molten weld pool.

4. You may have to flash the torch off the molten weld pool occasionally to control the heat and avoid dripping.

5. Inspect the bead for uniformity and for visual defects.

6. Repeat this practice until you can consistently make welds that are visually defect free.

7. Turn off the gas cylinders and regulators, and clean up your work area when you are finished.

INSTRUCTOR'S COMMENTS _____

_____

_____

■ PRACTICE 32-26

Name _____ Date _____

Base Metal Thickness _____ Filler Metal Diameter_____

Class _____ Instructor _____ Grade _____

OBJECTIVE: After completing this practice you should be able to make a lap joint in the overhead position.

EQUIPMENT AND MATERIALS NEEDED FOR THIS PRACTICE

1. Properly set-up oxyacetylene welding equipment.

2. Proper safety protection (welding goggles, safety glasses, spark lighter, and pliers). Refer to Chapter 2 in the text for more specific safety information.

3. Two or more pieces of 16-gauge carbon steel sheet, approximately 1 1/2 in. (38 mm) wide by 6 in. (152 mm) long.

4. 1/16 in. (2 mm), 3/32 in. (2.4 mm), and 1/8 in. (3 mm) diameter RG45 filler rods. Determine by trial and error the size of filler rod with which you are most comfortable.

INSTRUCTIONS

1. Tack the two pieces together in a lap joint configuration and position the workpiece in the overhead position.

2. Use the torch and add filler metal as you did in Practice 32-8.

3. Add the filler metal along the leading edge of the molten weld pool.

4. You may have to flash the torch off the molten weld pool occasionally to control the heat and avoid dripping.

5. Inspect the bead for uniformity and for visual defects.

6. Repeat this practice until you can consistently make welds that are visually defect free.

7. Turn off the gas cylinders and regulators, and clean up your work area when you are finished.

INSTRUCTOR'S COMMENTS _____

_____

_____

■ PRACTICE 32-27

Name _____ Date _____

Base Metal Thickness _____ Filler Metal Diameter_____

Class _____ Instructor _____ Grade _____

OBJECTIVE: After completing this practice you should be able to make a tee joint in the overhead position.

EQUIPMENT AND MATERIALS NEEDED FOR THIS PRACTICE

1. Properly set-up oxyacetylene welding equipment.

2. Proper safety protection (welding goggles, safety glasses, spark lighter, and pliers). Refer to Chapter 2 in the text for more specific safety information.

3. Two or more pieces of 16-gauge carbon steel sheet, approximately 1 1/2 in. (38 mm) wide by 6 in. (152 mm) long.

4. 1/16 in. (2 mm), 3/32 in. (2.4 mm), and 1/8 in. (3 mm) diameter RG45 filler rods. Determine by trial and error the size of filler rod with which you are most comfortable.

INSTRUCTIONS

1. Tack the two pieces together in a tee joint configuration and position the workpiece in the overhead position.

2. Use the torch and add filler metal as you did in Practice 32-9.

3. Add the filler metal along the leading edge of the molten weld pool.

4. You may have to flash the torch off the molten weld pool occasionally to control the heat and avoid dripping.

5. Inspect the bead for uniformity and for visual defects.

6. Repeat this practice until you can consistently make welds that are visually defect free.

7. Turn off the gas cylinders and regulators, and clean up your work area when you are finished.

INSTRUCTOR'S COMMENTS _____

_____

_____

■ PRACTICE 32-28

Name _____ Date _____

Base Metal Thickness _____ Filler Metal Diameter_____

Class _____ Instructor _____ Grade _____

OBJECTIVE: After completing this practice you should be able to make a stringer bead on a piece of pipe in the 1G position.

EQUIPMENT AND MATERIALS NEEDED FOR THIS PRACTICE

1. Properly set-up oxyacetylene welding equipment.

2. Proper safety protection (welding goggles, safety glasses, spark lighter, and pliers). Refer to Chapter 2 in the text for more specific safety information.

3. One or more pieces of schedule 40 mild steel pipe, approximately 2 in. (51 mm) in diameter by 6 in. (152 mm) long.

4. 1/16 in. (2 mm), 3/32 in. (2.4 mm), and 1/8 in. (3 mm) diameter RG45 filler rods. Determine by trial and error the size of filler rod with which you are most comfortable.

INSTRUCTIONS

1. Position the pipe in the 1G horizontal rolled position.

2. Start welding the bead at the 2 o'clock position weld up to the 12 o'clock position.

3. Stop and roll the pipe so that the crater is at the 2 o'clock position and weld again up to the 12 o'clock position.

4. Repeat this process until the weld bead extends all the way around the pipe.

5. Inspect the bead for straightness, uniform width and reinforcement, and for absence of visual defects.

6. Repeat this practice until you can consistently make welds that are defect free.

7. Turn off the gas cylinders and regulators, and clean up your work area when you are finished.

INSTRUCTOR'S COMMENTS _____

_____

_____

■ PRACTICE 32-29

Name _____ Date _____

Base Metal Thickness _____ Filler Metal Diameter_____

Class _____ Instructor _____ Grade _____

OBJECTIVE: After completing this practice you should be able to make a butt joint on pipe in the 1G position.

EQUIPMENT AND MATERIALS NEEDED FOR THIS PRACTICE

1. Properly set-up oxyacetylene welding equipment.

2. Proper safety protection (welding goggles, safety glasses, spark lighter, and pliers). Refer to Chapter 2 in the text for more specific safety information.

3. Two or more pieces of schedule 40 mild steel pipe, approximately 2 in. (51 mm) in diameter by 3 in. (76 mm) long with the ends beveled as shown in textbook Figure 32-64.

4. 1/16 in. (2 mm), 3/32 in. (2.4 mm), and 1/8 in. (3 mm) diameter RG45 filler rods. Determine by trial and error the size of filler rod with which you are most comfortable.

INSTRUCTIONS

1. Tack the beveled ends together and position the weldment in the 1G horizontal rolled position.

2. Start welding the joint at the 2 o'clock position weld up to the 12 o'clock position.

3. Stop and roll the pipe so that the crater is at the 2 o'clock position and weld again up to the 12 o'clock position.

4. Repeat this process until the weld bead extends all the way around the pipe.

5. Inspect the bead for straightness, uniform width and reinforcement, and for absence of visual defects.

6. Repeat this practice until you can consistently make welds that are defect free.

7. Turn off the gas cylinders and regulators, and clean up your work area when you are finished.

INSTRUCTOR'S COMMENTS _____

_____

_____

■ PRACTICE 32-30

Name _____ Date _____

Base Metal Thickness _____ Filler Metal Diameter_____

Class _____ Instructor _____ Grade _____

OBJECTIVE: After completing this practice you should be able to make a stringer bead on a piece of pipe in the 5G position.

EQUIPMENT AND MATERIALS NEEDED FOR THIS PRACTICE

1. Properly set-up oxyacetylene welding equipment.

2. Proper safety protection (welding goggles, safety glasses, spark lighter, and pliers). Refer to Chapter 2 in the text for more specific safety information.

3. One or more pieces of schedule 40 mild steel pipe, approximately 2 in. (51 mm) in diameter by 6 in. (152 mm) long.

4. 1/16 in. (2 mm), 3/32 in. (2.4 mm), and 1/8 in. (3 mm) diameter RG45 filler rods. Determine by trial and error the size of filler rod with which you are most comfortable.

INSTRUCTIONS

1. Position the pipe in the 5G horizontal fixed position.

2. Start welding the bead at the 6 o'clock position weld up to the 12 o'clock position.

3. Stop and begin welding again at the 6 o'clock position weld up to the 12 o'clock position around the other side of the pipe.

4. The weld bead should now extend all the way around the pipe.

5. Inspect the bead for straightness, uniform width and reinforcement, and for absence of visual defects.

6. Repeat this practice until you can consistently make welds that are defect free.

7. Turn off the gas cylinders and regulators, and clean up your work area when you are finished.

INSTRUCTOR'S COMMENTS _____

_____

_____

■ PRACTICE 32-31

Name _____ Date _____

Base Metal Thickness _____ Filler Metal Diameter_____

Class _____ Instructor _____ Grade _____

OBJECTIVE: After completing this practice you should be able to make a butt joint on pipe in the 5G position.

EQUIPMENT AND MATERIALS NEEDED FOR THIS PRACTICE

1. Properly set-up oxyacetylene welding equipment.

2. Proper safety protection (welding goggles, safety glasses, spark lighter, and pliers). Refer to Chapter 2 in the text for more specific safety information.

3. Two or more pieces of schedule 40 mild steel pipe, approximately 2 in. (51 mm) in diameter by 3 in. (76 mm) long with the ends beveled as shown in textbook Figure 32-64.

4. 1/16 in. (2 mm), 3/32 in. (2.4 mm), and 1/8 in. (3 mm) diameter RG45 filler rods. Determine by trial and error the size of filler rod with which you are most comfortable.

INSTRUCTIONS

1. Tack the beveled ends together and position the weldment in the 5G horizontal fixed position.

2. Start welding the bead at the 6 o'clock position weld up to the 12 o'clock position.

3. Stop and begin welding again at the 6 o'clock position weld up to the 12 o'clock position around the other side of the pipe.

4. The weld bead should now extend all the way around the pipe.

5. Inspect the bead for straightness, uniform width and reinforcement, and for absence of visual defects.

6. Repeat this practice until you can consistently make welds that are defect free.

7. Turn off the gas cylinders and regulators, and clean up your work area when you are finished.

INSTRUCTOR'S COMMENTS _____

_____

_____

■ PRACTICE 32-32

Name _____ Date _____

Base Metal Thickness _____ Filler Metal Diameter_____

Class _____ Instructor _____ Grade _____

OBJECTIVE: After completing this practice you should be able to make a stringer bead on a piece of pipe in the 2G position.

EQUIPMENT AND MATERIALS NEEDED FOR THIS PRACTICE

1. Properly set-up oxyacetylene welding equipment.

2. Proper safety protection (welding goggles, safety glasses, spark lighter, and pliers). Refer to Chapter 2 in the text for more specific safety information.

3. One or more pieces of schedule 40 mild steel pipe, approximately 2 in. (51 mm) in diameter by 6 in. (152 mm) long.

4. 1/16 in. (2 mm), 3/32 in. (2.4 mm), and 1/8 in. (3 mm) diameter RG45 filler rods. Determine by trial and error the size of filler rod with which you are most comfortable.

INSTRUCTIONS

1. Position the pipe in the 2G vertical position.

2. This bead is made similar to a horizontal stringer bead on a flat plate.

3. Start with a small bead and then increase the size. This will allow you to build a shelf to support the molten metal.

4. The weld bead should extend all the way around the pipe.

5. Inspect the bead for straightness, uniform width and reinforcement, and for absence of visual defects.

6. Repeat this practice until you can consistently make welds that are defect free.

7. Turn off the gas cylinders and regulators, and clean up your work area when you are finished.

INSTRUCTOR'S COMMENTS _____

_____

_____

■ PRACTICE 32-33

Name _____ Date _____

Base Metal Thickness _____ Filler Metal Diameter_____

Class _____ Instructor _____ Grade _____

OBJECTIVE: After completing this practice you should be able to make a butt joint on pipe in the 2G position.

EQUIPMENT AND MATERIALS NEEDED FOR THIS PRACTICE

1. Properly set-up oxyacetylene welding equipment.

2. Proper safety protection (welding goggles, safety glasses, spark lighter, and pliers). Refer to Chapter 2 in the text for more specific safety information.

3. Two or more pieces of schedule 40 mild steel pipe, approximately 2 in. (51 mm) in diameter by 3 in. (76 mm) long with the ends beveled as shown in textbook Figure 32-64.

4. 1/16 in. (2 mm), 3/32 in. (2.4 mm), and 1/8 in. (3 mm) diameter RG45 filler rods. Determine by trial and error the size of filler rod with which you are most comfortable.

INSTRUCTIONS

1. Tack the beveled ends together and position the weldment in the 2G vertical position.

2. This bead is made similar to a horizontal butt joint on flat plates.

3. Start with a small bead and then increase the size. This will allow you to build a shelf to support the molten metal.

4. The weld bead should extend all the way around the pipe.

5. Inspect the bead for uniform width and reinforcement and for absence of visual defects.

6. Repeat this practice until you can consistently make welds that are defect free.

7. Turn off the gas cylinders and regulators, and clean up your work area when you are finished.

INSTRUCTOR'S COMMENTS _____

_____

_____

■ PRACTICE 32-34

Name _____ Date _____

Base Metal Thickness _____ Filler Metal Diameter_____

Class _____ Instructor _____ Grade _____

OBJECTIVE: After completing this practice you should be able to make a stringer bead on a piece of pipe in the 6G 45° fixed inclined position.

EQUIPMENT AND MATERIALS NEEDED FOR THIS PRACTICE

1. Properly set-up oxyacetylene welding equipment.

2. Proper safety protection (welding goggles, safety glasses, spark lighter, and pliers). Refer to Chapter 2 in the text for more specific safety information.

3. One or more pieces of schedule 40 mild steel pipe, approximately 2 in. (51 mm) in diameter by 6 in. (152 mm) long.

4. 1/16 in. (2 mm), 3/32 in. (2.4 mm), and 1/8 in. (3 mm) diameter RG45 filler rods. Determine by trial and error the size of filler rod with which you are most comfortable.

INSTRUCTIONS

1. Position the pipe in the 6G 45° fixed inclined position.

2. Start welding the bead at the bottom and weld up to the top.

3. The bead shape will change as you move around the pipe.

4. Stop at the top and begin welding again at the bottom and weld up to the top around the other side of the pipe.

5. The weld bead should now extend all the way around the pipe.

6. Inspect the bead for straightness, uniform width and reinforcement, and for absence of visual defects.

7. Repeat this practice until you can consistently make welds that are defect free.

8. Turn off the gas cylinders and regulators, and clean up your work area when you are finished.

INSTRUCTOR'S COMMENTS _____

_____

_____

■ PRACTICE 32-35

Name _____ Date _____

Base Metal Thickness _____ Filler Metal Diameter_____

Class _____ Instructor _____ Grade _____

OBJECTIVE: After completing this practice you should be able to make a butt joint on pipe in the 6G 45° fixed inclined position.

EQUIPMENT AND MATERIALS NEEDED FOR THIS PRACTICE

1. Properly set-up oxyacetylene welding equipment.

2. Proper safety protection (welding goggles, safety glasses, spark lighter, and pliers). Refer to Chapter 2 in the text for more specific safety information.

3. Two or more pieces of schedule 40 mild steel pipe, approximately 2 in. (51 mm) in diameter by 3 in. (76 mm) long with the ends beveled as shown in textbook Figure 32-64.

4. 1/16 in. (2 mm), 3/32 in. (2.4 mm), and 1/8 in. (3 mm) diameter RG45 filler rods. Determine by trial and error the size of filler rod with which you are most comfortable.

INSTRUCTIONS

1. Tack the beveled ends together and position the weldment in the 6G 45° fixed inclined position.

2. Start welding the joint at the bottom and weld up to the top.

3. The bead shape will change as you move around the pipe.

4. Stop at the top and begin welding again at the bottom and weld up to the top around the other side of the pipe.

5. The weld bead should now extend all the way around the pipe.

6. Inspect the bead for uniform width and reinforcement and for absence of visual defects.

7. Repeat this practice until you can consistently make welds that are defect free.

8. Turn off the gas cylinders and regulators, and clean up your work area when you are finished.

INSTRUCTOR'S COMMENTS _____

_____

_____

CHAPTER 32: OXYACETYLENE WELDING QUIZ

Name _____ Date _____

Class _____ Instructor _____ Grade _____

Instructions: Carefully read Chapter 32 in the text and complete the following questions.

A. IDENTIFICATION

In the space provided, identify the items shown on the illustration.

1. If the reinforcement on the following welds is removed, which would be the stronger weld? Place an "x" next to the correct answer.

_____ (A) weld is the stronger

_____ (B) weld is the stronger

B. DRAW

In the space provided make a pencil drawing to illustrate the question.

2. Draw a picture of the following weld joints.

    a. corner joint

    b. butt joint

    c. tee joint

    d. lap joint

C. FILL IN THE BLANK

Fill in the blank with the correct word. Answers may be more than one word.

3. One of the arc welding processes is most often used today for welding metal thicker than _____ gauge.

4. Overheated tips will _____.

5. The ideal angle for the welding torch is _____°.

6. A _____ welding rod can be used to cool the molten weld pool, increase buildup, and reduce penetration.

7. To prevent burnout, the torch should be _____ or _____, keeping the outer flame envelope over the molten weld pool until it solidifies.

8. Small amounts of total penetration usually will not cause a _____.

9. OFW in the field or shop must often be done in positions that are less than _____, so the welder needs to be somewhat versatile.

10. Before you start oxyfuel welding, move the hoses so that there is no _____ of the torch.

11. Hold the torch at a 45° angle to the metal with the inner cone at about _____ in. (_____ mm) above the metal surface.

12. If the outside corner joint is tacked together properly, the addition of filler metal is _____.

13. To make the butt joint, place two clean pieces of metal flat on the table and tack weld _____ together.

14. When heating the two clean sheets of a lap joint, caution must be exercised to ensure that _____ at the same time.

15. It is important to hold the flame so that _____ tee joint sheets melt at the same time.

16. Welds made in the vertical, horizontal, or overhead position are _____.

17. To prevent the molten weld pool from dripping when making a vertical weld, the _____ of the molten weld pool must be watched.

18. Being able to control the weld bead this accurately can be very helpful for _____ welds and for filling gaps and holes for repair welding.

19. The vertical tee joint has a _____ and a _____ side.

20. When making a horizontal lap weld, if too large a molten weld pool is started, _____ does not have time to form properly.

21. With the overhead weld, the molten weld pool is held to the sheet by _____ in the same manner that a drop of water is held to the bottom of a glass sheet.

22. The OFW process for both pipe and tubing is usually the _____.

D. MATCHING

In the space provided to the left, write the letter from Column B that best answers the question or completes the statement in Column A.

Column A   Column B

_____ 23. Which position is the pipe held fixed at a 45° angle?

           a. 1G

_____ 24. Which position is the pipe rolled horizontally and the weld is made flat by depositing it near the top?

           b. 6G

_____ 25. Which position is the pipe fixed vertically and the weld made around the pipe horizontally?

           c. 5G

_____ 26. Which position is the pipe fixed horizontally and may not be rotated while welding?

           d. 2G

E. SHORT ANSWER

Write a brief answer in the space provided that will answer the question or complete the statement.

27. Put an X next to three reasons why filler metal is not added by melting the rod and dripping it into the pool.

      _____ a. The metal cannot always be added where it is needed.

      _____ b. The rod sticks to the base metal if it is added any other way.

      _____ c. The method works only in the flat position.

      _____ d. Important alloys are burned out.

      _____ e. It takes more skill.

28. What holds the molten pool in place when welding in the overhead position?

_____

_____

29. Why is it suggested that you might wish to cut the filler wire in two pieces for welding?

_____

_____

F. ESSAY

Provide complete answers for all of the following questions.

30. What effect does the torch angle have on the size of the pool?

_____

_____

31. What will happen to the molten weld pool if the protecting outer flame is quickly moved away?

_____

_____

32. How do welders know when they are close to burning through when welding?

_____

_____

33. How can changes in the sizes of the filler metal affect the size of the weld?

_____

_____

34. How is a vertical weld pool kept from dripping?

_____

_____

35. When practicing vertical welds, why should a welder start at a 45° angle before attempting a vertical sheet?

_____

_____

36. Why is only small-diameter pipe gas welded today?

_____

_____

37. How should a weld bead be started to prevent a cold lap?

_____

_____

38. How is the end of thin-wall piping prepared for welding?

_____

_____

39. How is the size and shape of the bead controlled as the weld moves from the overhead to the vertical position on a pipe held in the 5G position?

_____

_____

40. How will the weld bead change in shape if the welding manipulation is not changed as the weld progresses from the vertical to the flat position on a pipe in the 5G position?

_____

_____

41. Where should the filler rod be added to the weld pool on a pipe in the 2G position? Why?

_____

_____

42. Why is the 6G pipe welding position so difficult to do correctly?

_____

_____

43. What is a keyhole, and why is it used?

_____

_____

G. DEFINITIONS

44. Define the following terms:

    heat sink _____

    shelf _____

    tee joint _____

    flashing _____

    overlap _____

    keyhole _____

INSTRUCTOR'S COMMENTS _____

_____

_____

Chapter 33

Brazing, Braze Welding, and Soldering

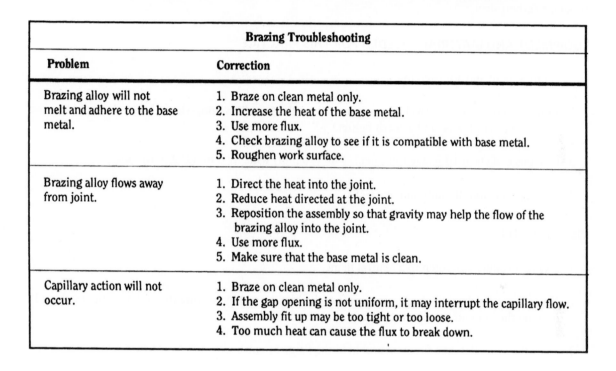

Brazing Troubleshooting	
Problem	**Correction**
Brazing alloy will not melt and adhere to the base metal.	1. Braze on clean metal only. 2. Increase the heat of the base metal. 3. Use more flux. 4. Check brazing alloy to see if it is compatible with base metal. 5. Roughen work surface.
Brazing alloy flows away from joint.	1. Direct the heat into the joint. 2. Reduce heat directed at the joint. 3. Reposition the assembly so that gravity may help the flow of the brazing alloy into the joint. 4. Use more flux. 5. Make sure that the base metal is clean.
Capillary action will not occur.	1. Braze on clean metal only. 2. If the gap opening is not uniform, it may interrupt the capillary flow. 3. Assembly fit up may be too tight or too loose. 4. Too much heat can cause the flux to break down.

■ PRACTICE 33-1

Name _____ Date _____

Class _____ Instructor _____ Grade _____

OBJECTIVE: After completing this practice, the student should be able to braze stringer beads on 16-gauge carbon steel.

EQUIPMENT AND MATERIALS NEEDED FOR THIS PRACTICE

1. A properly lit and adjusted torch.

2. Proper safety protection (welding goggles, safety glasses, pliers, long-sleeved shirt, long pants, leather shoes or boots, and a spark lighter).

3. One piece of clean 16-gauge mild steel, 3 in. (76 mm) wide by 6 in. (152 mm) long.

4. Brazing flux, and BRCuZn brazing rod.

INSTRUCTIONS

1. Light and adjust the torch to a neutral flame.

2. If a prefluxed rod is not used, you must flux the brazing rod by heating the end with the torch and dipping the hot end into the flux.

3. Place the sheet flat on a firebrick and hold the flame at one end.

4. Touch the flux-covered rod to the sheet and allow a small amount of brazing rod to melt onto the hot sheet.

5. Once the molten brazing metal wets the sheet, start moving the torch in a circular pattern while dipping the rod into the molten braze pool as you move along the sheet. If the size of the molten pool increases, you can control it by reducing the torch angle, raising the torch, traveling at a faster rate, or flashing the flame off the molten braze pool. Flashing the torch off a braze joint will not cause oxidation problems as it does when welding, because the molten metal is protected by a layer of flux.

6. As the braze bead progresses across the sheet, dip the end of the rod back in the flux, if a powdered flux is used, as often as needed to keep a small molten pool of flux ahead of the bead.

7. Repeat this practice until you can consistently produce brazed stringer beads that are straight with uniform width and buildup.

7. Turn off the gas cylinders and regulators, and clean up your work area when you are finished brazing.

INSTRUCTOR'S COMMENTS _____

_____

_____

■ PRACTICE 33-2 AND PRACTICE 33-3

Name _____ Date _____

Class _____ Instructor _____ Grade _____

OBJECTIVE: After completing Practice 33-2, you should be able to braze butt joints in the flat position on 16-gauge carbon steel; and after completing Practice 33-3, you should be able to make the same butt joint with 100% penetration.

EQUIPMENT AND MATERIALS NEEDED FOR THIS PRACTICE

1. A properly lit and adjusted torch.

2. Proper safety protection (welding goggles, safety glasses, pliers, long-sleeved shirt, long pants, leather shoes or boots, and a spark lighter).

3. Two pieces of clean 16-gauge mild steel, 1 1/2 in. (38 mm) wide by 6 in. (152 mm) long.

4. Brazing flux and BRCuZn brazing rod.

INSTRUCTIONS

1. Place the metal flat on a firebrick, hold the plates tightly together, and make a tack braze at both ends of the joint. If the plates become distorted, they can be bent back into shape with a hammer before making another tack weld in the center.

2. Align the sheets so that you can comfortably make a braze bead along the joint.

3. Starting as you did in Practice 33-1, make a uniform braze along the joint.

4. Repeat this practice until a uniform braze can be made without defects.

5. Repeat this practice using more heat so that 100% penetration occurs.

6. Turn off the gas cylinders and regulators, and clean up your work area when you are finished brazing.

INSTRUCTOR'S COMMENTS _____

_____

_____

■ PRACTICE 33-4

Name _____ Date _____

Class _____ Instructor _____ Grade _____

OBJECTIVE: After completing this practice, you should be able to braze tee and lap joints on two pieces of carbon steel of different thicknesses.

EQUIPMENT AND MATERIALS NEEDED FOR THIS PRACTICE

1. A properly lit and adjusted torch.

2. Proper safety protection (welding goggles, safety glasses, pliers, long-sleeved shirt, long pants, leather shoes or boots, and a spark lighter).

3. Two or more pieces of 16-gauge steel 1 1/2 in. (38 mm) wide by 6 in. (152 mm) long.

4. Brazing flux and BRCuZn brazing rod.

INSTRUCTIONS

1. Tack braze the pieces together in the lap and tee joint configurations being sure that they are held tightly together.

2. Place the metal on a firebrick with the thin metal up.

3. Apply heat to the exposed thick metal and more slowly to the overlapping thin metal so that conduction from the thin metal will heat the thick metal at the lap. If the braze is started before the thick metal is sufficiently heated, the filler metal will be chilled and a bond will not occur.

4. After the joint is completed and cooled, tap the joint with a hammer to see if there is a good bonded joint.

5. Repeat this practice until the joint can be made without defects.

6. Turn off the gas cylinders and regulators, and clean up your work area when you are finished brazing.

INSTRUCTOR'S COMMENTS _____

_____

_____

■ PRACTICE 33-5 AND PRACTICE 33-6

Name _____ Date _____

Class _____ Instructor _____ Grade _____

OBJECTIVE: After completing this practice you should be able to make a brazed lap joint in the flat position, Practice 33-5; and with 100% penetration, Practice 33-6.

EQUIPMENT AND MATERIALS NEEDED FOR THIS PRACTICE

1. Properly set-up, lit, and adjusted torch.

2. Proper safety protection (welding goggles, safety glasses, spark lighter, and pliers). Refer to Chapter 2 in the text for more specific safety information.

3. Two or more pieces of 16-gauge mild steel, 1 1/2 in. (38 mm) wide by 6 in. (152 mm) long.

4. Brazing flux and BRCuZn brazing rod.

INSTRUCTIONS

1. Braze tack the pieces together in a lap joint configuration.

2. Place the sheets in the flat position and align them so that you can comfortably make a braze bead along the joint.

3. Starting as you did in Practice 33-5, make a uniform braze bead along the joint.

4. Go slowly enough and apply enough heat so that you will achieve 100% penetration.

5. Inspect the bottom side of the bead to ensure that 100% penetration was achieved.

6. Repeat this practice until you can consistently make braze beads that are defect free.

7. Turn off the gas cylinders and regulators, and clean up your work area when you are finished.

INSTRUCTOR'S COMMENTS _____

_____

_____

■ PRACTICE 33-7

Name _____ Date _____

Class _____ Instructor _____ Grade _____

OBJECTIVE: After completing this practice you should be able to make a brazed tee joint in the flat position using thin to thick metal.

EQUIPMENT AND MATERIALS NEEDED FOR THIS PRACTICE

1. Properly set-up, lit, and adjusted torch.

2. Proper safety protection (welding goggles, safety glasses, spark lighter, and pliers). Refer to Chapter 2 in the text for more specific safety information.

3. One piece of 16-gauge mild steel plate, 1 1/2 in. (38 mm) wide by 6 in. (152 mm) long and one piece of 1/4 in. (6 mm) thick by 1 1/2 in. (38 mm) wide by 6 in. (152 mm) long.

4. Brazing flux and BRCuZn brazing rod.

INSTRUCTIONS

1. Braze tack the pieces together in a tee joint configuration with the thin plate in the vertical position.

2. The thin plate will heat up faster than the thick plate so direct the torch mostly on the thick plate. Both plates must be heated equally.

3. Direct the heat to the thicker plate while adding the brazing rod onto the thinner plate.

4. Make a braze along the joint that is uniform in appearance.

5. Inspect the bead for uniformity and for freedom from defects.

6. Repeat this practice until you can consistently make braze beads that are defect free.

7. Turn off the gas cylinders and regulators, and clean up your work area when you are finished.

INSTRUCTOR'S COMMENTS _____

_____

_____

■ PRACTICE 33-8

Name _____ Date _____

Class _____ Instructor _____ Grade _____

OBJECTIVE: After completing this practice you should be able to make a brazed lap joint in the flat position using thin to thick metal.

EQUIPMENT AND MATERIALS NEEDED FOR THIS PRACTICE

1. Properly set-up, lit, and adjusted torch.

2. Proper safety protection (welding goggles, safety glasses, spark lighter, and pliers). Refer to Chapter 2 in the text for more specific safety information.

3. One piece of 16-gauge mild steel plate, 1 1/2 in. (38 mm) wide by 6 in. (152 mm) long and one piece of 1/4 in. (6 mm) thick by 1 1/2 in. (38 mm) wide by 6 in. (152 mm) long.

4. Brazing flux and BRCuZn brazing rod.

INSTRUCTIONS

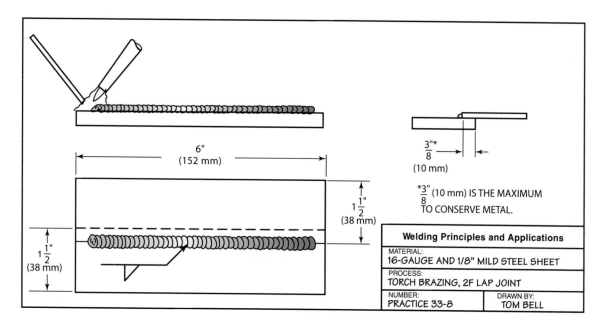

1. Braze tack the pieces tightly together in a lap joint configuration.

2. Place the tacked assembly on a firebrick in the flat position with the thin plate up.

3. The thin plate will heat up faster than the thick plate, so direct the torch mostly on the thick plate. Both plates must be heated equally.

4. Direct the heat to the thicker plate while adding the brazing rod.

5. Make a braze along the joint that is uniform in appearance.

6. Inspect the bead for uniformity and for freedom from defects.

7. Repeat this practice until you can consistently make braze beads that are defect free.

8. Turn off the gas cylinders and regulators, and clean up your work area when you are finished.

INSTRUCTOR'S COMMENTS _____

_____

_____

■ PRACTICE 33-9

Name _____ Date _____

Class _____ Instructor _____ Grade _____

OBJECTIVE: After completing this practice you should be able to make a braze welded butt joint in the flat position using thin to thick metal.

EQUIPMENT AND MATERIALS NEEDED FOR THIS PRACTICE

1. Properly set-up, lit, and adjusted torch.

2. Proper safety protection (welding goggles, safety glasses, spark lighter, and pliers). Refer to Chapter 2 in the text for more specific safety information.

3. One piece of 16-gauge mild steel plate, 1 1/2 in. (38 mm) wide by 6 in. (152 mm) long and one piece of 1/4 in. (6 mm) thick by 1 1/2 in. (38 mm) wide by 6 in. (152 mm) long. Grind the edge of the 1/4-in. (6 mm) plate, as shown in Figure 33-44 in the textbook.

4. Brazing flux and BRCuZn brazing rod.

INSTRUCTIONS

1. Braze tack the pieces together in a butt joint configuration. See Figure 33-44 in the textbook.

2. Place the tacked assembly on a firebrick in the flat position.

3. The thin plate will heat up faster than the thick plate so direct the torch mostly on the thick plate. Both plates must be heated equally.

4. The flame should be moved in a triangular motion so that the root is heated as well as the top of the bead.

5. Make a braze bead along the joint that is uniform in appearance.

6. Inspect the bead for uniformity and for freedom from defects.

7. When the assembly is cool, do a bend test, as shown in Figure 33-46 in the textbook.

8. Repeat this practice until you can consistently make braze beads that are defect free.

9. Turn off the gas cylinders and regulators, and clean up your work area when you are finished.

INSTRUCTOR'S COMMENTS _____

_____

_____

■ PRACTICE 33-10

Name _____ Date _____

Class _____ Instructor _____ Grade _____

OBJECTIVE: After completing this practice you should be able to make a braze welded tee joint in the flat position using thin to thick metal.

EQUIPMENT AND MATERIALS NEEDED FOR THIS PRACTICE

1. Properly set-up, lit, and adjusted torch.

2. Proper safety protection (welding goggles, safety glasses, spark lighter, and pliers). Refer to Chapter 2 in the text for more specific safety information.

3. One piece of 16-gauge mild steel plate, 1 1/2 in. (38 mm) wide by 6 in. (152 mm) long and one piece of 1/4 in. (6 mm) thick by 1 1/2 in. (38 mm) wide by 6 in. (152 mm) long.

4. Brazing flux and BRCuZn brazing rod.

INSTRUCTIONS

1. Braze tack the pieces together in a tee joint configuration with the thin plate in the vertical position.

2. Place the tacked assembly on a firebrick in the flat position.

3. The thin plate will heat up faster than the thick plate so direct the torch mostly on the thick plate. Both plates must be heated equally.

4. Direct the heat mostly at the thick plate and add the braze rod on the thin plate.

5. Make a braze bead along the joint that is uniform in appearance.

6. Inspect the bead for uniformity and for freedom from defects.

7. When the assembly is cool, do a bend test, as shown in Figure 33-47 in the textbook.

8. Repeat this practice until you can consistently make braze beads that are defect free.

9. Turn off the gas cylinders and regulators, and clean up your work area when you are finished.

INSTRUCTOR'S COMMENTS _____

_____

_____

■ PRACTICE 33-11

Name _____ Date _____

Class _____ Instructor _____ Grade _____

OBJECTIVE: After completing this practice you should be able to use braze welding to weld up a hole in the flat position.

EQUIPMENT AND MATERIALS NEEDED FOR THIS PRACTICE

1. Properly set-up, lit, and adjusted torch.

2. Proper safety protection (welding goggles, safety glasses, spark lighter, and pliers). Refer to Chapter 2 in the text for more specific safety information.

3. One piece of 16-gauge mild steel plate, 3 in. (76 mm) wide by 3 in. (76 mm) long having a 1-in. (25-mm) diameter hole drilled through it.

4. Brazing flux and BRCuZn brazing rod.

INSTRUCTIONS

1. Place the piece in the flat position on two firebricks so that the hole is between them.

2. Start by running a stringer bead around the hole. See Figure 33-48 in the textbook.

3. When the bead is complete, turn the torch at a very steep angle and point it at the edge of the hole nearest the torch. Hold the end of the brazing rod in the flame so that both the bead around the hole and the rod meet at the same time. See Figure 33-49 in the textbook.

4. Put the rod in the molten bead and then flash the torch off to allow the molten metal to cool.

5. Repeat this process around the hole until it is completely filled.

6. When the braze weld is complete, it should be fairly flat with the surrounding metal.

7. Inspect the braze weld for uniformity and for freedom from defects.

8. Repeat this practice until you can consistently make braze beads to fill up holes that are defect free.

9. Turn off the gas cylinders and regulators, and clean up your work area when you are finished.

INSTRUCTOR'S COMMENTS _____

_____

_____

■ PRACTICE 33-12

Name _____ Date _____

Class _____ Instructor _____ Grade _____

OBJECTIVE: After completing this practice you should be able to use braze welding to build up a flat surface.

EQUIPMENT AND MATERIALS NEEDED FOR THIS PRACTICE

1. Properly set-up, lit, and adjusted torch.

2. Proper safety protection (welding goggles, safety glasses, spark lighter, and pliers). Refer to Chapter 2 in the text for more specific safety information.

3. One piece of 1/4-in. (6-mm) mild steel plate, 3 in. (76 mm) wide by 3 in. (76 mm) long.

4. Brazing flux and BRCuZn brazing rod.

INSTRUCTIONS

1. Place the piece in the flat position on a firebrick.

2. Start along one side of the plate and make a braze weld down that side.

3. When you get to the end, turn the plate 180° and braze back alongside the first braze covering about one-half of the first braze. See Figure 33-50 in the textbook.

4. Repeat this process until the side is completely covered with braze metal.

5. Turn the plate 90° and repeat the process being careful to get good fusion with the layer beneath. See Figure 33-51 in the textbook.

6. Be sure that the edges are built up enough so that they could be cut back square.

7. Repeat this process until the buildup is at least 1/4 in. (6 mm) thick.

8. Inspect the braze weld for uniformity and for freedom from defects.

9. Repeat this practice until you can use brazing to build up flat surfaces that are consistently defect free.

10. Turn off the gas cylinders and regulators, and clean up your work area when you are finished.

INSTRUCTOR'S COMMENTS _____

_____

_____

■ PRACTICE 33-13

Name _____   Date _____

Class _____   Instructor _____   Grade _____

OBJECTIVE: After completing this practice you should be able to use braze welding to build up a rounded surface.

EQUIPMENT AND MATERIALS NEEDED FOR THIS PRACTICE

1. Properly set-up, lit, and adjusted torch.

2. Proper safety protection (welding goggles, safety glasses, spark lighter, and pliers). Refer to Chapter 2 in the text for more specific safety information.

3. One piece of mild steel bolt or rod, 1/2 in. (13 mm) in diameter by 3 in. (76 mm) long.

4. Brazing flux and BRCuZn brazing rod.

INSTRUCTIONS

1. Place the mild steel bolt or rod in the flat position on a firebrick.

2. Start along one end and make a braze weld 1 1/2 in. (38 mm) long.

3. Turn the rod 180° and braze back alongside the first braze covering about one-half of the first braze. See Figure 33-52 in the textbook.

4. Rotate the rod and repeat the process being careful to get good fusion with the layer beneath.

5. Repeat this process until the rod is completely covered with braze metal and is 1 in. (25 mm) in diameter. See Figure 33-53 in the textbook.

6. Inspect the braze weld for uniformity and for freedom from defects.

7. Repeat this practice until you can use brazing to build up rounded surfaces that are consistently defect free.

8. Turn off the gas cylinders and regulators, and clean up your work area when you are finished.

INSTRUCTOR'S COMMENTS _____

_____

_____

■ PRACTICE 33-14

Name _____ Date _____

Class _____ Instructor _____ Grade _____

OBJECTIVE: After completing this practice you should be able to silver braze copper pipe in the 2G position.

EQUIPMENT AND MATERIALS NEEDED FOR THIS PRACTICE

1. Properly set-up, lit, and adjusted air MAPP®, air propane, or any air fuel gas torch. Be sure that the regulator pressure is set to the manufacturer's specifications for your fuel type and tip size.

2. Proper safety protection (welding goggles, safety glasses, spark lighter, and pliers). Refer to Chapter 2 in the text for more specific safety information.

3. Two or more short random length pieces of 1/2-in. to 1-in. (13-mm to 25-mm) copper pipe with matching copper pipe fittings.

4. Brazing flux and BCuP-2 to BCuP-5 brazing rod.

5. Steel wool, sanding cloth, and/or a wire brush.

INSTRUCTIONS

1. Clean the pipe O.D. and the fitting I.D. using steel wool, sanding cloth, or a wire brush. Then slide the fitting onto the pipe. Be sure that the pipe is seated at the bottom of the fitting.

2. Heat the brazing rod and make a bend about 3/4 in. (19 mm) from the end. Use this bend as a gauge so that you do not put too much brazing metal into the joint.

3. Place the pipe in the 2G position. Heat the pipe first but not too much. When it is hot (but not glowing red) start heating the fitting. Keep the torch moving in order to uniformly heat the entire joint. The joint is at the correct temperature when the braze metal starts to wet the surface.

4. Move the flame to the back side of the pipe and feed the brazing rod into the joint. Move the torch and brazing rod all the way around the pipe and fitting so that the entire joint will be filled. There should be a slight fillet showing all the way around the joint. See Figure 33-55 in the textbook.

5. After cooling, hacksaw the joint apart for inspection. See Figures 33-56 to 33-60 in the textbook.

6. Repeat this practice until you can make defect free brazed pipe joints.

7. Turn off the gas cylinder and regulator, and clean up your work area when you are finished.

INSTRUCTOR'S COMMENTS _____

_____

_____

■ PRACTICE 33-15

Name _____ Date _____

Class _____ Instructor _____ Grade _____

OBJECTIVE: After completing this practice you should be able to silver braze copper pipe in the 5G horizontally fixed position.

EQUIPMENT AND MATERIALS NEEDED FOR THIS PRACTICE

1. Properly set-up, lit, and adjusted air MAPP®, air propane, or any air fuel gas torch. Be sure that the regulator pressure is set to the manufacturer's specifications for your fuel type and tip size.

2. Proper safety protection (welding goggles, safety glasses, spark lighter, and pliers). Refer to Chapter 2 in the text for more specific safety information.

3. Two or more short random length pieces of 1/2-in. to 1-in. (13-mm to 25-mm) copper pipe with matching copper pipe fittings.

4. Brazing flux and BCuP-2 to BCuP-5 brazing rod.

5. Steel wool, sanding cloth, and/or a wire brush.

INSTRUCTIONS

1. Clean the pipe O.D. and the fitting I.D. using steel wool, sanding cloth, or a wire brush. Then slide the fitting onto the pipe. Be sure that the pipe is seated at the bottom of the fitting.

2. Place the pipe in the 5G position. See Figure 33-61 in the textbook. Heat the pipe first but not too much. When it is hot (but not glowing red) start heating the fitting. Keep the torch moving in order to uniformly heat the entire joint. The joint is at the correct temperature when the braze metal starts to wet the surface.

3. Move the flame to the back side of the pipe and feed the brazing rod into the joint. Move the torch and brazing rod all the way around the pipe and fitting so that the entire joint will be filled. There should be a slight fillet showing all the way around the joint. See Figure 33-55 in the textbook.

4. After cooling, hacksaw the joint apart for inspection. See Figures 33-56 to 33-60 in the text-book.

5. Repeat this practice until you can make defect free brazed pipe joints.

6. Turn off the gas cylinder and regulator, and clean up your work area when you are finished.

INSTRUCTOR'S COMMENTS _____

_____

_____

■ PRACTICE 33-16

Name _____ Date _____

Class _____ Instructor _____ Grade _____

OBJECTIVE: After completing this practice you should be able to silver braze copper pipe in the 2G vertical up position.

EQUIPMENT AND MATERIALS NEEDED FOR THIS PRACTICE

1. Properly set-up, lit, and adjusted air MAPP®, air propane, or any air fuel gas torch. Be sure that the regulator pressure is set to the manufacturer's specifications for your fuel type and tip size.

2. Proper safety protection (welding goggles, safety glasses, spark lighter, and pliers). Refer to Chapter 2 in the text for more specific safety information.

3. Two or more short random length pieces of 1/2-in. to 1-in. (13-mm to 25-mm) copper pipe with matching copper pipe fittings.

4. Brazing flux and BCuP-2 to BCuP-5 brazing rod.

5. Steel wool, sanding cloth, and/or a wire brush.

INSTRUCTIONS

1. Clean the pipe O.D. and the fitting I.D. using steel wool, sanding cloth, or a wire brush. Then slide the fitting onto the pipe. Be sure that the pipe is seated at the bottom of the fitting.

2. Place the pipe in the 2G vertical position. See Figure 33-62 in the textbook. Heat the pipe first but not too much. When it is hot (but not glowing red) start heating the fitting. Keep the torch moving in order to uniformly heat the entire joint. The joint is at the correct temperature when the braze metal starts to wet the surface.

3. Move the flame to the top of the joint and feed the brazing rod into the joint. Move the torch and brazing rod all the way around the pipe and fitting so that the entire joint will be filled. There should be a slight fillet showing all the way around the joint. See Figure 33-55 in the textbook.

4. After cooling, hacksaw the joint apart for inspection. See Figures 33-56 to 33-60 in the textbook.

5. Repeat this practice until you can make defect free brazed pipe joints.

6. Turn off the gas cylinder and regulator, and clean up your work area when you are finished.

INSTRUCTOR'S COMMENTS _____

_____

_____

■ PRACTICE 33-17

Name _____ Date _____

Class _____ Instructor _____ Grade _____

OBJECTIVE: After completing this practice you should be able to make a soldered tee joint in the flat position.

EQUIPMENT AND MATERIALS NEEDED FOR THIS PRACTICE

1. Properly set-up, lit, and adjusted air MAPP®, air propane, or any air fuel gas torch. Be sure that the regulator pressure is set to the manufacturer's specifications for your fuel type and tip size.

2. Proper safety protection (welding goggles, safety glasses, spark lighter, and pliers). Refer to Chapter 2 in the text for more specific safety information.

3. Two or more pieces of 18- to 24-gauge mild steel sheet, 1 1/2 in. (38 mm) wide by 6 in. (152 mm) long.

4. Flux-cored tin-lead or tin-antimony wire solder or solid wire solder and a container of liquid flux.

5. Steel wool, sanding cloth, and/or a wire brush.

INSTRUCTIONS

1. Using the steel wool, sanding cloth, and/or a wire brush, clean the metal surface where the solder joint will be placed.

2. Hold one piece vertically on the other piece and spot solder both ends. If flux-cored solder is not being used, paint the liquid flux on at this time.

3. Place the piece in the flat position and point the torch flame in the same direction that you will be soldering. Flash the torch on and off of the joint while adding solder to the joint to control the heat and keep the molten pool small.

4. Continue flashing the torch on and off and adding solder until you reach the end of the joint. When the joint is completed, the solder bead should be uniform.

5. Inspect the bead for uniformity and visual defects.

6. Repeat this practice until you can make defect free soldered joints.

7. Turn off the gas cylinder and regulator, and clean up your work area when you are finished.

INSTRUCTOR'S COMMENTS _____

_____

_____

■ PRACTICE 33-18

Name _____ Date _____

Class _____ Instructor _____ Grade _____

OBJECTIVE: After completing this practice you should be able to make a soldered lap joint in the flat position.

EQUIPMENT AND MATERIALS NEEDED FOR THIS PRACTICE

1. Properly set-up, lit, and adjusted air MAPP®, air propane, or any air fuel gas torch. Be sure that the regulator pressure is set to the manufacturer's specifications for your fuel type and tip size.

2. Proper safety protection (welding goggles, safety glasses, spark lighter, and pliers). Refer to Chapter 2 in the text for more specific safety information.

3. Two or more pieces of 18- to 24-gauge mild steel sheet, 1 1/2 in. (38 mm) wide by 6 in. (152 mm) long.

4. Flux-cored tin-lead or tin-antimony wire solder or solid wire solder and a container of liquid flux.

5. Steel wool, sanding cloth, and/or a wire brush.

INSTRUCTIONS

1. Using the steel wool, sanding cloth, and/or a wire brush, clean the metal surface where the solder joint will be placed.

2. Place one piece tightly on the other piece in a lap joint configuration and spot solder both ends. If flux-cored solder is not being used, paint the liquid flux on at this time.

3. Place the piece in the flat position and point the torch flame in the same direction that you will be soldering. Flash the torch on and off of the joint while adding solder to the joint to control the heat and keep the molten pool small.

4. Continue flashing the torch on and off and adding solder until you reach the end of the joint. When the joint is completed, the solder bead should be uniform.

5. Inspect the bead for uniformity and visual defects.

6. Repeat this practice until you can make defect free soldered joints.

7. Turn off the gas cylinder and regulator, and clean up your work area when you are finished.

INSTRUCTOR'S COMMENTS _____

_____

_____

■ PRACTICE 33-19

Name _____ Date _____

Class _____ Instructor _____ Grade _____

OBJECTIVE: After completing this practice you should be able to solder copper pipe in the 2G vertical down position.

EQUIPMENT AND MATERIALS NEEDED FOR THIS PRACTICE

1. Properly set-up, lit, and adjusted air MAPP®, air propane, or any air fuel gas torch. Be sure that the regulator pressure is set to the manufacturer's specifications for your fuel type and tip size.

2. Proper safety protection (welding goggles, safety glasses, spark lighter, and pliers). Refer to Chapter 2 in the text for more specific safety information.

3. Two or more short random length pieces of 1/2-in. to 1-in. (13-mm to 25-mm) copper pipe with matching copper pipe fittings.

4. Solid tin-lead or tin-antimony wire solder and a container of liquid flux.

5. Steel wool, sanding cloth, and/or a wire brush.

INSTRUCTIONS

1. Clean the pipe O.D. and the fitting I.D. using steel wool, sanding cloth, or a wire brush. Then apply the liquid flux to the pipe and the fitting.

2. Slide the fitting onto the pipe. Be sure that the pipe is seated at the bottom of the fitting. Twist the pipe in the fitting to be sure that the flux is being applied completely around the inside of the joint.

3. Make a bend about 3/4 in. (19 mm) from the end of the wire solder. Use this bend as a gauge so that you do not put too much solder into the joint.

4. Place the pipe in the 2G vertical down position. Heat the pipe and fitting uniformly. The joint is at the correct temperature when the solder starts to wet the surface.

5. When the solder starts to wet, move the flame away from the pipe and feed the solder into the joint, wiping it all the way around the joint. See Figure 33-65 in the textbook.

6. After cooling, hacksaw and pry the joint apart for inspection. See Figure 33-66 in the textbook.

7. Repeat this practice until you can make defect free soldered pipe joints.

8. Turn off the gas cylinder and regulator, and clean up your work area when you are finished.

INSTRUCTOR'S COMMENTS _____

_____

_____

■ PRACTICE 33-20

Name _____   Date _____

Class _____   Instructor _____   Grade _____

OBJECTIVE: After completing this practice you should be able to solder copper pipe in the 1G horizontal rolled position.

EQUIPMENT AND MATERIALS NEEDED FOR THIS PRACTICE

1. Properly set-up, lit, and adjusted air MAPP®, air propane, or any air fuel gas torch. Be sure that the regulator pressure is set to the manufacturer's specifications for your fuel type and tip size.

2. Proper safety protection (welding goggles, safety glasses, spark lighter, and pliers). Refer to Chapter 2 in the text for more specific safety information.

3. Two or more short random length pieces of 1/2-in. to 1-in. (13-mm to 25-mm) copper pipe with matching copper pipe fittings.

4. Solid tin-lead or tin-antimony wire solder and a container of liquid flux.

5. Steel wool, sanding cloth, and/or a wire brush.

INSTRUCTIONS

1. Clean the pipe O.D. and the fitting I.D. using steel wool, sanding cloth, or a wire brush. Then apply the liquid flux to the pipe and the fitting.

2. Slide the fitting onto the pipe. Be sure that the pipe is seated at the bottom of the fitting. Twist the pipe in the fitting to be sure that the flux is being applied completely around the inside of the joint.

3. Make a bend about 3/4 in. (19 mm) from the end of the wire solder. Use this bend as a gauge so that you do not put too much solder into the joint.

4. Place the pipe in the 1G horizontal rolled position. Heat the pipe and fitting uniformly. The joint is at the correct temperature when the solder starts to wet the surface.

5. When the solder starts to wet, move the flame away from the pipe and feed the solder into the joint, wiping it from the 10 o'clock to the 2 o'clock position. Roll the pipe 90° and repeat. See Figure 33-67 in the textbook.

6. After cooling, hacksaw and pry the joint apart for inspection. See Figure 33-66 in the textbook.

7. Repeat this practice until you can make defect free soldered pipe joints.

8 Turn off the gas cylinder and regulator, and clean up your work area when you are finished.

INSTRUCTOR'S COMMENTS _____

_____

_____

■ PRACTICE 33-21

Name _____ Date _____

Class _____ Instructor _____ Grade _____

OBJECTIVE: After completing this practice you should be able to solder copper pipe in the 4G vertical up position.

EQUIPMENT AND MATERIALS NEEDED FOR THIS PRACTICE

1. Properly set-up, lit, and adjusted air MAPP®, air propane, or any air fuel gas torch. Be sure that the regulator pressure is set to the manufacturer's specifications for your fuel type and tip size.

2. Proper safety protection (welding goggles/safety glasses, spark lighter, and pliers). Refer to Chapter 2 in the text for more specific safety information.

3. Two or more short random length pieces of 1/2-in. to 1-in. (13-mm to 25-mm) copper pipe with matching copper pipe fittings.

4. Solid tin-lead or tin-antimony wire solder and a container of liquid flux.

5. Steel wool, sanding cloth, and/or a wire brush.

INSTRUCTIONS

1. Clean the pipe O.D. and the fitting I.D. using steel wool, sanding cloth, or a wire brush. Then apply the liquid flux to the pipe and the fitting.

2. Slide the fitting onto the pipe. Be sure that the pipe is seated at the bottom of the fitting. Twist the pipe in the fitting to be sure that the flux is being applied completely around the inside of the joint.

3. Make a bend about 3/4 in. (19 mm) from the end of the wire solder. Use this bend as a gauge so that you do not put too much solder into the joint.

4. Place the pipe in the 4G vertical up position. Heat the pipe and fitting uniformly. The joint is at the correct temperature when the solder starts to wet the surface.

5. When the solder starts to wet, move the flame away from the pipe and feed the solder into the joint, wiping it all the way around the joint.

6. After cooling, hacksaw and pry the joint apart for inspection. See Figure 33-66 in the textbook.

7. Repeat this practice until you can make defect free soldered pipe joints.

8. Turn off the gas cylinder and regulator, and clean up your work area when you are finished.

INSTRUCTOR'S COMMENTS _____

_____

_____

■ PRACTICE 33-22

Name _____ Date _____

Class _____ Instructor _____ Grade _____

OBJECTIVE: After completing this practice you should be able to solder aluminum to copper.

EQUIPMENT AND MATERIALS NEEDED FOR THIS PRACTICE

1. Properly set-up, lit, and adjusted air MAPP®, air propane, or any air fuel gas torch. Be sure that the regulator pressure is set to the manufacturer's specifications for your fuel type and tip size.

2. Proper safety protection (welding goggles/safety glasses, spark lighter, and pliers). Refer to Chapter 2 in the text for more specific safety information.

3. One or more pieces of 1/8-in. (3-mm) by 1 1/2-in. (38-mm) square aluminum plate and a copper penny.

4. Solid tin-lead or tin-antimony wire solder and a container of liquid flux.

5. Steel wool, sanding cloth, and/or a wire brush.

INSTRUCTIONS

1. Clean the surface of the aluminum plate and the penny until all coatings and/or oxides have been completely removed.

2. Tin the aluminum by heating it slightly and then dripping molten solder on it. Do this by heating the solder in the flame. Use no flux. Rub the molten solder around on the aluminum using a piece of steel wool. Continue to heat the plate and rub the solder with the steel wool until the surface is tinned. See Figure 33-68 in the textbook.

3. Repeat step 2 with the copper penny. Flux should be used on the penny.

4. Place the tinned surface of the penny in contact with the tinned surface of the aluminum. See Figure 33-69 in the textbook.

5. Heat the two until the solder melts and flows out from between the penny and the plate.

6. After cooling, check the bond by trying to break the joint.

7. Repeat this practice until you can make defect free aluminum to copper soldered joints.

8. Turn off the gas cylinder and regulator, and clean up your work area when you are finished.

INSTRUCTOR'S COMMENTS _____

_____

_____

CHAPTER 33: SOLDERING, BRAZING, AND BRAZE WELDING QUIZ

Name _____ Date _____

Class _____ Instructor _____ Grade _____

Instructions: Carefully read Chapter 33 in the text and answer the following questions.

A. SHORT ANSWER

Write a brief answer in the space provided that will answer the question or complete the statement.

1. Select five of the following that are advantages of soldering and brazing. (Place an X by the correct examples.)

   _____ a. It is impossible to make inclusions by soldering or brazing.

   _____ b. Dissimilar materials can be joined.

   _____ c. Parts of varying thickness can be joined.

   _____ d. Parts can be temporarily joined without damaging them.

   _____ e. You do not have to worry about overheating the parts.

   _____ f. The internal stresses caused by rapid temperature changes can be reduced.

Define the following terms associated with soldering, brazing, and braze welding.

2. Liquid-solid phase bonding process.

   _____
   _____
   _____

3. Flux.

   _____
   _____
   _____

4. Paste range.

   _____
   _____
   _____

5. Capillary action.

_____

_____

_____

B. MATCHING

In the space provided to the left, write the letter from Column B that best answers the question or completes the statement in Column A.

Column A

Column B

_____ 6. What is the ability to withstand being pulled apart?

a. flux

_____ 7. What is the ability to bend without failing?

b. cadmium-silver

_____ 8. What is the ability to be bent repeatedly without exceeding the elastic limit and without failure?

c. induction heating

d. shear strength

_____ 9. What removes light surface oxides, promotes wetting, and aids in capillary action?

e. cadmium-zinc

_____ 10. What is a method of soldering or brazing that uses a high-frequency electrical current to establish a corresponding current on the surface of the part?

f. corrosion resistance

g. fatigue resistance

_____ 11. What is the temperature range in which a metal is partly liquid and partly solid as it is heated and cooled?

h. ductility

i. paste range

_____ 12. What is the solder alloy that is often used on food-handling equipment and copper pipes?

j. tensile strength

_____ 13. Which solder or brazing alloy has a good wetting action and corrosion resistance on aluminum and aluminum alloys?

_____ 14. What has the ability to withstand a force parallel to the joint?

_____ 15. What has the ability to resist chemical attack?

C. FILL IN THE BLANK

Fill in the blank with the correct word. Answers may be more than one word.

16. The only difference between soldering and brazing is that soldering is the _____ at which each process takes place.

17. This small spacing allows _____ to draw the filler metal into the joint when the parts reach the proper phase temperature.

18. With brazing and soldering, it is easy to join _____.

19. With brazing and soldering, very _____ parts or a _____
    part and a _____ part can be joined without burning or overheating them.

20. As the joint spacing _____, the surface tension increases the tensile strength
    of the joint.

21. The _____ the area that is overlapped, the greater is the strength.

22. For most soldered or brazed joints, fatigue resistance is usually _____.

23. The flux, when heated to its reacting temperature, must be _____ and
    _____ provided at the joint.

24. Paste and liquid fluxes can be injected into a joint from tubes using a _____.

25. The flux is picked up by the fuel gas as it is _____ and is then carried to the
    torch where it becomes part of the flame.

26. The use of fluxes does _____ eliminate the need for good joint cleaning.

27. Care should be taken to avoid _____ fluxes.

28. Fuel gases such as MAPP, propane, butane, and natural gas have a flame that will heat parts
    _____.

29. The furnace temperature can be _____ controlled to ensure that the parts will
    not overheat.

30. The very localized heating of the induction brazing and soldering process may result in
    _____.

31. With resistance brazing and soldering, the heat can be localized so that the entire part may not have
    _____.

32. As the joined part cools through the paste range, it is important that the part not be
    _____.

33. Most health and construction codes will _____ allow tin-lead solders for use
    on water or food-handling equipment.

34. Tin-antimony is the most common solder used in plumbing because it is _____.

35. Cadmium-silver solder alloys can be used to join aluminum to _____ or
    _____ metals.

36. The next series of letters in the classification indicates the atomic symbol of metals used to make
    up the alloy, such as CuZn (_____ and _____).

37. Copper-zinc alloys are the most popular _____ alloys.

38. Class BRCuZn-A is commonly referred to as _____ brass and can be used to fuse weld naval brass.

39. Phos-copper has good _____ and _____ on copper and copper alloys.

40. Copper-phosphorus-silver alloy is sometimes referred to as _____.

41. Some applications of nickel alloys include joining _____, _____, _____, and _____.

42. Since copper oxides can cause porosity, Class BCu-2 ties up the oxides with the _____ compounds reduces the porosity.

43. Some joints can be designed so that the flux and filler metal may be _____.

44. Braze metal is ideal for parts that receive limited _____ because the buildup is easily machinable.

45. To prevent overheating with an oxyacetylene flame, _____ and hold the flame so that the inner cone is about 1 in. (25 mm) from the surface.

D. ESSAY

Provide complete answers for all of the following questions.

46. Soldering is performed below what temperature? Brazing is performed above what temperature?

_____

_____

47. What is capillary action?

_____

_____

48. List four advantages of soldering and brazing as compared to other methods of joining.

_____

_____

49. Why are MAPP®, propane, or natural gas better fuel gases than acetylene for brazing or soldering?

_____

_____

50. How does an induction brazing machine work?

_____

_____

51. Why is it important that both pieces of metal being joined be heated at the same rate to the same temperature?

_____

_____

52. How are the edges of a plate shaped for braze welding? Why?

_____

_____

E. DEFINITIONS

53. Define the following terms:

liquid-solid phase bonding _____

tensile strength _____

shear strength _____

ductility _____

fatigue resistance _____

corrosion resistance _____

paste range _____

eutectic _____

INSTRUCTOR'S COMMENTS _____

_____

_____

NOTES